Fundamentals of Refractory Technology

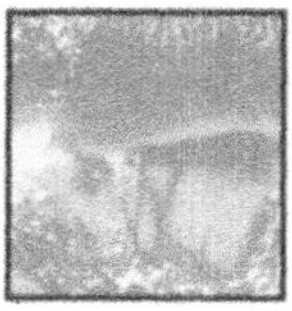

Ceramic Transactions
Volume 125

Fundamentals of Refractory Technology

Proceedings of the Refractory Ceramics Division Focused Lecture Series presented at the 101st and 102nd Annual Meetings held April 25–28, 1999, in Indianapolis, Indiana, and April 30–May 3, 2000, in St. Louis, Missouri, respectively.

Edited by
James P. Bennett
U.S. Department of Energy
Albany Research Center

Jeffery D. Smith
University of Missouri—Rolla

Published by
The American Ceramic Society
735 Ceramic Place
Westerville, Ohio 43081
www.ceramics.org

Proceedings of the Refractory Ceramics Division Focused Lecture Series presented at the 101st and 102nd Annual Meetings held April 25–28, 1999, in Indianapolis, Indiana, and April 30–May 3, 2000, in St. Louis, Missouri, respectively.

***Cover photo:* "Dense MgO layer formed in the used MgO-C refractory with Al-Mg alloy," is courtesy of A. Yamaguchi, and appears as figure 10 in the paper "Application of Thermochemistry to Refractories," which begins on page 157.**

Library of Congress Cataloging-in-Publication Data

A CIP record for this book is available from the Library of Congress.

For information on ordering titles published by The American Ceramic Society, or to request a publications catalog, please call 614-794-5890.

ISSN 1042-1122
ISBN 1-57498-133-1

Contents

2000 Focused Sessions—St. Louis, Missouri, May 1, 2000

Preface

This volume contains papers presented in focussed sessions of two Refractory Ceramics Division (RCD) meetings held in conjunction with the 101st Annual Meetings & Expositions (Indianapolis, Indiana—April 25–28, 1999) and the 102nd Annual Meeting and Exposition (St. Louis, Missouri—April 30–May 3, 2000). The focussed sessions were organized by RCD and the editors of this volume. Refractory scientists from throughout the world were invited to provide overviews of the scientific principles related to refractory manufacturing and performance. The sessions were well received, with at times as many as 100 scientists and engineers from industry, university, and government laboratories in attendance.

This book contains seven of eight papers presented during the 1999 Focused Sessions and nine of the 11 papers presented during the 2000 Focused Sessions. One paper for the 2000 Focused Session ("A Novel Rhoemeter for Refractory Castables," by Pandolfelli, Pileggi, and Paiva) was published in the *American Ceramic Society Bulletin.* The papers were reviewed by committee. The editors would like to recognize the balance of the committee and thank them for their assistance. They include Dr. Richard C. Bradt, University of Alabama, Dr. Robert E. Moore, University of Missouri—Rolla, and Dr. Michel A. Rigaud, École Polytechnique.

The editors are especially grateful for the efforts of the authors who presented in the focused sessions and were diligent in completing their manuscripts. In addition, the editors would like to thank all those individuals who attended the technical sessions. It is our hope that the information presented in the focused sessions and contained in this book will help advance knowledge in refractory science.

Jeffery D. Smith
James P. Bennett

1999 Focused Sessions

Indianapolis, IN, April 26, 1999

Dr. Jeffery D. Smith, Program Chair
Department of Ceramic Engineering
University of Missouri - Rolla
Rolla, MO, USA

Dr. Michel A. Rigaud, Session Chair
École Polytechnique
Montreal, Canada

Dr. Richard C. Bradt, Session Chair
University of Alabama
Tuscaloosa, AL, USA

PARTICLE SIZE DISTRIBUTION AS A PREDICTOR OF SUSPENSION FLOW BEHAVIOR

Randall M. German
Brush Chair Professor in Materials
Center for Innovative Sintered Products
P/M Lab, 147 Research West
The Pennsylvania State University
University Park, PA 16802-6809, USA

ABSTRACT

Component fabrication from powder suspensions has many variants and they all have magic particle size distributions that give improved behavior. Distributions that give the highest packing density and lowest forming viscosity have been known for decades. Size distributions with high packing densities and low flow viscosities characteristically are wide with an abundance of large particles. This presentation shows examples from several powder mixtures to illustrate links between particle packing and mixture viscosity. Operative principles and distribution parameters provide insight on suspension viscosity and yield strength. Homogeneity is the largest source of error between theory and practice, so principles are introduced for measuring homogeneity.

INTRODUCTION

An understanding of particle packing characteristics is important in many common ceramic forming operations, ranging from slip casting to reaction boding. Most important, an understanding of packing density is an invaluable basis for understanding rheology in powder processing. The study of particle packing is well documented in three books [1-3]. Packing density describes the volume fraction of a container that is filled with powder and is measured by fractional density. The rheology of a powder suspension is measured by the stress needed to initiate flow and the resistance to sustained flow as measured by the viscosity. The goal is to induce a high packing density via mixed particle sizes or continuous particle size distributions, recognizing this results in low viscosity suspensions.

Many industrial situations have proprietary links between the particle size, particle shape, distribution width, packing density, and processing. Detailed relations are beyond the scope of this article; however, it is important to recognize that packing density is an easily measured attribute that provides much insight into powder flow [4], viscosity [5], elasticity [6], compaction [7], infiltration [8], sintering [9], permeation [10], molding [11], melting [12], hot isostatic compaction [13], and other attributes. Indeed, the problem of optimizing particle size distributions to improve packing density and flow is encountered in many fields, including chocolate, ice cream, filled polymer molding, asphalt, cement, solder pastes, paints, slip casting, thixomolding, powder metallurgy, tape casting, and castable refractories.

Although refractory systems are the focus of the seminar series, major interest arises from studies on the flow behavior of powders used in powder injection molding [14]. A powder and thermoplastic polymer mixture are mixed to form a viscous slurry that can be molded into a cold die cavity. The polymer freezes in the die to hold the particles in the cavity shape. The particle size distribution and particle shape determine the solids loading (ratio of powder volume to total volume) in the feedstock, which controls the viscosity and subsequent processing steps. The lower the packing density of the powder, the greater the binder content needed for flow. Generally, in selecting a powder for a process, the desire is to attain a high solids loading, recognizing that such powders require less polymer for molding and sintering shrinkage for densification.

Packing structures are categorized as either random or ordered. As shown in Figure 1, a random packing is constructed by a sequence of events that are not correlated with one another. When a powder is poured into a container, the structure is random. On the other hand, an ordered structure occurs when objects are placed systematically into periodic positions, such as the stacking of bricks to form a wall. Random structures lack long-range repetition, and typically exhibit lower packing densities.

For monosized spheres the maximum packing density occurs in an ordered close-packed array with a coordination number of 12 and a density of 74%. With respect to most ceramic powders, the highest packing density is random and the structure forms with a density less than that for ideal, monosized spheres at 60 to 64%. Tap density, the highest density random packing, occurs when the particles have been vibrated without introducing long range order or deformation. The random loose packing that results when particles are poured into a container without agitation or vibration is commonly called the apparent density.

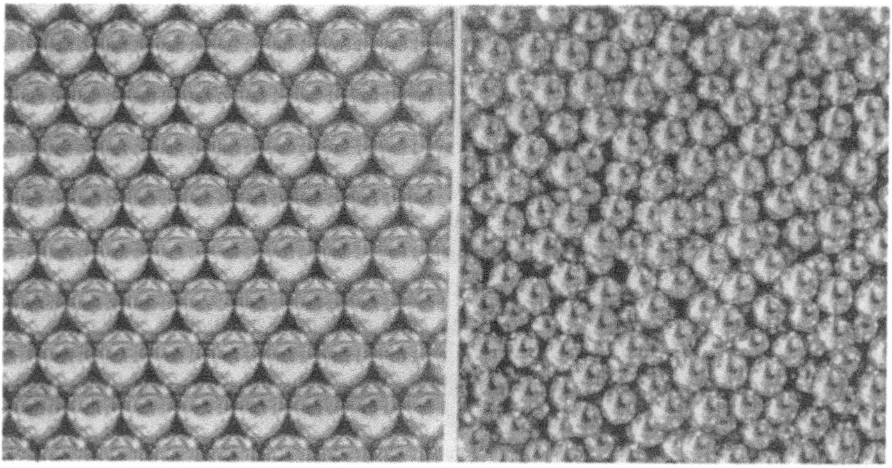

Figure 1. Pictures contrasting the structure for a random and ordered array of monosized spheres.

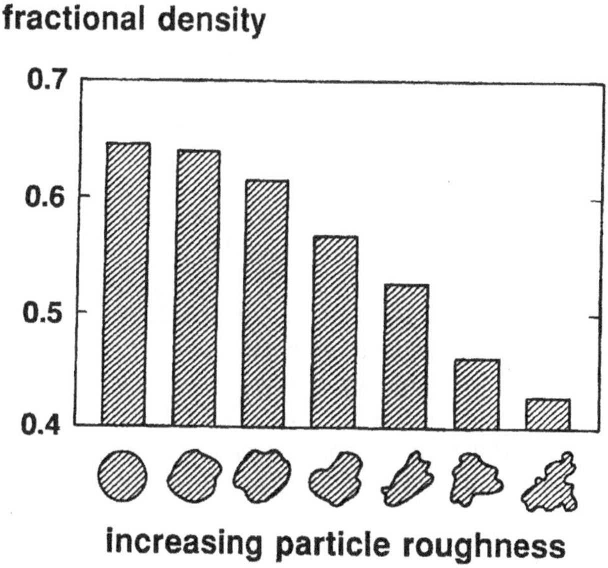

Figure 2. Packing density for random monosized particles as a function of the particle shape and surface roughness, showing the benefit of equiaxed and smooth particles.

HISTORICAL INTEREST IN PACKING AND SUSPENSION FLOW

Concern over particle packing and flow traces back at least 300 years. A detailed history on particle packing is available in reference [1-3] and the implications with respect to rheology are covered in references [14-19]. Today we recognize that a coordination number of 12 is the highest possible for monosized spheres (the close-packed structure evident in crystallography). However, Newton in 1694 speculated on the creation of a high packing density structure with 13-fold coordination. Hales in 1727 examine the packing of peas, using dried peas that swelled in water to form flat contact faces, allowing first determination of the relation between packing density and coordination number. Einstein treated the problem of dilute suspension flow in 1906, proposing a linear relation between viscosity and solids loading. About this same time, Fuller and Thompson in 1907 were constructing earthen dams for the New York City water system and explored relations between rock-gravel-sand grain size distribution and the mixture density and strength. Their findings are valuable today, showing four rules for creation of high strength dams (which happen to be those with high packing densities):

- ❑ use the largest mean particle size possible
- ❑ use rounded particles
- ❑ skew the size distribution to use a high fraction of large particles
- ❑ rely on a low volume of intermediate sized particles.

Subsequent studies showed these concepts were very accurate and extended to continuous particle size distributions. Important conceptualizations were provided by Furnas in 1928 and Andreasen in 1930, both suggesting that optimal distributions could be envisioned based on filling space first with the largest particles. Subsequently, smaller particles are selected to fill the voids, and at each subsequent level smaller particles are used to fill remaining void space. From a demonstration view, McGeary in 1961 showed a five-mode particle mixture with a packing density of 95%. About this same time, Scott experimentally determined the maximum packing density obtainable with monosized spheres. Meanwhile, interest in viscosity was related to the maximum packing density, and the work of Money in 1950 proved most accurate in showing that viscosity of a suspension is directly linked to the maximum packing density.

It is on this base that today we look at the effect of particle size distribution on packing as a precursor problem to a study of suspension rheology. Fundamentally, high packing density powders also have low suspension viscosities for a given powder content.

KEY FACTORS

The particle characteristics important to ceramic processing are packing density, strength, surface area, permeability, and pore size. A critical characteristic is packing density. For example, in powder injection molding, the required quantity of binder approximately equals the void space between particles [20]. Optimal quantities of binder ensure defect-free molding. Likewise, many other processing factors, including sintering shrinkage, debinding rate, and compact strength depend on the packing density.

Particles of differing materials, but equal size and shape, will pack to the same fractional density in spite of differing theoretical densities. However, several other factors cause differences in packing densities. For a mean particle size below approximately 100 μm there is more interparticle friction to inhibit packing; thus, for small particles the loose random packing density is a function of the mean particle diameter [3,21]. Part of the problem arises because of agglomeration, since surface forces are large when compared to the particle mass. This problem is most apparent as the particle size decreases below approximately 1 μm. One option is to create repulsive forces between particles by using thin surface coatings of polar molecules or control of the powder surface charge via pH adjustment. The surface repulsive forces contribute to high packing densities by reducing the interparticle friction. Thus, attention to surface chemistry is beneficial in formulating powder mixtures requiring high packing densities.

Another form of interparticle friction arises from surface irregularities on the particles. The greater the surface roughness or the more irregular the particle shape, the lower the packing density. Data for the particle shape effect on packing density are scattered, yet some general patterns are apparent. A higher packing density is associated with spherical particles with smooth surfaces. For particles of the same size but different shapes, the packing density will decrease as the shape departs from spherical or equiaxed [3,22].

When two or more powders are mixed together, there will be opportunities to improve the packing density. However, this assumes the structure is homogeneous, with the small particles ideally situated in the voids between the large particles. Although the ideal is simple to describe, there is still difficulty in attaining the desired homogeneity. The best symptom of inhomogeneity is the viscosity and instability of the viscosity [23]. A related problem with wide particle size distributions is segregation based on particle size. Size segregation is more of a problem with large mean particle sizes and large differences in particle size. One consequence of size segregation is a decrease in the overall mixture density and point-to-point density variations. In well-mixed powders, size segregation is not a serious problem. Organic processing aids are used to minimized segregation in handling, especially for smaller powders that exhibit poor flow and packing.

For highest densities, it is appropriate to vibrate the powder to eliminate bridging, large voids, or other defects. For this reason, the tap density provides a best first measure of particle packing and proves relevant to many forming operations. The measurement depends on the material, vibration amplitude, vibration direction, applied pressure, vibration frequency, particle density, shear, and test apparatus [24,25]. During vibration the density varies with the number of vibrations by an exponential function,

$$f(N) = f_i + (f_f - f_i)\ \exp\left(-\frac{K}{N}\right) \tag{1}$$

where K is a constant that depends on the device, height of fall, and velocity, N is the number of vibration cycles, f_f is the final solids density, $f(N)$ is the solids density after N cycles, and f_i is the initial solids density. Generally, the more irregular the particle shape, the greater the packing benefit from vibration. Powders will reach the dense random packing limit more rapidly as the particle size increases.

IDEAL PACKING

As a starting point, consider the packing of an ideal powder, where each particle is the same size and spherical. This is probably the best studied packing problem, yet it is far from the typical powders. It is the basis for most models, partly because of similarities to hard sphere models for crystalline atomic structure - body-centered cubic, face-centered cubic, and simple cubic packings.

An ordered packing consists of perfectly placed spheres. Many common examples are known in crystallography tables; for example the body-centered cubic structure is 68% dense. There is a systematic increase in packing density as the number of contacts per sphere increases. The highest packing for monosized spheres is a coordination number of 12 with a 74% density. On the other hand, the lowest possible packing density in a gravitationally stable structure formed from nonadhesive spheres is a coordination number of 4 with a packing density near 34%.

Various models link the coordination number and packing density. Unfortunately there is no exact relation between packing density and coordination number, but a simple model is,

$$N_C = 2\ \exp(2.4\ f) \tag{2}$$

where N_C is the packing coordination number and f is the fractional density.

An important packing with respect to ceramic processing is the random dense structure. This is formed by vibrating a powder into a high packing density without the application of external stress, other than gravity. For the monosized spheres, this packing density is 63.7% [26], which equals $2/\pi$. As a point of comparison, the packing density of a random loose array of monosized spheres is 60%, with approximately 6 contacts per sphere. Such a structure forms when uniform spheres are poured into a container, but not vibrated.

The random dense packing and random loose packing are fairly similar for large spherical particles. Consequently, spherical particles of sizes greater than approximately 100 μm undergo little densification during vibration. In contrast, smaller spheres and nonspherical particles exhibit a greater difference between the apparent and tap densities. These particles undergo a larger packing density increase with vibration and exhibit higher packing densities in the presence of fluids, surfactants, and pressure.

The packing density and coordination number decrease as the particle shape departs from that of a sphere. Figure 2 shows the fractional packing density for various monosized irregular particle shapes. Powders with highly irregular particle shapes do not match the packing density for spheres. As the particle shape becomes more rounded (spherical) the packing density increases. The difference between random and ordered packing densities increases as the particle shape becomes nonspherical.

Spherical or rounded particles are desirable in applications that require a high packing density and easy flow. But if the particle has a regular polygonal shape, then a decrement in packing density does not always occur. Anisotropic particles can be packed to high densities if they are ordered. For example, cubic particles can be packed to 100% density when placed in an ordered packing. The highest packing densities and most isotropic structures are observed with equiaxed particles, such as spheres and cubes. Fibers in random packing exhibit a decrease in density as the length-to-diameter ratio increases [27]. Fractional densities below 10% result from random packing of fibers with large length-to-diameter ratios.

BIMODAL MIXTURES OF SPHERES

Basic Structure - Bimodal particle mixtures pack to higher densities than do monosized particles. The key to improved packing rests with the particle size ratio. Small particles are selected such that they fit into the interstices between large particles without forcing the large particles apart. In turn, even smaller particles can be selected to fit into the remaining pores, giving a corresponding improvement in packing density.

To improve the packing density, the added powder must fill the void spaces without dilating the overall volume. For a random dense packing, the basic behavior is sketched in Figure 3. The packing volume, termed the specific volume (volume-to-mass ratio), is plotted as a function of the composition for a mixture of large and small spheres. There is a composition of maximum packing density that has a majority of large particles. The relative improvement in packing density depends on the particle size ratio of the large and small particles. Within a limited range, the greater the size ratio, the higher the maximum packing density. This is true up to a limiting size ratio of approximately 20:1, but requires at least a 20% difference in particle sizes [28,29].

Optimal Packing - The optimal packing density for a bimodal mixture of spheres corresponds to the minimum specific volume. At this optimal point the large particles are in point contact with one another and all of the interstitial voids are filled with small particles. A mathematical description of bimodal packing starts by designating the large particles with the subscript "L" and the small particles with the subscript "S". Calculation of the optimal composition in terms of the weight fraction of large particles X_L^* is an obvious goal. In general, the weight fraction of large particles at any composition X_L depends on the following equation:

$$X_L = \frac{W_L}{V_L + W_S} \tag{3}$$

with W indicating the weight. The weight of large particles is calculated from the theoretical density ρ_L, fractional packing density f_L, and container volume v as $W_L = f_L \rho_L v$.

For the condition of maximum packing density the desire is to add sufficient small particles to just fill the void space between the large particles, without forcing the large particles apart. The amount of void space equals $1 - f_L$, and the fractional packing density for the small particles times the volume and the theoretical density of the small particles gives its weight faction as

$$W_L = F_L \rho_L v \tag{4}$$

Thus, the optimal packing X_L* is given as follows:

$$X_L^* = \frac{f_L \rho_L}{f_L \rho_L + (1 - f_L) f_S \rho_S} \tag{5}$$

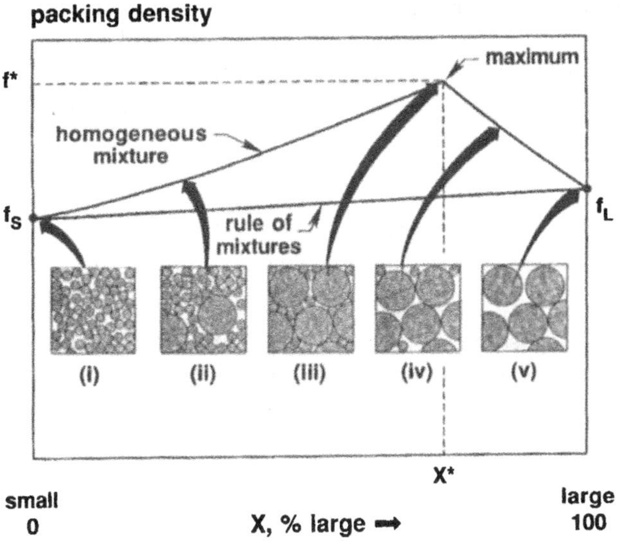

Figure 3. Packing with a bimodal size involves selection of smaller particles that fill the interstitial voids between the larger particles.

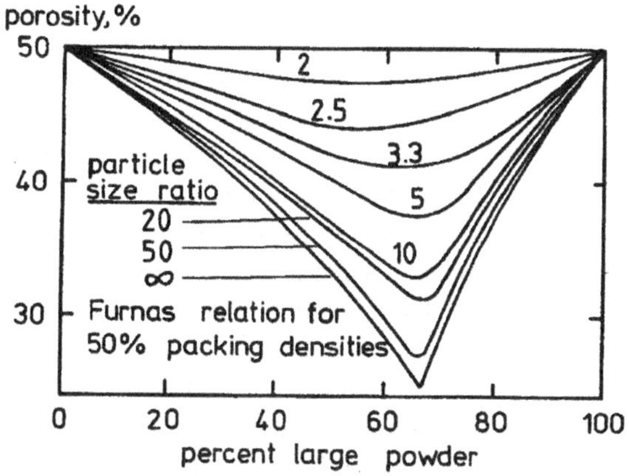

Figure 4. Porosity of various mixtures versus composition (percent large particles) and particle size ratio (large:small) to show how maximum packing density depends on both features.

In the case where the two particles composing the bimodal mixture are of the same composition, then the density of the small and large particles will be the same ($\rho_L = \rho_S$). A further simplification occurs for the case where the fractional packing densities for the two powders are the same. This case gives the optimal packing composition by the simple relation,

$$X_L^* = \frac{1}{2 - F_L} \tag{6}$$

Equations 5 and 6 give the compositions in terms of the proportional weight of large particles for the general case and for the specific case of equal theoretical densities and equal fractional packing densities.

Particle Size Ratio - Besides composition, the packing density is dependent on the particle size ratio up to a limit. The larger the particle size ratio, the higher the packing density at all compositions. For high packing densities, avoid porous and agglomerated particles. Consider two powders with a large difference in particle size, exhibiting random dense packing with the usual density of 63.7%. The corresponding weight fraction of large particles for maximum packing is 0.734, or 26.6 wt.% small particles.

For the more general cases, Figure 4 plots the porosity of mixed powders versus the percentage of the large powder. This plot shows the simultaneous effects of composition and particle size ratio (large divided by small) on the mixture density [30]. A minimum porosity occurs for each mixture at a composition rich in the large particles. The minimum porosity is larger and shifts toward the small particle axis as the particle size ratio decreases.

Figure 5 gives the relation between the inherent packing density of the large particles and the amount of small particles needed to attain maximum packing density as calculated by Furnas [29]. There is no gain in packing density if the particles have the same size, whereas the packing benefit is maximized by a large difference in particle sizes. Figure 6 shows the measurements by McGeary [24] for random dense packings. The fractional packing density increases as the particle size ratio increases up to a ratio of 15:1 or so. Note the change in behavior at the particle size ratio corresponding to a particle just filling the triangular pores between the large particles at roughly a 7:1 size ratio. The packing density is unchanged by particle size ratios larger than 20:1. In the size ratio range from 1:1 to 20:1 for D_L/D_S there is a major change in packing density. Figure 7 compares the prediction of Furnas [29] with that of Fedors and Landel [30]. The porosity of the mixture is expressed as a function of the inverse particle size ratio. Both models show that major packing benefits are associated with large

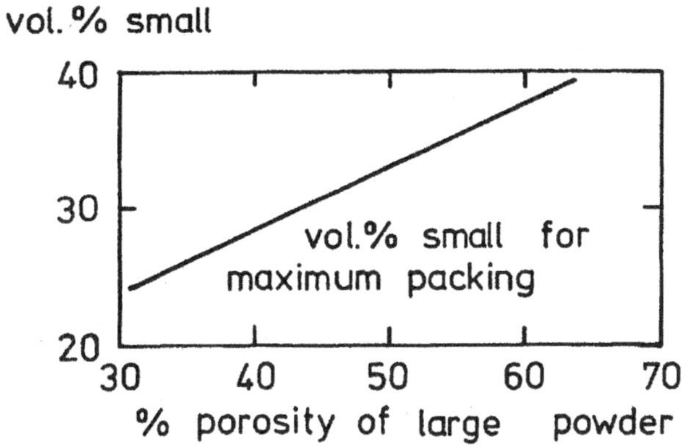

Figure 5. The amount of small powder in a bimodal mixture at the optimal packing condition as a function of the packing density of the large powder.

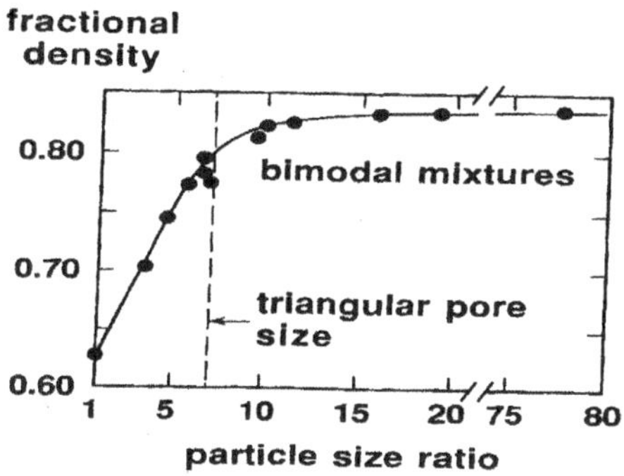

Figure 6. Experimental results from McGeary showing how the particle size ratio (large:small) impacts on the packing density for bimodal mixtures of spheres at the maximum packing density.

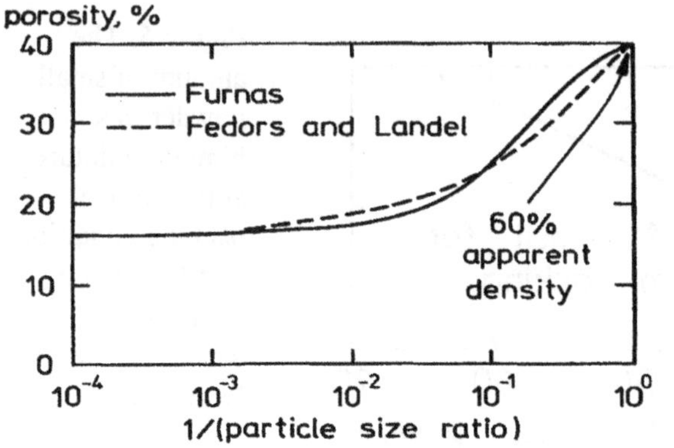

Figure 7. A comparison of the predicted bimodal powder packing density behavior for powders that pack to 60% density, showing the mixture porosity versus inverse particle size ratio for the models by Furnas and Fedors and Landel.

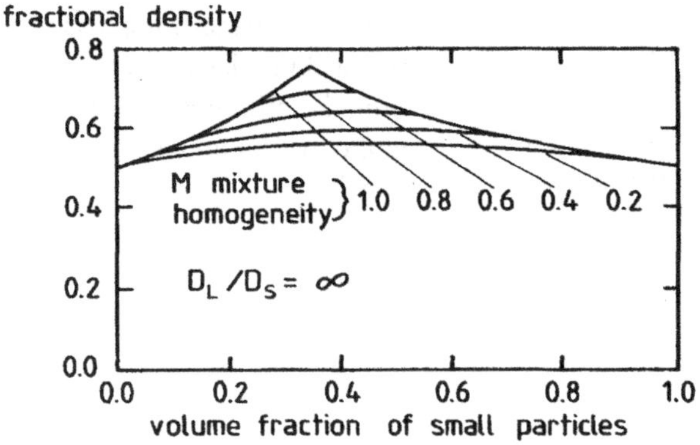

Figure 8. Homogeneity has impact on the fraction packing density in the same manner as the large:small particle size ratio. This plot shows density variation with composition for a large size ratio, but varying homogeneity.

differences in sphere sizes, but very large differences have no further benefit.

Mixture Homogeneity - Since the packing density varies with composition for mixtures of differing particle sizes, mixture homogeneity has a major effect on the packing density [23]. In an inhomogeneous packing, some regions will have more large particles than others. Consequently, the fractional density will vary from point-to-point in proportion to the local composition. Most models assume orderly positioning of the particles. In reality this is a difficult state to attain. Accordingly, the predicted density for a random bimodal mixture often overestimates that attainable in actual mixtures.

The packing density will increase with the homogeneity of the mixture. Figure 8 shows how composition and the degree of mixing will influence the packing density. This plots the fractional packing density versus composition for several levels of mixture homogeneity for an infinite particle size ratio. The density improves with the degree of mixing. The upper limit is based on perfect mixing, while the lower limit assumes a totally segregated structure. Randomly mixed systems will range between the unmixed and fully mixed limits, depending on the mixture homogeneity. The mixture homogeneity *M* is defined as follows:

$$M = 1 - \frac{\sigma}{\sigma_o} \tag{7}$$

where σ is the standard deviation in the composition between random samples taken from the mixture and σ_o is the standard deviation in composition for the unmixed system,

$$\sigma_o = \left(X_L\,(1 - X_L)\right)^{1/2} \tag{8}$$

Most random mixtures exhibit a homogeneity near 0.7. However, in well mixed ceramic powder systems an effective polymer can prevent segregation. Thus, slurry processing generally gives a higher homogeneity when compared to dry processing and binders contribute to a higher packing density.

A high packing density for the individual component particles is helpful in attaining a high packing density for the mixture. In the case of randomly packed monosized spheres with a density of 63.7%, a mixture of 73% large and 27% small will give a mixture packing density approaching 86%. The larger the particle size ratio and the more uniform the mixture homogeneity, then the closer the observed packing density will approach the model limits. All of the concepts addressed here for spheres carry over into mixtures of nonspherical particles, but often the starting densities are lower because of inhibited packing.

MULTIPLE MODE DISTRIBUTIONS

The ideas developed for bimodal spherical packings have been extended to multiple mode systems. The first step is to consider trimodal mixtures, where there are three particle sizes. Subsequently, the trimodal systems form the basis for considering multiple mode mixtures (mixtures of several different monosized powders) and continuous particle size distributions. The continuous particle size distribution is of great interest since it is a practical approach to creating high packing density systems. As with bimodal packing, mixture homogeneity influences packing density for multiple mode or continuous distributions.

A few comments are in order about the practical aspects of multiple mode mixtures. It is often claimed that at least a 7:1 particle size ratio is needed for optimal packing. For a trimodal packing, this corresponds to a size ratio of 49:7:1. Assuming the smallest particles are 1 μm in size, then the largest particles will be 49 μm. Such a size is within a realistic working range. However, the classification of particles into sizes of 1, 7, and 49 μm is difficult. Obviously, as the size ratio or number of size classes becomes larger the practical problems increase. Consequently, continuous particle size distributions are more typical in practice.

The maximum packing density for bimodal mixtures of spheres is probably near 86%, whereas that of monosized spheres is 64%. This corresponds to a 34% improvement in packing. For a trimodal mixture, the maximum packing density is probably 95%. This is slightly more than a 10% improvement in packing over the bimodal situation. The addition of more components will return a decreasing benefit in packing, but the practical problems in preparing such mixtures may be insurmountable. Hence, a clear case can be made for bimodal mixtures in certain instances of ceramic processing, because of the large packing gain and relative ease of finding adequately different sizes. Mixtures formed from more components may not be fruitful.

The trimodal mixture of monosized spheres is very sensitive to the particle size ratio. For this discussion, the ternary sizes will be termed large, medium, and small. Depending on the ratio of sizes, a ternary mixture may have a higher packing density than the large-small binary. It is a key characteristic of all multiple mode powder mixtures that a large difference in particle sizes aids packing. Furnas [29] calculated, based on the particle size ratio, whether optimal packing would occur in the binary, ternary, or higher-order system; the results are shown in Figure 9. For a given particle size ratio and 60% inherent packing density, this plot indicates the number of components corresponding to optimal packing. For example, with a small particle size ratio such as 2, optimal packing

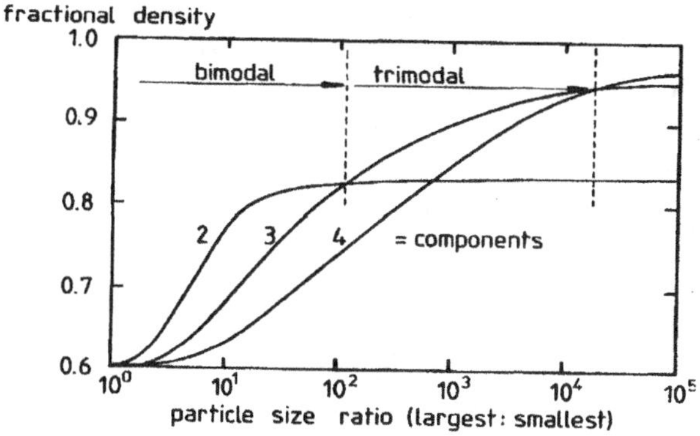

Figure 9. The effect of particle size ratio (large:small) on the packing density, showing how the selection of bimodal versus trimodal compositions depends on the available range of powder sizes.

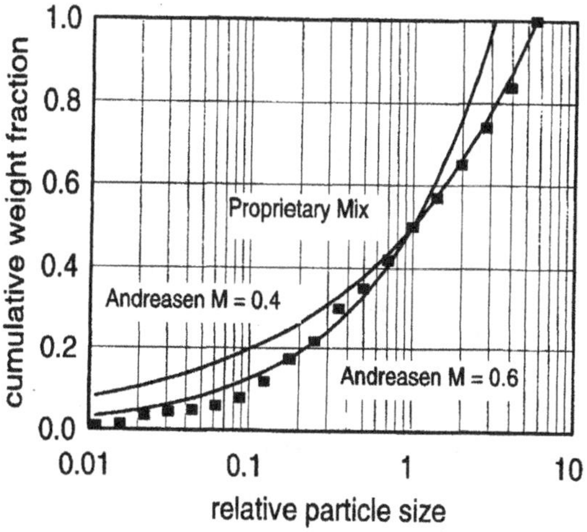

Figure 10. The Andreasen continuous particle size distribution with high packing densities has a skewed character with a high proportion of large particles and a long tail of small particles. Three distributions are included in the log size plot, showing a proprietary mixture for high solids loading as compared with the distributions identified by Andreasen.

occurs with two components. Alternatively, with larger particle size differences, like 1000:1, better packing would be attained with a ternary mixture.

It is noteworthy to examine the number of particles corresponding to optimal multiple mode packings. For a particle size ratio of 1:7:77, the volume percentages for each powder at optimal packing are 10, 23, and 67, respectively. The approximate number of particles corresponds to 69000, 490, and 1 for the small, medium, and large size classes. This means that a few large particles (but a high weight fraction) go a long way toward filling space. In the limiting case of an infinite particle size ratio, Lee [31] calculates a maximum packing density of 92.6% at a composition (by volume) of 8.7% small, 24.2% medium, and 67.1% large particles. For spheres with an inherent random packing density of 64%, the maximum density available with a trimodal mixture at an infinite particle size ratio is at 95% [32].

Further consideration of mixed particle sizes leads to four, five, and more components. As the number of components increases, the experimentation necessary to find an optimal packing composition becomes more difficult. One of the early treatments of multiple mode packing was performed by Horsfield [33]. For a mixture of 4 or 5 components, packing densities near 85% were demonstrated with modest size differences. At the extreme, a 6 component mixture with 96% density is possible. McGeary [24] experimentally produced a packing density of 95% in a random packing.

In multiple mode models, the structure takes its basic form from the largest particles, being sensitive to their size, shape and distribution [34]. The remaining void volume is then filled by the smaller particles. These in turn leave smaller pores that are filled by smaller particles. The effect of a small particle size ratio is a decrease in the maximum packing fraction. Using this concept, Furnas predicts maximum packing densities of 87, 95, 98, and 99% for 2, 3, 4, and 5 component systems.

CONTINUOUS PARTICLE SIZE DISTRIBUTIONS

From multiple mode packings it is obvious to extend the concepts to continuous distributions. The packing density of a multiple mode particle mixture increases as the number of components increases, as long as the particles are very different in size and an optimal composition is maintained. From this concept, a wide particle size distribution gives a higher packing density than with a narrow particle size distribution.

Sohn and Moreland [35] used Gaussian distributions to show the packing density increased with wider particle size distributions, independent of the mean particle size. However, there was a particle shape effect, with irregular particles giving a lower packing density.

For limited size widths, the packing density increases as the distribution becomes broader. With wide size distributions the packing density approaches a limiting value. For an infinitely wide particle size distribution, the projected packing density has been estimated to range between 82 and 96%. These values are comparable with the densities attainable with bimodal and trimodal mixtures with large particle size ratios. It is speculated that an upper density of 96% might be obtained with a very wide particle size distribution.

Furnas [29] showed a wide size distribution provided a high packing density and he describes a mixture process to secure a high density. His distribution was similar to the empirical findings of Fuller and Thompson [36]. Many subsequent measurements confirm these ideas [3]. More commonly, high packing densities are fit with a particle size distribution described by the Andreasen equation [37]:

$$W(D) = A_o + A\,D^M \tag{9}$$

where W is the weight fraction of particles less than size D, and A_o, A and M are empirical constants used to fit the particle size distribution. The maximum apparent density occurs for M values between 0.5 and 0.67, while M values between 0.33 and 0.5 give the highest tap density. Figure 10 plots the Andreasen distribution, showing the characteristic long tail of small particles and rapid rise in larger particles. For comparison, a proprietary high packing density mixture with low viscosity is included to show agreement with this simple model. A high proportion of small particles helps fill the voids between the nearly-continuous matrix of large particles.

Funk and Dinger [15] suggest a similar continuous particle size distribution function which traces directly to the Andreasen equation, but requires knowledge on the largest D_L and smallest D_S particle sizes,

$$W(D) = \frac{D^N - D_S^N}{D_L^N - D_S^N} \tag{10}$$

Again W represents the weight fraction of particles smaller than size D, and N is an adjustable constant to fit the distribution. Rearranging Equation 10 gives a form identical to Equation 9,

$$W(D) = A_1 + A_2\,D^N \tag{11}$$

where the constants are defined by characteristics particle sizes such that $A_1 = 1/(D_L^N - D_S^N)$ and $A_2 = - D_S^N/(D_L^N - D_S^N)$. Further, by definition $W = 0.5$ at D_{50} (the median particle size), thus $D_{50}^N = (D_L^N + D_S^N)/2$, which defines N as a function of the three particle sizes. As demonstrated in Figure 11, this distribution function gives the same result as the Andreasen model.

Continuous particle size distributions can be mixed to improve the packing density if the mean sizes of the two distributions are very different. Lewis and Goldman [38] successfully argue that the mixture that gives optimal mixing corresponds to the widest particle size distribution. Thus, with continuous particle size distributions the key is to seek wide distributions to increase packing density and to form low viscosity suspensions.

RHEOLOGY LINKS

Most powder forming technology relies on particle lubrication from a fluid phase to assist in shaping. The fluid phase is divided into two partitions - that portion needed to fills all voids when the powder is at the maximum packing density (termed the immobile fluid) and the excess fluid that provides lubrication (termed the mobile fluid). These are saturated suspensions with all pore space filled with fluid. It is the extra fluid from dilation to a lower packing density than the maximum packing density that ensures lubricated particle flow. Schematically, this concept is sketched in Figure 12. It is the thickness of the lubricating layer between particles that determines suspension viscosity [39].

A viscous suspension as consists of discrete particles with all of the pores between the particles filled with fluid. It is the lubricating layer thickness that determines viscosity; hence, a linkage is needed between the layer thickness and the solid content in the suspension, known as the solids loading. As long was the particles are relatively large, typically not colloids, then at the critical solids loading the particles are in point contact, reflecting the same condition as the maximum packing density. If the solids loading is below the critical level, meaning that the powder is below the maximum packing density, then the excess fluid lubricates the particles.

Associated with this powder-fluid structure are concerns with strength and viscosity. Strength measures the stress needed to initiate flow and viscosity measures the stress needed to sustain flow. The higher the desired flow rate, typically the greater the stress. Particles have a natural tendency to adhere and agglomerate, leading to a low yield strength. Flow is initiated once the applied stress exceeds the yield strength, which increases with shear strain rate and solids loading. Typically this strength is low (in the range of 100 Pa or less) and can be ignored, but it is beneficial in holding shape after forming.

It is the excess fluid, termed the mobile fluid, that provides lubricity and

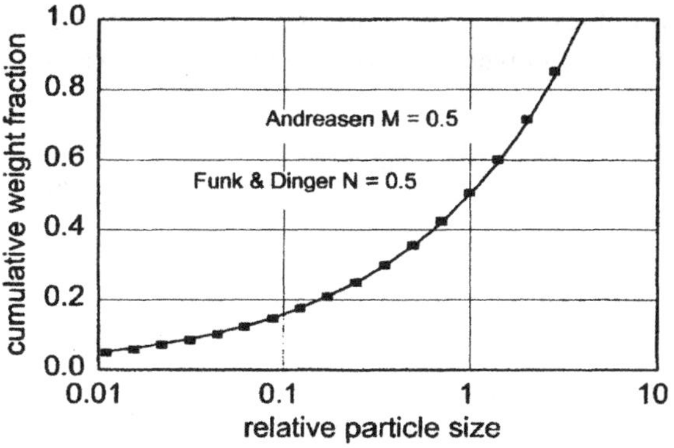

Figure 11. A comparison of the Funk and Dinger particle size distribution and the Andreasen distribution, showing they are identical.

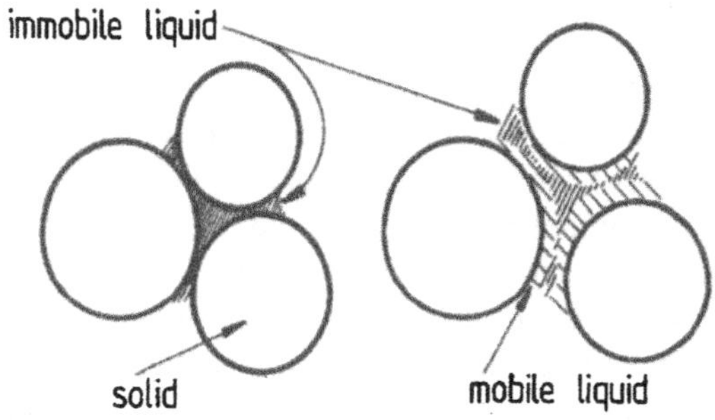

Figure 12. The concept of a lubricating layer, where the voids are filled between the particles with the immobile fluid and the flow between particles is lubricated by the mobile or excess fluid over that required to fill the voids at the critical solids loading which corresponds with the maximum packing density.

lowers the suspension viscosity. The thickness of the lubrication layer δ can be estimated based on the solids loading and powder surface area on a volume basis S_v,

$$\delta = \frac{4\,(1 - \Phi)}{S_V} = \frac{2}{3}\,D\left(\frac{1 - \Phi}{\Phi}\right) \tag{12}$$

The second form assumes the particles are monosized spheres. If we further assume a homogeneous mixture, then the lubricant layer thickness can be expressed as a function of the particle size D, critical solids loading or maximum packing density Φ_C, and the actual solids loading Φ as follows [39]:

$$\delta = D\left[\left(1 + \frac{1}{\Phi} - \frac{1}{\Phi_C}\right)^{1/3} - 1\right] \tag{13}$$

As the lubricating layer increases in thickness, the viscosity and strength decrease, desirable attributes in high solids loading suspensions. As repeated throughout this presentation, it is the particle packing density that determines the system behavior. Low viscosities are possible in high solids loadings if the particle size distribution is adjusted for a high packing density.

Mooney [40] provided the desired link between viscosity and solids loading. Today the Mooney equation is the most common expression linking viscosity of crowed suspensions of nonagglomerating particles to the solids loading. His idea was to relate the mixture viscosity η_M to the liquid viscosity η_L and the ratio of the solids loading to the maximum packing density. The solids loading Φ is defined as the solid volume of powder (not the apparent volume, but the true volume) divided by the total volume. Thus, the maximum solids loading, also known as the critical solids loading Φ_C, is the inverse of the maximum packing density.

$$\eta_M = \frac{\eta_L}{\left(1 - \dfrac{\Phi}{\Phi_C}\right)^N} \tag{14}$$

In many powder systems the exponent N is in the range from 2 to 3.5, and for most larger particle sizes (over 10 μm) $N = 2$ is most appropriate. For a fixed solids content Φ, the higher the maximum solids loading Φ_C the lower the mixture viscosity. In all suspensions, the viscosity declines as the solids loading decreases [3,18]. If the underlying powder used in the suspension has a high maximum packing density, then the mixture viscosity is lower for all solids loadings. Particle suspensions decrease viscosity in inverse proportion to the shear rate [16,19]. Further, when the particles are anisotropic, then a hysteresis problem arises. Sharp changes in strength and viscosity are observed with angular particles.

As a demonstration of viscosity behavior, Figure 13 plots the relative viscosity for mixtures of spheres at a constant 55% solids loading. Since bimodal powders have a maximum packing density (critical solids loading) that depends on composition, then according to Equation 14 the viscosity will shift. As is evident, the lowest viscosity occurs at the same composition expected to give the highest packing density. Further confirmation of the link between critical solids loading and maximum packing density captured in Figure 14. This plot superimposes results from powders with differing maximum packing densities, showing convergence to a normalized curve of relative viscosity versus relative solids loading (actual solids loading divided by the maximum packing density or critical solids loading).

The principle of using high solids loading particle size distributions to lower viscosity is used in many aspects of human endeavor. Examples are found in solder paste, inks, ice cream, chocolate, asphalt, and castable refractories. To obtain lower fat contents, lower caloric contents, higher strengths, or easier flow, these suspensions build on Andreasen particle size distributions and then tailor the ratio of powder to polymer, fluid, fat, oil, or water to adjust the strength and viscosity.

SUMMARY

Castable refractories build on the same particle packing and viscosity principles encountered in many fields ranging from "light chocolate" to "superstrong" asphalt. They all start with powder size distributions with high packing densities. A spherical particle helps packing, so rounded shapes are beneficial. Mixtures of differing particle sizes increase the packing density, especially when the particles are quite different in size. For every ratio of particle sizes, there is a composition that will optimize the packing density and lower the viscosity. Conceptually, optimal packing starts with the largest particles and forms a skeleton so that successively smaller particles fill the voids. Each of the remaining smaller voids is then filled by even smaller particles. Inhomogeneities in the structure degrade the packing density.

The optimal number of component particle sizes for high packing densities and low viscosities depends on the inherent packing density of the

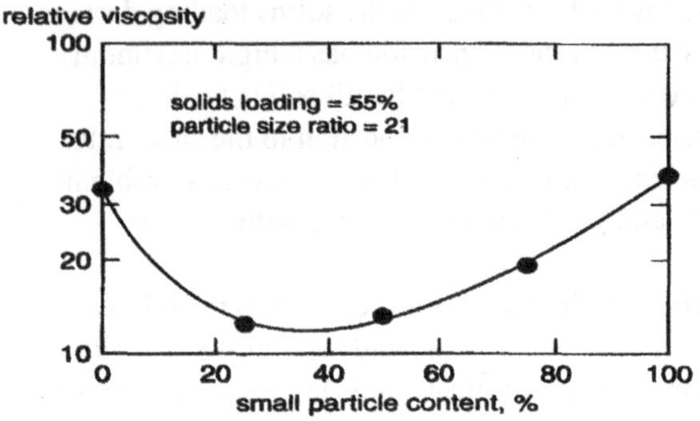

Figure 13. Relative viscosity versus amount of small particles for bimodal mixtures of spheres with a size ratio of 21 (large:small) tested at a fixed 55 vol.% solids loading, showing the low viscosity associated with the maximum packing density.

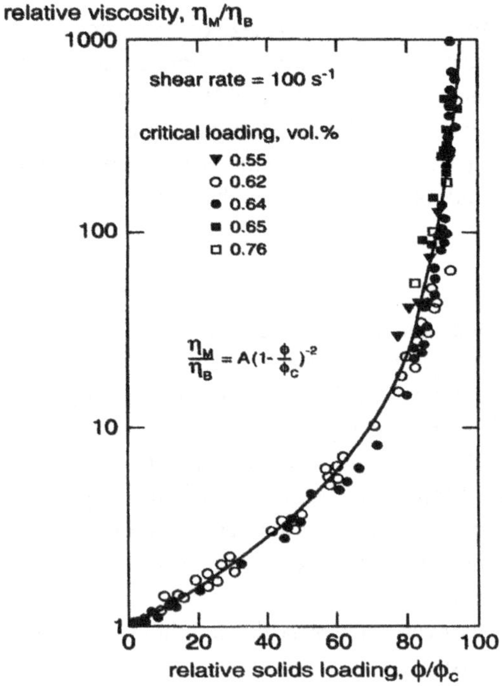

Figure 14. Relative viscosity versus the Mooney equation, showing how differences in the maximum packing density (critical solids loading Φ_C) can be rationalized by examining viscosity based on the solids loading Φ divided by the critical value.

powders and the particle size ratio involved in the mixture. Bimodal mixtures give higher packing densities than trimodal mixtures for smaller particle size ratios. In all cases, the optimal packing condition has a high proportion (by volume or weight) of the largest particles.

For a continuous particle size distribution the packing density increases as the width of the distribution increases. Additionally, the more equiaxed and spherical the particles, the higher the packing density. But there is a limitation to the maximum density attainable. For continuous particle size distributions, various theories predict maximum densities as high as 96%. An experimental density of 95% has been demonstrated. Near-optimal distributions have been empirically discovered in several industries. These have a characteristic wide range of small particles. In this regard, the early research by Fuller and Thompson [36] is most relevant. They found that an optimal particle size distribution for maximum density was obtained with 1) use of the largest mean particle size, 2) use of round particles as opposed to angular particles, 3) a high proportion by volume of large particles, and 4) a low proportion of intermediate sized particles.

Accordingly, it is possible to identify ideal powders for various ceramic forming processes [3,11,14,15]. Such a powder will have a tailored particle size distribution for a high packing density, yet low cost. It will consist of dense, discrete particles free from agglomeration with smooth surfaces and a nearly spherical shape. The size distribution should be wide, but skewed to follow the Andreasen equation. Generally the liquid lubrication film thickness should be sufficient to allow uninhibited particle flow, meaning the actual solids loading should slightly below the critical solids loading. By attending to selecting high packing density particle size distributions it is possible to form low viscosity suspensions with excellent flow.

ACKNOWLEDGMENTS

For the past several years the effort in particle packing and rheology and the application of these principles to powder injection molding has been supported by several companies, primary funding is from 90 firms that are members of the Center for Innovative Sintered Products at Penn State. Their support in these studies is greatly appreciated.

REFERENCES

1. W. A. Gray, *The Packing of Solid Particles*, Chapman and Hall, London, UK, 1968.
2. D. J. Cumberland and R. J. Crawford, *The Packing of Particles*, Elsevier Science, Amsterdam, Netherlands, 1987.
3. R. M. German, *Particle Packing Characteristics*, Metal Powder Industries

Federation, Princeton, NJ, 1989.

4. P. A. Smith and R. A. Haber, "Reformulation of an Aqueous Alumina Slip Based on Modification of Particle-Size Distribution and Particle Packing," *J. Am. Ceram. Soc.*, **75**, 290-294 (1992).
5. P. A. Smith and A. G. Haerle, "Particle Crowding Analysis of Slip Casting," *J. Am. Ceram. Soc.*, **78**, 809-812 (1995).
6. R. D. Sudduth, "A Generalized Model to Predict the Effects of Voids on Modulus in Ceramics," *J. Mater. Sci.*, **30**, 4451-4462 (1995).
7. A. Jagota and P. R. Dawson, "Micromechanical Modeling of Powder Compacts - I. Unit Problems for Sintering and Traction Induced Deformation," *Acta Metall.*, **36**, 2551-2561 (1988).
8. J. Narciso, A. Alonso, A. Pamies, C. Garcia-Cordovilla, and E. Louis, "Factors Affecting Pressure Infiltration of Packed SiC Particulates by Liquid Aluminum," *Metall. Mater. Trans.*, **26A**, 983-990 (1995).
9. R. M. German, "Sintering Densification for Powder Mixtures of Varying Distribution Widths," *Acta Metall. Mater.*, **40**, 2085-2089 (1992).
10. J. P. Fitzpatrick, R. B. Malt, and F. Spaepen, "Percolation Theory and the Conductivity of Random Close Packed Mixtures of Hard Spheres," *Physics Lett.*, **47A**, 207-208 (1974).
11. R. M. German, "The Importance of Particle Characteristics in Powder Injection Molding," pp. 109-160 in *Reviews in Particulate Materials,* vol. 1. Edited by A. Bose, R. M. German and A. Lawley. Metal Powder Industries Federation, Princeton, NJ, 1993,
12. O. A. Plumb, "Convective Melting of Packed Beds," *Inter. J. Heat Mass Trans.*, **37**, 829-836. (1994).
13. E. K. H. Li and P. D. Funkenbusch, "Hot Isostatic Pressing (HIP) of Powder Mixtures and Composites: Packing, Densification and Microstructural Effects," *Metall. Trans.*, **24A**, 1345-1354 (1993).
14. R. M. German, and A. Bose, *Injection Molding of Metals and Ceramics*, Metal Powder Industries Federation, Princeton, NJ, 1997.
15. J. E. Funk and D. R. Dinger, *Predictive Process Control of Crowded Particulate Suspensions*, Kluwer Academic., Norwell, MA, 1994.
16. T. A. Ring, *Fundamentals of Ceramic Powder Processing and Synthesis*, Academic Press, San Diego, CA, 1996.
17. R. M. German, *Powder Injection Molding*, Metal Powder Industries Federation, Princeton, NJ, 1990.
18. C. W. Macosko, *Rheology Principles, Measurements, and Applications*, VHC Publ., New York, NY, 1994.
19. R. J. Hunter, *Introduction to Modern Colloid Science*, Oxford University Press, Oxford, UK, 1993.

20. J. Warren, and R. M. German, "The Effect of Powder Characteristics on Binder Incorporation for Injection Molding Feedstock"; pp. 391-402 in *Modern Developments in Powder Metallurgy*, vol. 18. Edited by P. U. Gummeson and D. A. Gustafson. Metal Powder Industries Federation, Princeton, NJ, 1988.
21. J. R. Parrish, "Packing of Spheres," *Nature*, **190**, 800 (1961)
22. H. E. White and S. F. Walton, "Particle Packing and Particle Shape," *J. Am. Ceram. Soc.*, **20**, 155-166 (1937).
24. R. M. German, "Homogeneity Effects on Feedstock Viscosity in Powder Injection Molding," *J. Am. Ceram. Soc.*, **77**, 283-285 (1994).
24. R. K. McGeary, "Mechanical Packing of Spherical Particles," *J. Am. Ceram. Soc.*, **44**, 513-522 (1961).
25. M. Takahashi and S. Suzuki, "Numerical Analysis of Tapping Behavior of Ceramic Powders," *Ceram. Bull.*, **65**, 1587-1590 (1986).
26. G. D. Scott, "Packing of Equal Spheres," *Nature*, **188**, 908-909 (1960).
27. J. V. Milewski, "Packing Concepts in the Utilization of Filler and Reinforcement Combinations"; pp. 66-78 in *Handbook of Fillers and Reinforcements for Plastics*. Edited by H. S. Katz and J. V. Milewski. Van Nostrand Reinhold, New York, NY, 1978.
28. K. Ridgway and K. J. Tarbuck, "Particulate Mixture Bulk Densities," *Chem. Proc. Eng.*, **49** [2], 103-105 (1968).
29. C. C. Furnas, "Grading Aggregates I - Mathematical Relations for Beds of Broken Solids of Maximum Density," *Indust. Eng. Chem.*, **23**, 1052-1058 (1931).
30. R. F. Fedors and R. F. Landel, "An Empirical Method of Estimating the Void Fraction in Mixtures of Uniform Particles of Different Size," *Powder Tech.*, **23**, 225-231 (1979).
31. D. I. Lee, "Packing of Spheres and Its Effect on the Viscosity of Suspensions," *J. Paint Tech.*, **42**, 579-587 (1970).
32. A. R. Dexter and D. W. Tanner, "Packing Density of Ternary Mixtures of Spheres," *Nature Phys. Sci.*, **230**, 177-179 (1971).
33. H. T. Horsfield, "The Strength of Asphalt Mixtures," *J. Soc. Chem. Ind.*, **53**, 107T-115T (1934).
34. A. B. Yu and R. P. Zou, "Prediction of the Porosity of Particle Mixtures," *Kona*, **16**, 68-81 (1998).
35. H. Y. Sohn and C. Moreland, "The Effect of Particle Size Distribution on Packing Density," *Can. J. Chem. Eng.*, **46**, 162-167 (1968).
36. W. B. Fuller and S. E. Thompson, "The Laws of Proportioning Concrete," *Am. Soc. Civil Eng. Trans.*, **59**, 67-143 (1907).
37. A. H. M. Andreasen, "Ueber die Beziehung Zwischen Kornabstufung und

Zwischenraum in Produkten aus losen Kornern (mit einigen Experimenten)," *Kolloid Z.*, **50**, 217-228 (1930).
38. H. D. Lewis and A. Goldman, "Theorems for Calculation of Weight Ratios to Produce Maximum Packing Density of Powder Mixtures," *J. Am. Ceram. Soc.*, **49**, 323-327 (1966).
39. T. S. Shivashankar, *Study of Flow and Deformation Behavior of Concentrated Metal Powder Suspensions with Relevance to Noninteracting Systems*, Ph. D. Thesis, Department of Engineering Science and Mechanics, The Pennsylvania State University, University Park, PA, 1998.
40. M. Mooney, "The Viscosity of Concentrated Suspensions of Spherical Particles," *J. Coll. Sci.*, **3**, 162-195 (1951).

RHEOLOGY AND PLASTICITY FOR CERAMIC PROCESSING

William M. Carty
Whiteware Research Center
School of Ceramic Engineering and Materials Science
New York State College of Ceramics at Alfred University
Alfred, NY 14802

ABSTRACT

Rheology and plasticity in ceramic processing is dependent on five factors: particle-particle interactions (colloidal behavior), particle concentration (i.e., water content), particle size and distribution, particle morphology, and the rheology of the suspension medium. This presentation will focus on the factors that control rheology and plasticity through the use of traditional and advanced ceramic processing examples, ranging from extensive work on the dispersion character of clays to the processing of sub-micron alumina.

From the plasticity perspective, an overview of a direct shear plasticity measurement will be given. In addition, it will be demonstrated that the colloidal behavior concepts apply to plastic bodies in a similar manner to the behavior of particle suspensions.

I. INTRODUCTION

Rheology, the study of the deformation and flow of matter, provides a means of quantitatively describing fluid flows, and is well suited to the characterization of colloidal particle suspensions commonly used in industrial practices. The interaction between colloidal particles in suspension significantly effects the rheological properties. If the suspension concentration is low and the particle-particle interactions are weak (i.e., dispersed*), the suspension can exhibit Newtonian behavior; if the interaction is strong, the suspension may act like a solid. Suspension rheology is frequently used to provide an indication of the state of dispersion or, conversely, agglomeration.

* The terms dispersed and flocculated are preferred over the terms stable or unstable. The definition of stability varies with industry and usually denotes the relative tendency of particles to remain in suspension. For example, in a clay-based suspension, as used in the whiteware industry, a "stable" suspension is one in which the tendency for segregation by particle size is avoided by preparing the suspension at a dispersant concentration below that necessary to obtain the minimum viscosity. From an advanced ceramic processing perspective, the "stable" case is at the dispersant level that produces the minimum viscosity. Both suspensions would be considered stable, however.

In particle suspensions there is often a dramatic dependence of viscosity on shear rate. If the shear rate is extremely low, the suspension may act like a Newtonian fluid regardless of the nature of the interactions between particles. In most cases, however, particle suspensions exhibit shear thinning and, sometimes, thixotropic effects. If the particle concentration is high or the shear rate exceeds a certain critical value, shear thickening behavior is observed. There are five factors that control suspension rheology but, before addressing these factors in detail, it is necessary to define rheology and to understand the various types of rheological behaviors.

II. RHEOLOGICAL THEORY AND MEASUREMENT METHODS.

A. General definition of viscosity.

Viscosity, which is also referred to as internal friction or resistance to flow, is a proportionality constant which relates the shear stress (τ), to the shear rate (dV_X/dy or $\dot{\gamma}$), and is written:[1]

$$\tau = \eta\left(\frac{dV_x}{dy}\right) = \eta\dot{\gamma} \tag{1}$$

where $\dot{\gamma}$ ("gamma dot") is used to indicate shear strain rate, or shear rate (γ would denote shear strain). Fluids which exhibit Newtonian behavior, such as water, alcohols, some oils, gases, molten glasses, etc., are relatively easy to characterize provided the geometry of the flow situation can be modeled mathematically. Unfortunately, most fluids of interest in the area of ceramic and polymer processing are non-Newtonian, making such analysis considerably more difficult and in some cases impossible.

B. Generalized Newtonian fluid behavior and shear dependent viscosity.

The simplest category for non-Newtonian fluids is that of generalized Newtonian or power-law fluid in which the apparent viscosity varies with shear rate. The power-law expression is[1,2]

$$\eta = m\dot{\gamma}^{n-1} \quad \text{or} \quad \log\eta = \log m + (n-1)\log\dot{\gamma} \tag{2}$$

in which m is the "intercept" (the viscosity value at $\dot{\gamma}$ = 1.0/s), and n-1 is the slope of the log η versus log $\dot{\gamma}$ curve. Most non-Newtonian fluids exhibit shear thinning behavior, i.e., n<1, in which the viscosity decreases with increasing shear rate, sometimes as much as several orders of magnitude. Less common is the case of shear thickening in which the viscosity increases with increasing shear rate, i.e., n>1 (although this is an oversimplification of dilatency, which may exhibit discontinuous viscosity at a critical shear rate[3,4]). The primary problem with the power-law model is that it is unable to accurately predict the shear rate independent regions, η_0 and η_∞, frequently observed in the viscosity versus shear rate curves for particle suspensions (and sometimes in polymeric solutions).

Shear thinning behavior occurs because the particles form an ordered structure in the presence of a shear field. As the shear rate increases (and the shear field becomes greater) the structure breaks up into smaller and smaller elements, and the shear stress required to maintain a specific shear rate decreases. In essence, the apparent viscosity decreases with increasing shear rate due to a reduction in the resistance to flow as the particle packets become more ordered and the shear bands get closer together.[3,5]

Non-Newtonian fluids rarely (if ever) exhibit shear thinning effects throughout extremely large ranges of shear rate. It is commonly observed that as the shear rate approaches zero, or conversely as the shear rate approaches infinity, the apparent viscosity becomes independent of shear rate. To characterize this behavior, the Carreau-Yasuda model was developed in which the viscosities as the shear rate approached zero (η_0) and infinity (η_∞) were identified:[1,2]

$$\frac{\eta - \eta_\infty}{\eta_0 - \eta_\infty} = \left[1 + (\lambda'\dot{\gamma})^a\right]^{\frac{n-1}{a}} \tag{3}$$

The other parameters of the model are: **n**, the power-law exponent; λ', a time constant; and **a**, a dimensionless parameter which describes the transition from shear rate independent to shear rate dependent viscosity. Figure 1 illustrates both the simple power-law model over the shear rate range of 10^0 to 10^2 s^{-1}, and overall the Carreau-Yasuda model.

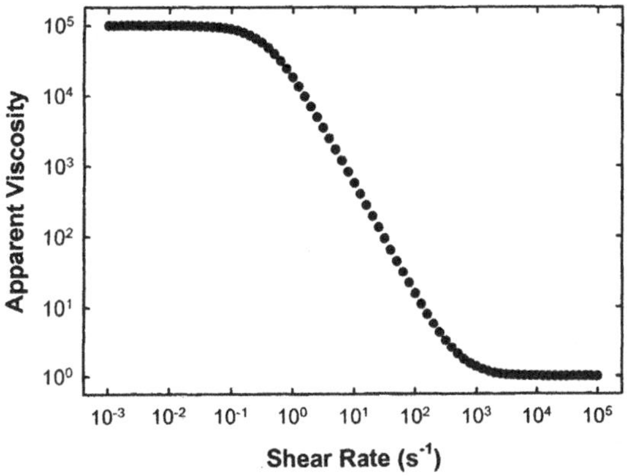

Figure 1. A schematic example of the dependence of apparent viscosity on shear rate using the Carreau-Yasuda model with parameters $\eta_0 = 10^5$, $\eta_\infty = 10^0$, a = 1.5, λ' = 2.5, and n = -0.6. Note that power-law behavior is illustrated over the shear rate range of 10^0-10^2 s^{-1}.

C. Rheological measurement.

Rotational rheometers are the most commonly used, and there are two types: 1) the stress-controlled rheometer (the most common); and 2) the strain-controlled rheometer (usually reserved for research measurements). In stress-controlled rheometry, the measurement head is also the driving head – the stress required to move the measurement head is measured. In the strain-controlled instrument, torque generation is maintained separate from the torque measurement transducer. The commonly used "Brookfield" viscometer is essentially a stress controlled instrument in which the torque required to maintain a specific rotation rate is measured. The shear rate can only be calculated when the test geometry is well known. In the case of Brookfield viscometers, the concentric cylinder geometry is assumed with the gap between the inner and outer cylinder assumed to be infinite and, therefore, the RPM value approximates shear rate.

With the exception of Newtonian fluids, true viscosity is almost never measured due to the shear rate dependence of the fluid. Also, a viscometer measures a discrete shear stress at a given shear rate, without knowledge of the previously measured viscosity (at a different shear rate). Since viscosity is defined as the proportionality constant between shear stress and shear rate, each data pair produces a viscosity value that is the slope of a line assumed to pass through the origin. Therefore, all viscometers measure apparent viscosity, as illustrated in Figure 2.

D. General types of rheological behavior.

As illustrated in Figure 3, there are four basic types of steady-state rheological flows under shearing conditions. A graphical example of the time independent shear rheological behaviors is presented in. These types of behavior are directly dictated by the five factors that control suspension rheology; therefore, it is necessary to consider the particle contributions to the rheological behavior.

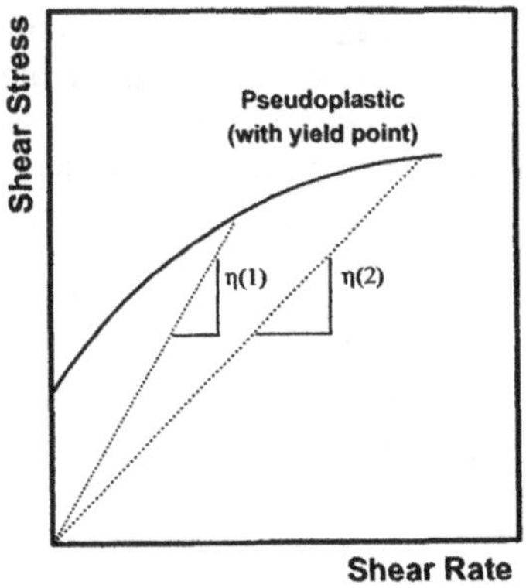

Figure 2. Schematic illustration of different apparent viscosities due to shear rate dependence of a fluid. The apparent viscosity is the slope through the origin for a shear stress/shear rate data pair. ($\eta_{(1)} > \eta_{(2)}$: shear thinning)

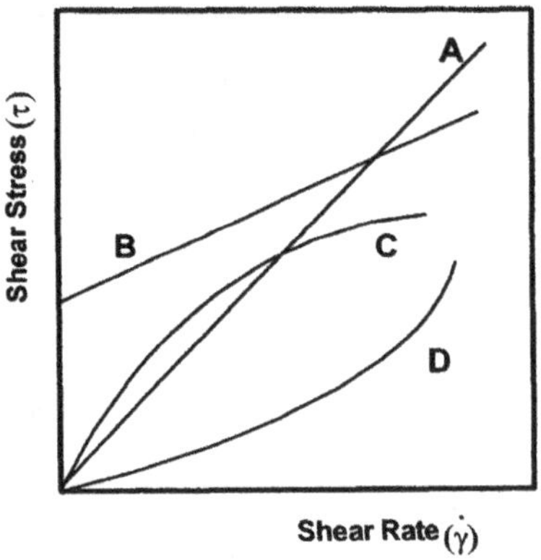

Figure 3. Four major types of time independent rheological behavior exhibited by particle suspensions: A) Newtonian; B) Bingham (ideal); C) pseudoplastic; D) dilatent. The axes are linear.

E. Rheology of Colloidal Particle Suspensions.

The previous discussion addressed time independent viscosity of fluids. When working with particle suspensions, particularly those containing nonspherical particles, such as clay platelets, it is common to observe a measurable time dependence. The most common phenomena is thixotropy, where the viscosity decreases with time at a constant shear rate. Conversely, when the shear is removed, the viscosity will slowly increase with time, often requiring hours or days to reach the viscosity exhibited before the fluid was sheared. If these suspensions are allowed to remain at rest for a sufficient period of time, the particles form a three dimensional network. When a shear stress is applied, the structure begins to break down, and the particles will begin to align with the shear flow, reducing the resistance to flow and hence lowering the viscosity.[3,6] As the shearing flow continues, more particles become aligned, further reducing flow resistance and therefore viscosity. When the shearing action is removed, the particles immediately begin slowly to rearrange due to Brownian motion and eventually develop a similar particle network to that present prior to shearing.

III. THE FIVE FACTORS CONTROLLING SUSPENSION RHEOLOGY.

Colloidal suspension rheology is controlled by five factors: 1) particle-particle interactions; 2) particle concentration; 3) particle size and distribution; 4) particle morphology; and 5) rheology of the suspension medium. Approaching the suspension rheology field from the perspective of these five controlling factors allows the contribution of each to be evaluated categorically.

A. Particle-particle interactions.

Particle-particle interactions are by far the most important factor, providing the greatest control over the suspension rheology. As illustrated in Figure 4 for the electrostatic and electrosteric stabilization of colloidal alumina particles, the viscosity of a suspension can be reduced by nearly a factor of 10^4 simply by moving from a pH level where the suspension is flocculated (≈7) to one in which the suspension is dispersed (≈4). In this case, the difference between a dispersed and a flocculated suspension occurs over a very narrow pH range.

The suspension structure is also heavily dependent on the degree of interaction between the particles. In a dispersed suspension the particles essentially act independently, while in a flocculated suspension the particles form a three-dimensional network. In the flocculated case, the particle structure creates voids which trap water causing the system to behave as if at a much higher particle concentration. It should also be noted that the degree of particle-particle interaction can be specifically controlled and that dispersed and flocculated behaviors are really two ends of a broad spectrum of possible behaviors.

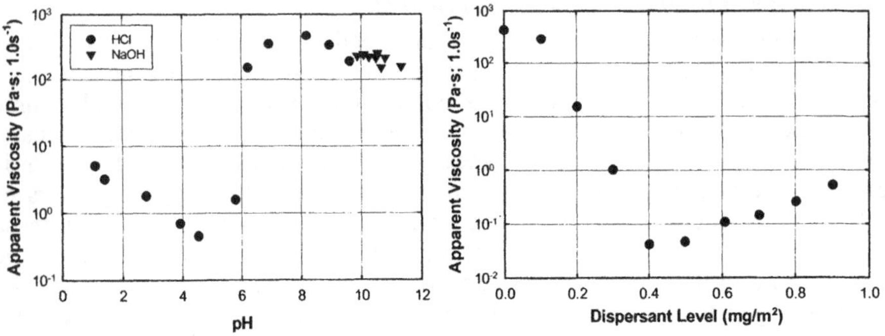

Figure 4. The viscosity of aqueous alumina (A-16 S.G., Alcoa Chemicals, 0.34 μm mean particle size, 9.7 m^2/g, 35 v/o) suspensions showing electrostatic stabilization (left) and electrosteric stabilization (right) by the adjustment of pH (HCl and NaOH) and the addition of NH_4-PMAA (Darvan C, R.T. Vanderbilt), respectively. The apparent viscosity values are extrapolated from the viscosity versus RPM data to a RPM value of 1.0.[7]

B. Particle concentration.

At the same level of dispersion, increasing the particle concentration increases the viscosity. This increase is attributed to increased particle-particle collisions and a decrease in the excess fluid necessary for the particles in move past one another in shear flow, as illustrated in Figure 5.

C. Particle size and particle size distribution.

While it could be argued that the third rheological factor of particle size and particle size distribution could be separated into two categories, for this discussion, they will be addressed together because there is no clear means to distinguish the effects of particle size from particle size distribution. Figure 6 shows the decrease in viscosity obtained by blending two particle sizes, keeping the overall concentration of particles constant. As the particle size decreases, the specific surface area increases and the number of particles in suspension increases dramatically. There are two contributions to the increase in viscosity as particle size decreases, 1) the surface area increase reduces the amount of free water, and 2) the increased collision frequency caused by the increase in particle number density (at an identical solids loading in the suspension).

The particle size distribution also plays an important, complementary role. A broader particle size distribution leads to better packing so that the same solids loading, the particles pack more efficiently, providing the impression of a lower overall solids loading (i.e., small particles fill voids between larger particles). In general, it is still unclear exactly how to predict the effects of particle size distribution on the rheology of particle suspensions, although the impact of the distribution can be readily demonstrated as shown in Figure 6.

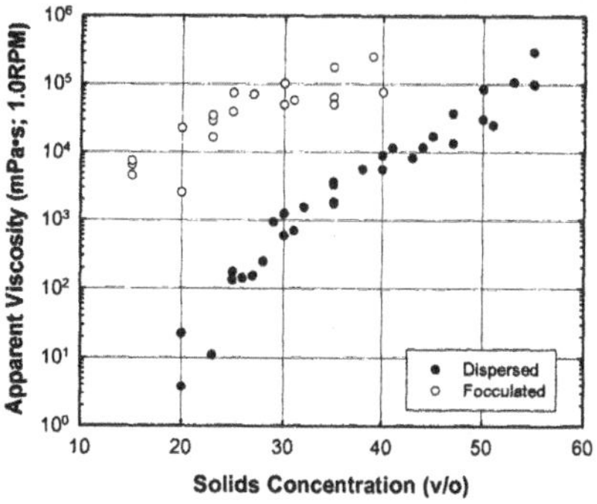

Figure 5. The effect of particle concentration on viscosity of aqueous alumina (A-16 S.G.) suspensions. Two cases are illustrated — one dispersed (using PMAA at a level of 0.5 mg/m^2), and one flocculated (no dispersant, no pH adjustment). The viscosity data becomes unreliable below 100 mPa·s due to measurement limitations.[7] (1 mPa·s = 1 centipoise)

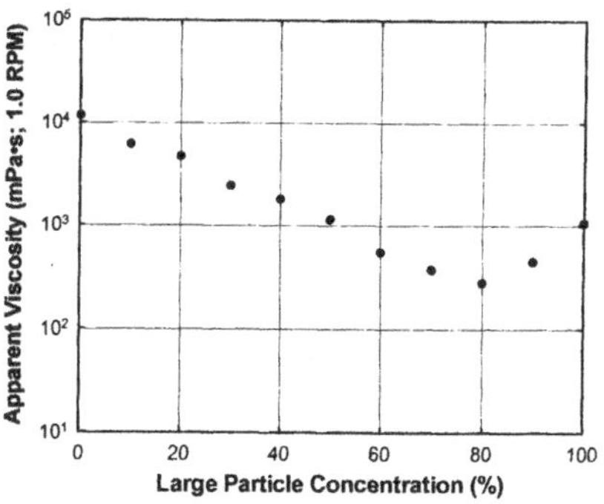

Figure 6. The impact of particle size and particle size distribution on the viscosity of a dispersed suspension. The x-axis is the fraction of A-10 (Alcoa Chemicals, 5.0 μm mean particle size, 1.5 m^2/g), the balance is A-16 S.G. The total concentration is kept constant at 45 v/o, and the system is dispersed with 0.5 mg/m^2 NH_4-PMAA.[7]

D. Particle morphology.

Spherical particles are "best behaved" in terms of suspension rheology and are the most commonly addressed from both experimental and theoretical perspectives. As the particles deviate from the spherical shape, the suspension viscosity increases, because the volume necessary for a particle to remain in suspension increases, primarily due to rotational Brownian motion. Essentially, higher suspension concentrations are possible with spherical particles compared to the concentrations possible with other particle morphologies. Figure 7 illustrates the differences between alumina (A-16 S.G.) and clay (EPK, Zemex Minerals, Edgar, FL) at the same concentration and roughly the same mean particle size. (The alumina actually has a similar mean particle size but a lower specific surface area than the clay.)

Because a sphere represents the lowest surface to volume ratio of any shape, a non-spherical particle of similar size would have a larger specific surface area, thus requiring more water to create the adsorbed water layer. The reduction of excess water is considered a minimal contribution, however, and if a spherical particle size is assumed for the sake of calculation, the surface area effected is somewhat overestimated, leading to the potential for erroneous evaluation of the data. The random orientation of the particles, and the consequent difficulty the particles have moving past each other, is the primary cause of the viscosity increase. Thixotropy is a direct result of particle shape. As the system is sheared, the particles align in the shear field, causing a reduction in the drag which translates to a reduction in viscosity. At higher shear rates, the particles align more rapidly so that the thixotropic effect may be less noticeable; at lower shear rates, this effect can be dramatic. Once the shear field is removed, the particles revert to a random orientation due to Brownian motion.

E. Rheology of the suspension medium.

In aqueous systems, the contribution of the suspension medium is generally small. If excess polymer is present, however, the contribution can be substantial. This concept is illustrated in Figure 8. An important example wherein the rheology of the suspension medium plays an important role is in injection molding systems. Usually, an injection molding system requires a suspension medium which exhibits an acute temperature sensitivity, where the viscosity of the system may change several orders of magnitude with a few degrees decrease in temperature.

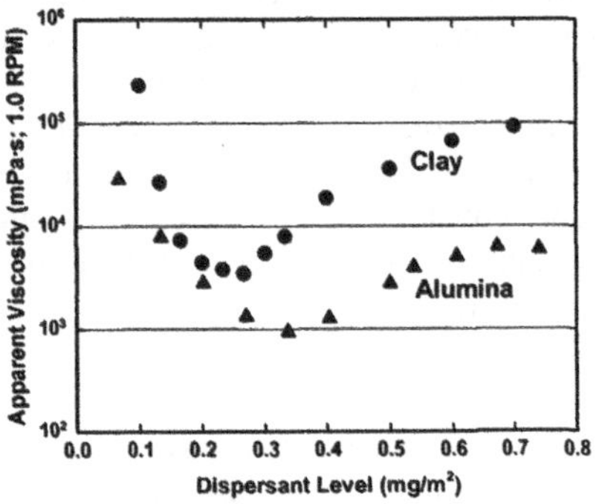

Figure 7. The contribution of morphology on the rheology of Na-PAA-dispersed (pH ~ 9.0)clay and alumina suspensions at 35 v/o. Both have a mean particle size of ~0.5μm; the alumina particles are roughly spherical and the clay particles possess a plate-like morphology.[8,9]

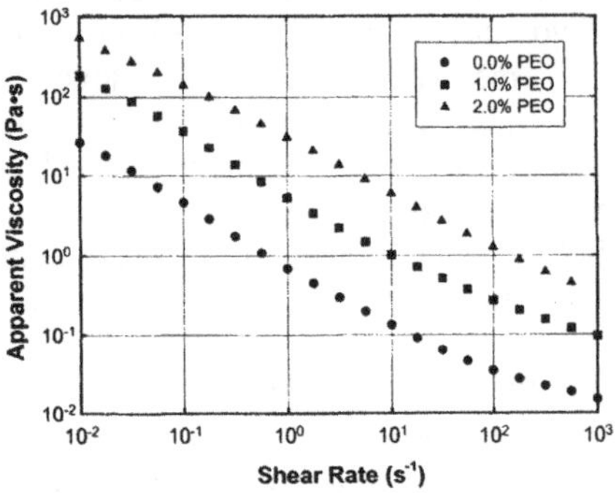

Figure 8. The increase in viscosity as a result of high molecular weight ($4x10^6$) polymer additions to a 35 v/o alumina suspension. The polymer is poly (ethylene oxide) at levels of 0.0%, 1.0%, and 2.0%. At the 1.5% level, the viscosity of the suspension is increased by a factor of ~8.[10]

IV. COLLOIDAL BEHAVIOR OF CLAY AND ALUMINA.

There is a great amount of confusion, both in industry and academia, regarding the colloidal nature of kaolinite particles in aqueous suspension stemming largely from a generally incorrect "picture" of the charge distribution on a kaolinite particle. The commonly held view that kaolinite possesses negatively charged basal-plane surfaces and positively charged edges is inconsistent with the mineralogy of kaolinite particles – a 1:1 sheet silicate composed of a $[Si_2O_5]^{2-}$

tetrahedral layer and an $[Al_2(OH)_4]^{2+}$ octahedral layer. Based on the colloidal behavior of silica (isoelectric point of 2.3-3.5) and hydrated alumina (isoelectric point 8.5-9.5),[11] in the pH range of 3.5 to 8.5 silica surfaces must be negatively charged and alumina surfaces positively charged. 2:1 sheet silicates (i.e., two $[Si_2O_5]^{2-}$ surfaces sandwiching an octahedral layer), such as talc, pyrophyllite, mica, bentonite, etc., do possess the negatively charged basal-plane surfaces and presumably neutral (or possibly slightly positively-charged) edges, consistent with the mineralogy and the charge character of silicate surfaces in water. Details on the nature of kaolinite is available elsewhere.[12,13]

Another problem obstructing the understanding of kaolin* suspension behavior is the myth regarding the role of sodium. It is incorrect, although widely accepted, that Na^+ functions as a dispersant and is responsible for the reduction in viscosity of clay suspensions. This misconception may stem from the conclusions of Johnson and Norton:[14] "...*(a) the charge on the kaolinite particles controls the degree of deflocculation and is governed by the type of cation and (b) the stability of system is controlled by the anion of the medium and is governed by the type of anion preferentially adsorbed*" (emphasis existing). The intent of these two conclusions is not entirely clear, and while mostly correct, their study focused on salts of Na and Ca, the common interpretation is that Na^+ *causes* stability. The large net negative charge on kaolinite particles means that Na^+ is a counterion, and as such, serves to compress the ionic double layer surrounding a colloidal particle, promoting coagulation of the particles.

This research demonstrates that the colloidal nature of kaolinitic clays is consistent with a particle composed of a silica-like and alumina-like surface. The adsorption of dispersants is consistent with that observed for alumina for a large variety of commercial dispersants, and when corrected for specific surface area, supports the dual nature argument. In addition, the impact of ionic strength, specifically altered by the addition of chlorides and sulfates of Na^+, Ca^{2+}, and Mg^{2+}, clearly indicates that the behavior of Na^+ is consistent with the divalent cations, demonstrating that sodium cannot be viewed as a dispersant for clay-based suspensions.

To demonstrate the dispersion behavior of clay and alumina, three separate sets of experiments were performed: 1) dispersion behavior of kaolin suspensions (composed solely of EPK); 2) dispersion of a typical porcelain batch composition; and 3) dispersion of sub-micron alumina suspensions. The raw materials and batch compositions are listed in Table I; the dispersants are listed in Table II. Details on the experimental procedures are presented elsewhere.[8,9,15,16] All dispersant additions were corrected for the specific surface area of the powders to correlate with surface coverage. The results indicated that the route taken to create the suspensions did not measurably impact the experimental results.[16]

---

* Kaolinite is a mineral. Kaolin refers to a rock, or in this case, a clay powder, composed of at least 50% kaolinite.

Although previous experiments have shown little change in rheological behavior with time, the clay samples were stored after mixing for at least 14 days prior to rheology measurement. The alumina suspensions were initially prepared at the desired solids concentration and not aged prior to testing.

Table I. Raw materials and the porcelain batch composition (dry weight basis, d.w.b.) with specific surface area and solids loading information.

Raw Materials	Source of Materials	Wt. % (d.w.b.)	SSA* (m^2/g)	Initial (vol%)	Final (vol%)
Kaolin	EPK; Zemex Minerals (Feldspar Corp.), Edgar, FL	29.0	26.9	45	40
Ball Clay	Todd Light; Kentucky-Tennessee Clay Co., Mayfield, KY	7.0	25.9		
Nepheline Syenite	A200; Unimin Canada Ltd., Nephton/Blue Mountain, Ontario	22.0	1.1		
Quartz	Oglebay Norton Industrial Sands, Inc., Glenford, OH (325 mesh)	29.5	0.9		
Alumina	A10; Alcoa, Pittsburgh, PA	12.5	1.0		
Alumina Suspensions					
A-16 S.G.	Alcoa, Pittsburgh, PA	100.0	8.9		35
APA-0.5	Ceralox Corp., Tucson, AZ	100.0	7.8		35

* N_2-BET; Gemini III 2375 Surface Area Analyzer, Micromeritics, Norcross, Georgia.

Table II: The dispersants evaluated in the suspensions are listed along with the corresponding chemical formula and the common abbreviations.

Dispersant	Chemical Formula	Abbreviation
Na-Poly Acrylic Acid	$H\text{-}(NaC_3O_2H_3)_n\text{-}H$	Na-PAA
Na-Poly Methacrylic Acid	$H\text{-}(NaC_4O_2H_5)_n\text{-}H$	Na-PMAA
Na-Hexametaphosphate	$(NaPO_3)_6$	SHMP
Na-Silicate (x=0.22)**	$xNa_2O\bullet(1\text{-}x)SiO_2$	Na-silicate
Na-Ash	$Na_2CO_3\bullet 10H_2O$	Na-ash
Na-Silicate:Na-Ash Blend (1:1)	*as noted above*	1:1

** Ratio via ICP analysis of the Na-Silicate solution, (Acme Analytical Laboratories Ltd., Vancouver, British Columbia, Canada).

A. Alumina Suspension Behavior – establishing a baseline for comparison.

If the basal-plane surfaces of a kaolinite particle are silica-like and alumina-like, it should only be necessary to create a negatively charged alumina surface and thus impart colloidal stability. This can be most easily accomplished through the specific adsorption of an anionic species, such as the polyelectrolytes PAA and PMAA, or the polyanions of SiO_3^{2-} or phosphate. The affinity of PMAA for alumina has been clearly demonstrated by Cesarano, *et al*,[17] but to establish a benchmark, the behavior of alumina was evaluated under similar conditions to those used to evaluate clay suspensions. Examples of the alumina dispersion results for the five primary dispersants (Na-PAA, Na-PMAA, Na-Silicate, SHMP, and Na-ash) are presented in Figure 9 for APA; A-16 S.G. results were similar.[9]

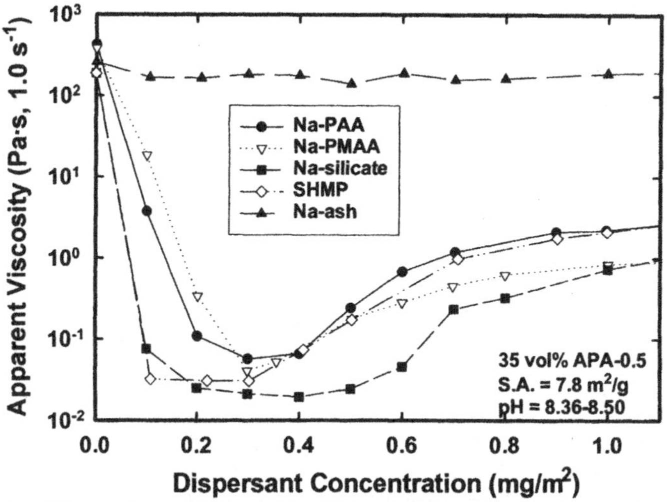

Figure 9. Dispersion of 35 vol% APA alumina suspensions using the five Na-based dispersants. Note the amount of dispersant necessary to reach the minimum in the viscosity curve. It is also clear that Na-ash is an ineffective dispersant for alumina, but that Na-silicate and SHMP are slightly more effective than Na-PAA or Na-PMAA.[9]

The amount of dispersant is corrected for the surface area of the suspended particles. To properly evaluate the dispersion effectiveness with comparison to clays, these results should be compared at equivalent pH levels because the dissociation behavior of PAA and PMAA is highly dependent on pH. Na-silicate and SHMP require almost exactly one-half as much dispersant to disperse clay as to disperse alumina, as discussed in the next section. Previous work has demonstrated that PAA and PMAA have a limited affinity for silica surfaces, such as the silica-like basal plane of kaolinite particles.[18] Therefore it should only be necessary to coat the alumina-like side of a kaolinite particle, which accounts for approximately one half of the total specific surface area of the suspension, thus requiring roughly one-half of the amount necessary to disperse colloidal alumina.

B. Clay Suspension Behavior.

The effectiveness of the dispersants was similar for the clay and batch suspensions,[8,16] and, as shown in Figure 10, Na-PAA, Na-PMAA, Na-silicate, and SHMP are all efficient dispersants. Even though the batch is composed of only 36% (d.w.b.) clay, the clay accounts for 93.7% of the available surface area and therefore dominates the batch dispersion behavior. Industry commonly uses a blend of Na-silicate and Na-ash. It is clearly evident that Na-ash is a poor dispersant, and that when mixed with Na-silicate, the viscosity of the suspension follows the behavior of Na-ash, then behaves at low concentrations as an intermediate, and finally as Na-silicate.

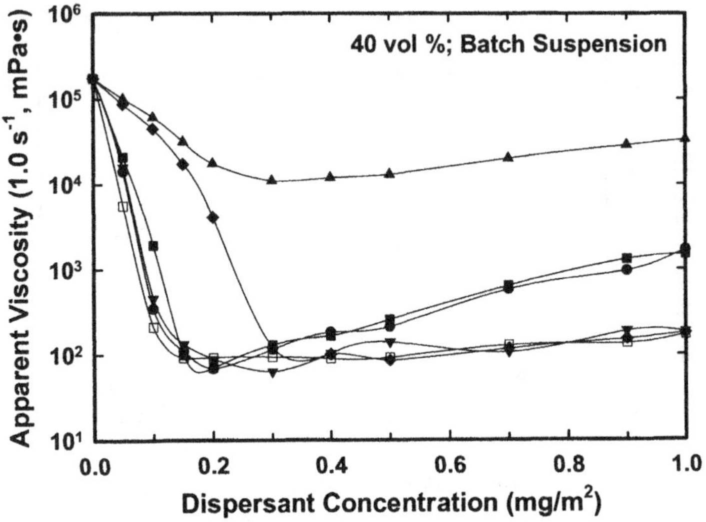

Figure 10. The apparent viscosity in the batch suspension at 1.0 s^{-1} as a function of dispersant concentration is shown for all six dispersants: Na-PAA (●), Na-PMAA (■), Na-silicate (▼), SHMP (□), Na-ash (▲), and 1:1 Na-Ash:Na-Silicate (♦). Similar results were measured in the kaolin suspensions.[8,16]

C. Comparison of Clay and Alumina Dispersion: The Effect of pH.

If it was only necessary to coat the alumina surface of kaolinite particles to cause dispersion, it should only require roughly one-half as much dispersant when corrected for the surface area of the powders (as in Figure 9 and 10). It seems clear from Figures 9 and 10 that the amount needed to disperse clay is roughly similar to that necessary for alumina. This is due to suspension pH. In the case of PAA and PMAA, the polyelectrolytes become 95% dissociated at pH = 8.5.[17,18] Clay suspensions tend to maintain a lower pH, around pH = 6.0, presumably due to the relatively high concentration of silica-like surfaces.[8,16] PAA and PMAA are 50% dissociated at pH = 6, meaning that it requires twice as much dispersant at pH = 6 than is necessary at pH = 9.0. When the data are corrected for pH (or when the pH of the suspension is controlled during the experiment), then the dispersant amounts can be evaluated on an equivalent basis, as is illustrated in Figure 7 for pH = 9.0. Note that in Figure 7 it requires 58% as much dispersant for clay as for alumina to reach the minimum in the viscosity curve.[13]

D. Effect of Ionic Strength on Suspension Rheology

Increasing ionic strength compresses the electrical double-layer surrounding a colloidal particle in suspension. It is generally accepted that the counter-ion, e.g., the ion oppositely charged to the net particle surface charge, is the more important species. The co-ion can usually be ignored, except in cases in which the co-ion has a specific affinity for the particle surface. The ζ-potential data demonstrates that the kaolin particles are strongly negatively charged, and that charge increases until surface coverage is achieved. These experiments had two goals: 1) to demonstrate that sodium is not a dispersant; and 2) to determine the critical coagulation concentration (CCC) (the ionic concentration necessary to cause a noticeable change in suspension rheology). Six salts were evaluated (NaCl, Na_2SO_4, $CaCl_2$, $CaSO_4$, $MgCl_2$, $MgSO_4$) over a broad concentration range. (The salts were added in solution form; experimental details are provided elsewhere.[15])

As illustrated in Figure 11, the effect of ionic strength on viscosity is dramatic. Na-PAA was added at three levels to impart varying degrees of stability. Under the most dispersed condition (0.05 mg/m^2), the suspension viscosity increases by a factor of nearly 1000 as the ionic concentration exceeds the CCC. Even in samples to which no dispersant has been added, increasing ionic concentration causes an increase in suspension viscosity. In the case of the Na_2SO_4 additions, it is clear that even at small sodium additions, stability is not improved.

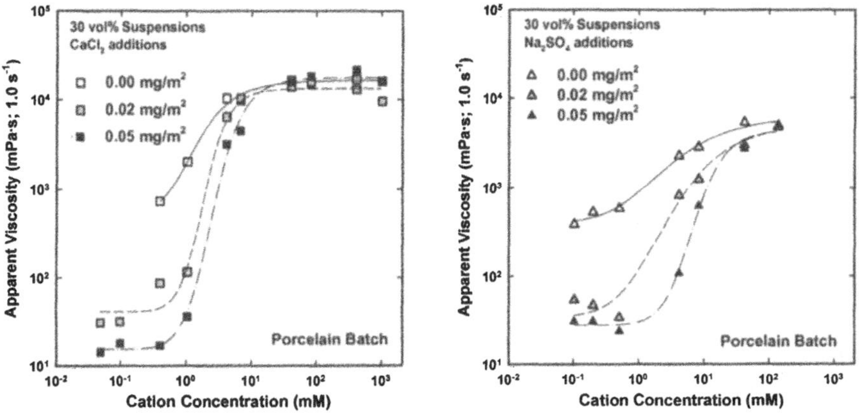

Figure 11. Effect of $CaCl_2$ (left) and Na_2SO_4 (right) salt additions on the viscosity of the porcelain batch suspensions. As the dispersant level increases, the amount of salt necessary to induce coagulation increases (albeit slightly), consistent with colloidal theory. In the suspensions containing $CaCl_2$, a plateau is reached; greater salt additions do not further increase viscosity. In the Na_2SO_4 samples, the plateau has not yet been achieved. Note that even small amount of sodium does not impart suspension stability, further supporting the idea that Na^+ is not responsible for dispersion.[15]

To demonstrate that the CCC is independent of the anionic species involved, $CaCl_2$ and $CaSO_4$ additions were evaluated, as shown in Figure 12. The limited solubility of $CaSO_4$ prevents further addition without adding solid. Figure 13 indicates that the CCC for Ca^{2+} and Mg^{2+} are identical in samples in which chloride salts have been added, and are approximately eight times lower than the CCC for NaCl. It is also evident that above a certain ionic strength level, the viscosity reaches a stable plateau, that appears to be at an intrinsic limit for this suspension. (It has been suggested that this may be an instrumental limitation, however, the data in Figure 10 clearly shows viscosity values above 10^5 mPa·s.)

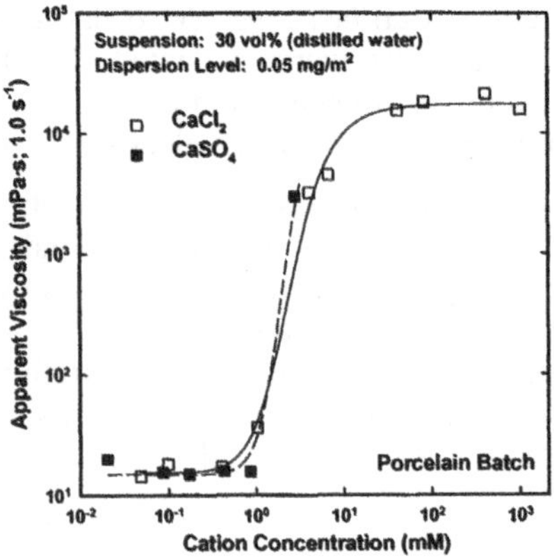

Figure 12. The concentration of Ca^{2+} required to reach the CCC is the same for both $CaCl_2$ and $CaSO_4$ at a dispersant level of 0.05 mg/m^2. Both curves appear to be approaching the same plateau, but $CaSO_4$ concentration level is limited by the solubility of $CaSO_4$.[15]

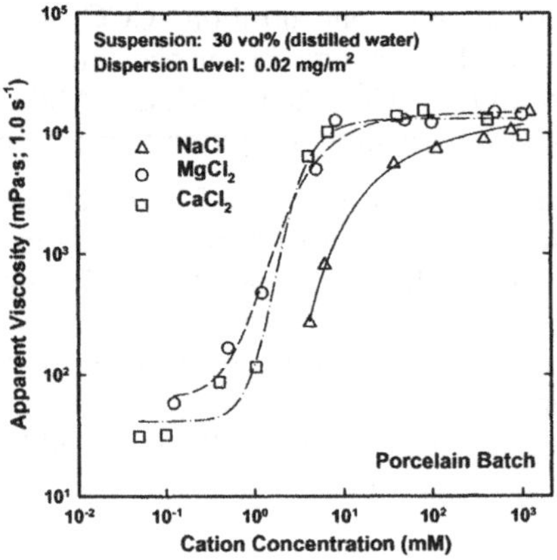

Figure 13. Additions of $CaCl_2$ and $MgCl_2$ produce identical CCC values at levels approximately eight times lower than that required for NaCl. All suspensions were prepared at a dispersant level of 0.02 mg/m^2, but reached a similar viscosity plateau above the CCC.[15]

V. PLASTICITY MEASUREMENT AND PLASTIC BODY BEHAVIOR

The High Pressure Annular Shear Cell (HPASC) was proposed as a plasticity characterization instrument based on the direct shear testing concepts derived from soil mechanics. The HPASC was constructed in 1993 and allows direct measurement of shear forces within a plastic body under an externally applied pressure, mimicking the environment that a plastic body would experience in a forming process such as jiggering or extrusion.[19-21] Test results over the past several years indicate that divalent cation concentration can profoundly change the plasticity of a whiteware body, and as such, may be the root cause of aging in plastic bodies. In addition, there is evidence that ionic strength may have important implications for drying ceramic objects.

A. The High Pressure Annular Shear Cell Concept.

The underlying concept of the HPASC is that plastic bodies behave as Bingham fluids, that is, Newtonian-like with a yield stress. As the pressure increases, the yield stress increases linearly with the applied normal pressure, as shown in Figure 14. Figure 15 shows the proposed dependence of the slope (pressure dependence) and intercept (cohesion) of the yield stress versus applied pressure for three hypothetical samples. Note that in this example, ***A*** and ***B*** have the same cohesion and ***B*** and ***C*** have the same pressure dependence.

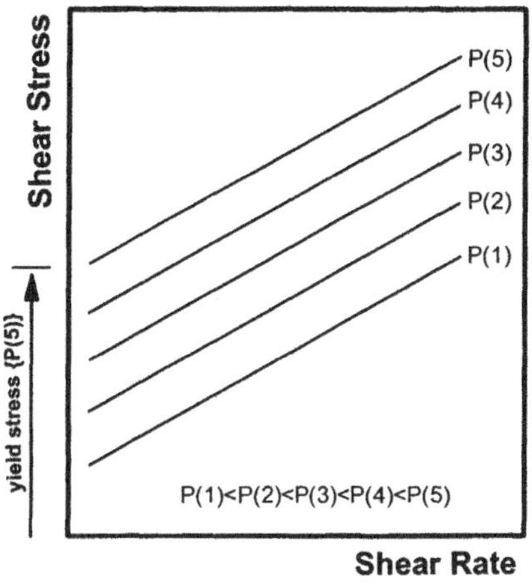

Figure 14. Schematic illustration of the effect of pressure on a Bingham fluid. Note that yield stress scales with increasing pressure.

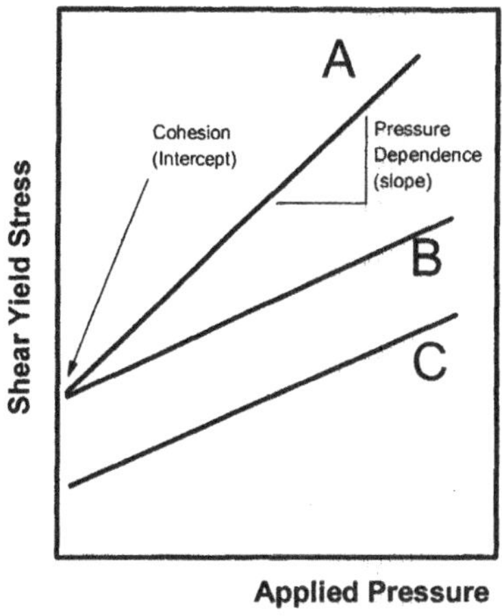

Figure 15. The definition of cohesion and the pressure dependence (the intercept and the slope, respectively) for the examples, A, B, and C.

The HPASC is represented schematically in Figure 16, and consists of a donut-shaped test cell that holds 75-100 grams of sample. Pressure is applied via a dead-weight loading system with a hydraulic amplifier at the top of the test fixture. The sample is allowed to compact during testing. Rotation rate and applied pressure can be altered independently, allowing a broad range of testing conditions. Figure 17 is a photograph of the test cell, showing the torque transfer and hydraulic loading setup. The torque required to prevent the top from rotating is measured and used to calculate the shear stress within the sample. The surfaces of the test fixture have been grooved to prevent sample slippage at the platens.

In the most general sense, three simple categories of behavior can be imagined: dry, plastic, and wet. The shear yield stress versus pressure behavior of these materials is illustrated in Figure 18. The yield stress of the dry sample would be highly dependent on the applied normal pressure in contrast to the wet sample, in which the shear stress is nearly independent of the applied pressure. The plastic sample behavior ideally lies somewhere between wet and dry. Figure 19 illustrates the behavior of a series of clay samples with differing water contents and an example of a typical porcelain body formulation.

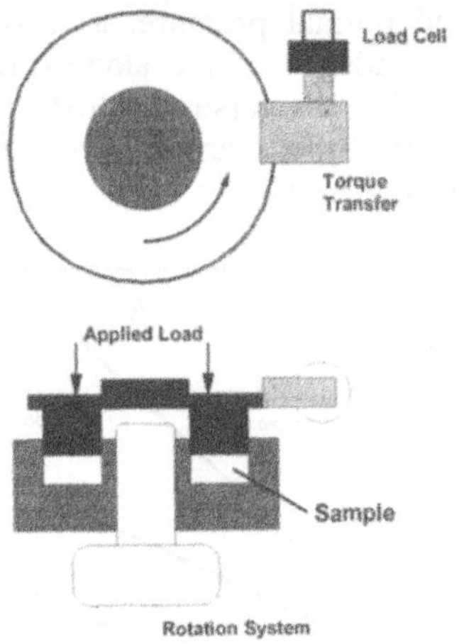

Figure 16. HPASC schematic. Pressure is applied to the top of the test fixture and the sample can compact during the test.

Figure 17. A photograph of the HPASC testing fixture with the testing chamber as illustrated schematically in Figure 16.

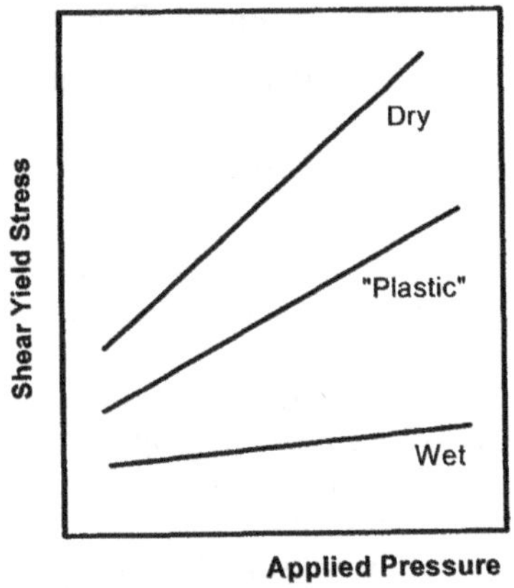

Figure 18. Schematic illustration of the change in slope and intercept of yield stress versus applied pressure for wet, plastic, and dry samples.

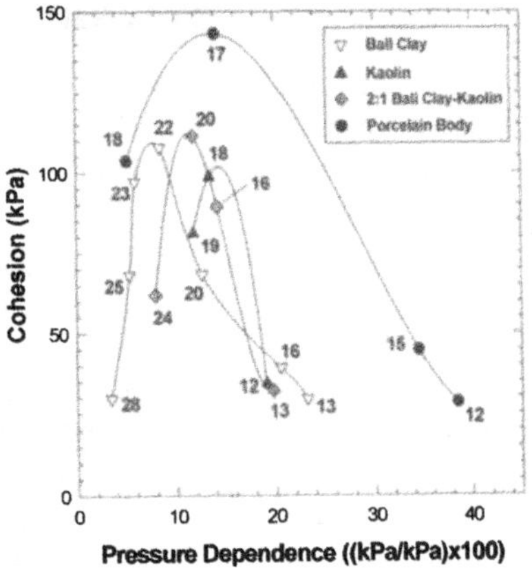

Figure 19. HPASC results of clays, and a standard porcelain body[1] showing the dependence of rheology on water content (noted next to data points).[19]

B. Experimental Approach.

The plasticity of two whiteware bodies was evaluated. Body A is a generic triaxial porcelain composed of clay, quartz, and feldspar; Body B was an industrial porcelain body, prepared in an industrial setting. Body samples identified '*As-Received*' were tested within hours of filter pressing. The '*Aged*' samples were stored for two weeks in a sealed container prior to testing. Other details on the compositions, the body preparation techniques, and the HPASC testing protocols, are presented elsewhere.[22]

C. HPASC Results and Discussion.

A typical example of the effect of water content on HPASC results of cohesion and pressure dependence is shown in Figure 20 for Body A. The maximum cohesion occurs at a water content of approximately 15%. The pressure dependence decreases linearly with increasing water content, up to the water content at which the maximum cohesion occurs, at which point the pressure dependence changes. It is proposed that the maximum cohesion obtained is a function of particle packing and a combination of capillary (due to partially filled pores) and colloidal forces (van der Waals attraction).

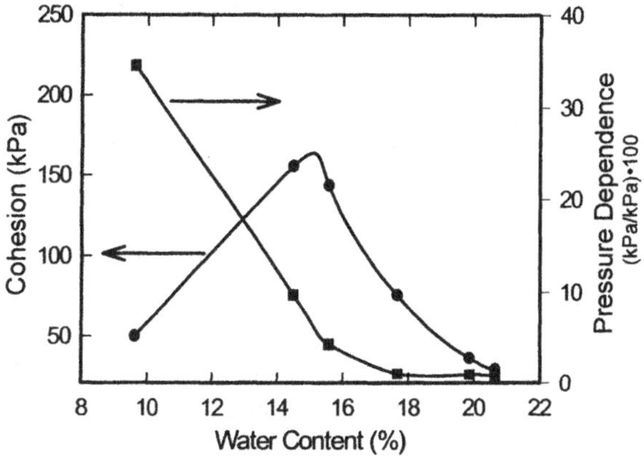

Figure 20. Typical HPASC behavior as a function of water content for *Body A* in the as-received condition. The circles (●) represent the cohesion; squares (■) represent the pressure dependence.

As the body is aged, the maximum cohesion strength decreases by nearly 25%, as shown in Figure 21. In addition, *Body B* exhibits a much greater maximum cohesion stress, than either the 'as received' or the 'aged' *Body A*.

In a typical plastic forming process, the water contents are in the 18-22% range. During drying, the water content decreases therefore changing the cohesive strength of the body, following the curve in Figure 21. It is proposed that drying stresses must be below this cohesion to prevent cracking. In the high water content regime, the body possesses low cohesion strength, allowing particles rearrangement to accommodate drying stresses. As cohesive strength increases particle rearrangement is more difficult allowing drying stresses to develop.

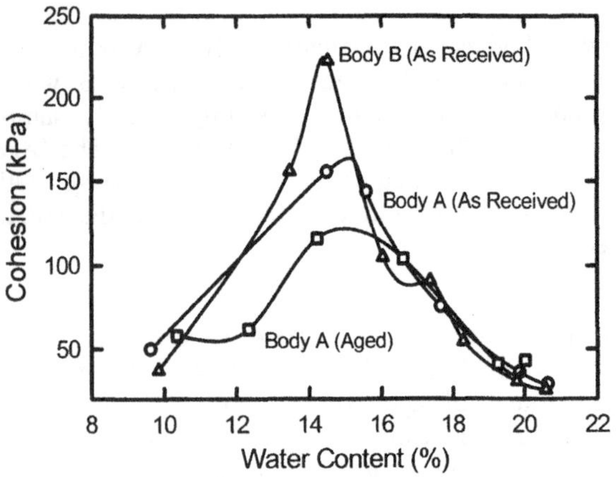

Figure 21. Cohesion with changing water content showing the difference between *Body A (as received)*, *Body A (aged)* and *Body B (as received)*.

It is proposed that drying performance and green strength are inversely related to the peak cohesion values. A higher cohesion value creates a strength differential in the body associated with the drying front. A moisture content difference of approximately 2% could create a strength difference of over 100 kPa, potentially resulting in localized shear cracking. (Anecdotal evidence supports the correlation of green strength with cohesion strength – industrial observations are that *Body B* exhibited lower strength than *Body A*; specific strength values are not available.) The change in the peak cohesion for the two "as-received" bodies can be partly attributed to changes in chemistry, but more probably, and with greater importance, to changes in particle packing. It has long been argued that particle packing is essential for body development,[23] and the substitution of one clay for another will certainly result in a change in particle packing.

To demonstrate that the changes in plasticity associated with time (aging) were both reasonable and repeatable, data is presented in Figure 22 that shows the effect of aging time of a ball clay-water sample. Cohesion decreases with time while pressure dependence shows a slight increase. When the data in Figure 22 was first generated, it was unclear what caused the reduction in cohesion with time. In addition, the time frame for a ball clay sample to achieve a steady-state HPASC behavior was nearly two weeks—about three to four times longer than required for a porcelain body composition.[19,20] Subsequent work indicated that cation dissolution was a function of composition and closely related to raw material selection and mixing intensity.[24] It became clear that water chemistry played an important role, as dissolved salts in the water could "jump-start" the aging process, in addition to the cations dissolved from the raw materials.

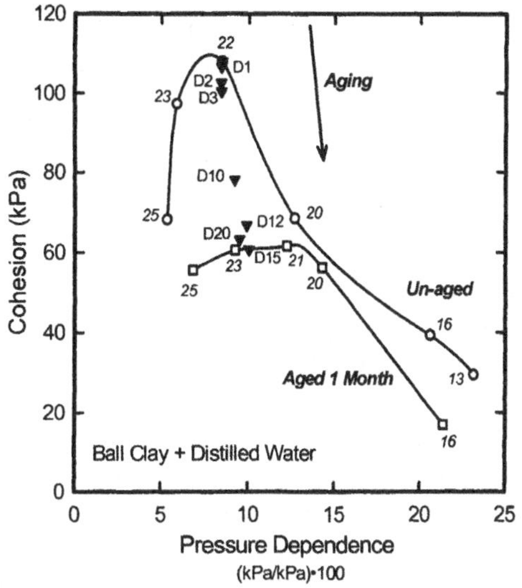

Figure 22. The plasticity of a ball clay-water body showing the change with time.[20] The water content levels are noted in italics (d.w.b.) and the change in plasticity for the 22% water content sample over 20 days is denoted by ▼. After one month, the maximum cohesion has decreased by approximately 45%.

D. Proposed Correlation with Drying Behavior.

The impact of initial water quality and the suspension solids loading on divalent cation concentration generated by the dissolution of the raw materials, is presented in Figure 23. The cation dissolution level after an initial relatively short (<24 hours) incubation period becomes constant and is equivalent to approximately 1.5% of the bulk cation concentration level in the batch.[24] (The deviation of the measured cation dissolution level from the predicted appears to be due to cation exchange.[25])

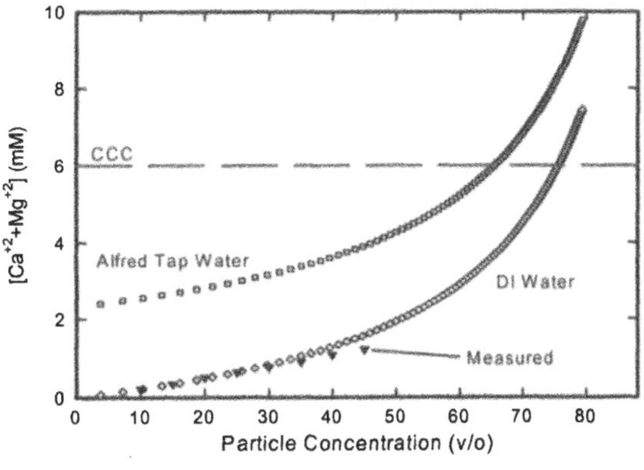

Figure 23. The effect of particle concentration on the equilibrium cation dissolution level for a porcelain body slip prepared with de-ionized (DI) water and with Alfred tap water (containing 2.5 mM dissolved divalent cations initially). The (CCC) is denoted by the dashed line.

At a constant raw material dissolution level the dissolved cation concentration increases with increasing particle concentration and is non-linear with solids concentration because the water content decreases with increasing solids loading.

The dissolution of cations from ball clay in water, without kaolin or feldspar, was demonstrated to be slower, thus indicating a possible explanation for the differences in aging behavior. Bodies prepared with distilled and Alfred water reach the CCC at a solids loading of ~65v/o (~20% d.w.b.) and ~75v/o (~13% d.w.b.), respectively, as shown in Figure 23. In this water concentration range, these samples will exhibit plastic character, but depending on the cation concentration, the plastic character differs. Based on the results from the plasticity measurements and the cation dissolution levels, it is proposed that raw material dissolution is responsible for aging by changing the dissolved cation levels in plastic bodies. Furthermore, as indicated by Figure 23, it is clear that suspensions may not exhibit aging effects if the cation concentration due to dissolution is far below the CCC, a condition common at moderate solids loadings commonly used for suspension preparation.

In an industrial process, however, the solids loading is not changed in a continuous manner. Any process which employs de-watering, such as slip casting or filter pressing, causes an abrupt shift in the solids loading at a constant cation level (Figure 24). The cation concentration in the expressed water is identical to the cation concentration remaining in the plastic body, thus providing an opportunity to characterize the chemistry of the water within the plastic body. Also, in slip casting operations, changes in chemistry may not be obvious in the relatively low concentration casting slip, but become apparent in the cast piece which possesses a solids loading similar to that of a plastic body. Once the de-watering step is complete, drying causes an increase in dissolved ions due to the removal of water by evaporation.

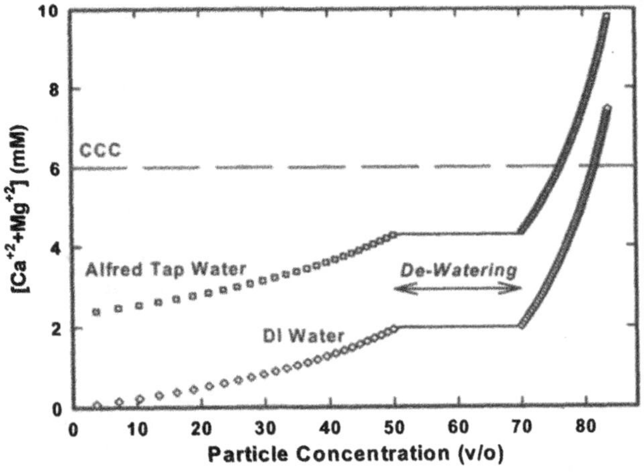

Figure 24. The effect of de-watering on the cation concentration versus solids loading for a porcelain body, as illustrated in Figure 23. De-watering simply shifts the curve to the right, but does not change the dissolved cation levels.

E. Effect of Ionic Strength on Plasticity.

If plasticity is dictated by dissolved cations, it should be possible to change plasticity by the addition of salts to a plastic body. Figure 25 clearly shows the effect of divalent cation (Ca^{2+}) concentration on the cohesion of Body A. To demonstrate that the decrease in cohesion occurs at a reasonable cation concentration, the results from the ionic strength experiments (Figures 11-13) are superimposed on Figure 25 to generate Figure 26. Both systems exhibit a dramatic effect of Ca^{2+} concentration at the same ionic concentration level*, indicating that the colloidal concepts are applicable even in highly loaded systems, such as in plastic bodies.

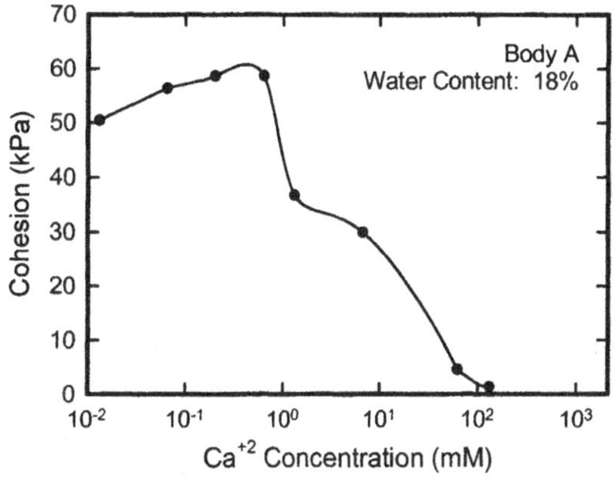

Figure 25. The decrease in cohesion above 1.0 mM Ca^{2+} concentration (via $CaCl_2$ additions) on cohesion of *Body A* measured using the HPASC.

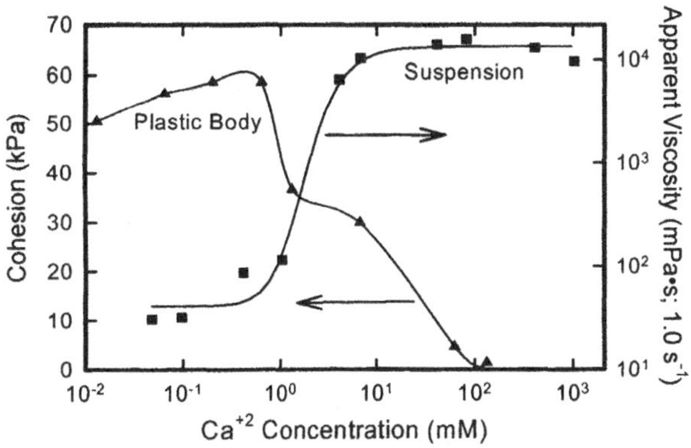

Figure 26. Comparison of plastic body behavior as cohesion via the HPASC and apparent viscosity from porcelain body suspensions showing that the critical coagulation is nearly identical and therefore independent of the water content.

* Consistent with colloidal theory, cation concentration is represented as the concentration of ions in the suspending medium, not in the total suspension volume. Therefore, the cation concentration is independent of solids loading and solely determined by water content.

Figure 26 shows that when the cation level increases in a suspension, the viscosity increases, but in a plastic body, the cohesion decreases. If cohesion is considered to be parallel to deformability, higher cation levels would produce softer clay, and in some processes, easier forming. It is also proposed from this data that drying can be assisted by increased cation levels, leading to a reduction in cohesion stress, and ultimately drying stresses in the body. Cation levels can be artificially changed by specific salt additions, or can be elevated in the body during the aging process, causing a transition from a dispersed condition below the CCC to a coagulated condition above the CCC. This may be the critical step in the drying process. If higher ionic concentrations are present in the bodies in the suspension preparation stage, higher cation concentrations remain after de-watering. Higher cation levels in the body after de-watering reduce the amount of water that needs to be evaporated during drying to reach the CCC level, thus allowing the dispersed-coagulated transition to occur earlier in the drying process.

VI. SUMMARY AND CONCLUSIONS

Of the five factors that control suspension rheology, the most important is particle-particle interactions as described by colloidal theory. Clay suspensions behave similarly to colloidal alumina suspensions, with the dispersant demand for clays roughly one-half that necessary for alumina, supporting the two-basal plane model for kaolinite and consistent with mineralogy. It is also clear that Na^+ functions as a coagulant rather than as a dispersant, with the polyanionic species specifically adsorbing to the particle surface. Salt additions increase the cation concentration, leading to coagulation, and divalent cations (such as Ca^{2+} and Mg^{2+}) are more efficient than monovalent cations (such as Na^+).

The cation level necessary to reduce the cohesion stress in the plastic body is consistent with the measured critical coagulation concentration in suspensions, demonstrating that chemistry can also used to control plasticity. In addition, it is proposed that achieving a critical cation concentration, through a process that appears to be directly related to raw material dissolution, may significantly improve drying behavior by reducing the peak cohesion stress in the body.

VI. REFERENCES

1. H. A. Barnes, J. F. Hutton, and K. Walters, *An Introduction to Rheology*, Chapters 1 (1-10), 2 (11-35), and 7 (115-139), Elsevier Science Publishers, B.V., Amsterdam, 1989.

2. R. B. Bird, R. C. Armstrong, and O. Hassager, *Dynamics of Polymeric Liquids, Volume 1, Fluid Mechanics*, 2nd Edition, John Wiley and Sons, New York (1987).

3. R. L. Hoffman, "Interrelationships of Particle Structure and Flow in Concentrated Suspensions," *MRS Bulletin*, **16** [8], 32-37, (1991).

4. R. L. Hoffman, "Discontinuous and Dilatent Viscosity Behavior in Concentrated Suspensions; I. Observations of a Flow Instability," *Transactions of the Society of Rheology*, **16** [1], 155-173, (1967).

5. R. H. Ottewill, "Concentrated Dispersions," *Colloidal Dispersions*, Ed. J. W. Goodwin, Special Publication #43, The Royal Society of Chemistry, London, 197-217, (1982).

6. R. J. Hunter, *Introduction to Modern Colloid Science*, Oxford Science Publications, New York, 1993.

7. Unpublished data from the undergraduate lab class "Introduction to Ceramic Processing (CES 205), Alfred University, collected 1994-1997.

8. K. R. Rossington, U. Senapati, and W. M. Carty, "A Critical Evaluation of Dispersants for Clay-Based Systems," *Ceram. Eng. and Sci. Proc.*, **19** [2], 77-86, (1998).

9. B. R. Sundlof and W. M. Carty, "Organic and Inorganic Dispersion of Alumina," *Ceram. Eng. and Sci. Proc.*, **20** [2], 151-166, (1999).

10. W. Carty, *Processing of Ceramic Fibers from Particle Suspensions*, Ph.D. Dissertation, University of Washington, 1992.

11. R. O. James, "Characterization of Colloids in Aqueous Systems," *Ceramic Powder Science; Advances in Ceramics, Vol. 21*, Eds., G. L. Messing, K. S. Mazdiyasni, J. W. McCauley, and R. A. Haber, Amer. Ceramic Soc., Westerville, OH, pp. 401-403, 1987.

12. N. Güven, "Molecular Aspects of Clay-Water Interactions," *Clay-Water Interface and its Rheological Implications*, CMS Workshop Lectures **4**, Eds. N. Güven and R. M. Pollastro, The Clay Minerals Society, Boulder, CO, 2-12, 1992.

13. W. M. Carty, "The Colloidal Nature of Kaolinite," *Amer. Cer. Soc. Bull.*, **78** [8], 72-76, (1999).

14. A. L. Johnson and F. H. Norton, "Fundamental Study of Clay: II, Mechanisms of Deflocculation in the Clay-Water System," *J. Amer. Ceram. Soc.*, **24** [6], 189-203 (1941).

15. K. R. O'Connor and W. M. Carty, "The Effect of Ionic Concentration on the Viscosity of Clay-Based Systems," *Ceramic Engineering and Science Proceedings*, **19** [2], 65-76, (1998).

16. K. R. Rossington, U. Senapati, and W. M. Carty, "A Critical Evaluation of Dispersants: Part II. Effects on Rheology, pH, and Specific Adsorption," *Ceram. Eng. and Sci. Proc.*, **20** [2], 119-131, (1999).

17. J. Cesarano III, I. A. Aksay, and A. Bleier, "Stability of Aqueous α-Al_2O_3 Suspensions with Poly(methacrylic acid) Polyelectrolyte," *J. Amer. Ceram. Soc.*, **71** [4], 250-255 (1988).

18. G. G. Hong, H. Lee, B. R. Sundlof, and W. M. Carty, "Evaluation of Kaolin Surface by Rheological Behavior of Suspensions in the Al_2O_3-SiO_2 System," manuscript in preparation.

19. W. M. Carty and C. Lee, "The Characterization of Plasticity," *Science of Whitewares (Proc. Sci. Whitewares Conf.)*, Eds., V. Henkes, G. Onoda, and W. Carty, 89-101 (1996).

20. C. Lee, *The Characterization of Plasticity in Clay-Based Systems*, M.S. Thesis, Alfred University, Alfred, NY, 1995.

21. P. A. Nowak, *Correlation of Extrusion Behavior with High Pressure Shear Rheometry*, M.S. Thesis, Alfred University, Alfred, NY, 1995.

22. W. M. Carty, K. R. Rossington, and D. S. Schuckers, "Plasticity Revisited," *Science of Whitewares II (Proc. Sci. Whitewares II Conf.)*, Eds., W. Carty and C. W. Sinton, 225-236, (2001).

23. D. R. Dinger, "Influences of Particle Size Distribution on Whiteware Properties and Processing," *Science of Whitewares (Proc. Sci. Whitewares Conf.)*, Eds., V. Henkes, G. Onoda, and W. Carty, 105-115 (1996).

24. H. Lee, *The Effect of Mixing Route on the Properties of Whiteware Suspensions*, M.S. Thesis, Alfred University, Alfred, NY, 1997.

25. P. Sillapachai, *Factors Controlling Aging in Whiteware Bodies*, M.S. Thesis, Alfred University, Alfred, NY (expected completion date: 2001).

ACKNOWLEDGEMENTS

Portions of this work were funded by the Whiteware Research Center, the New York State Energy Research and Development Authority, Buffalo China, Inc., Victor Insulators, Inc., and Syracuse China Co. The author would also like to thank Katherine Rossington, Brian Sundlof, Hyojin Lee, Peter Kupinksi, Pattarin Sillapachai, Nik Ninos (Buffalo China), Mike Dempsey (Victor Insulators), and Joe Benoit (Syracuse China) for their assistance.

THE NATURE OF CHEMICAL REACTIONS THAT OCCUR DURING CASTABLE INSTALLATION AND ANALYTICAL TECHNIQUES USED TO FOLLOW THESE REACTIONS.

C.D. Parr, C Revais, H. Fryda*

Lafarge Aluminates
28 rue Emile Meunier
Paris
France
F75792

*Lafarge Central Research Laboratory
95, rue du Montmurier
St. Quentin Fallavier Cedex
France

1 ABSTRACT

The control of refractory castable placing characteristics has become a major development theme among refractory producers. An understanding of the underlying mechanisms which control castable placing characteristics is a pre-requisite to the further development of such castable systems

This paper presents a review of the chemical reactions, which have been identified within various deflocculated calcium aluminate cement based castable systems and their impact upon the castable placing characteristics. These reactions will be illustrated through analytical techniques such as rheology, calorimetry and conductivity.

The conclusions will show the need for the continuing development of analytical techniques and they will also highlight future challenges in terms of castable formulation control.

2 INTRODUCTION

In order to understand the role of calcium aluminate cement (CAC) in modern refractory castable technology, it is first necessary to consider the reactions that take place within CAC during the hydration process.

The classical approach to investigate CAC hydration has been to use the well-established techniques of mortar/castable based physical testing[1]. With these techniques, rheological properties are measured through the use of flow values at different time intervals. Cement reactivity is measured indirectly through setting times. Mechanical strength development is measured through destructive testing. While this program of testing is well adapted for measuring quality and assessing industrial production, it is not as useful for evaluating the hydration process and, more specifically, the elaboration of mechanisms. The latter is necessary for the development and optimisation of new systems.

In the case of CAC within refractory systems, it is necessary to investigate in more detail the reactions that occur from the point that water is added at the beginning of mixing. These reactions and their kinetics will be influenced by the composition of the mineralogical phases present in a given cement. A general comparison of the reactivity of the anhydrous phases is shown in table 1 taken from George[2]. It can be seen that as the phases become more lime rich, i.e. as the C/A ratio increases, then so does their reactivity.

Table 1 : The reactivity of the various calcium aluminate phases

	C_3A	$C_{12}A_7$	CA	CA2	CA6
C/A	3	1,7	1	0,5	0,2
Reactivity at 20°C	Very rapid	Fast	Slow	Very slow	None

Note C = CaO, A = Al_2O_3

The mechanism of hydration of calcium aluminate is via solution, where an anhydrous phase dissolves and is followed by the precipitation of the hydrates from solution[3]. Three distinct phases can be identified; dissolution, nucleation and precipitation. The hydration process[4] is initiated by the hydroxylation of the cement surface followed by dissolution of cement in water and the liberation of calcium and aluminium ions. A small amount of hydrates will form at this point if the solution concentration rises over the super saturation limit level of the hydrates C_2AH_8* and AH_3. The dissolution will continue, with a consequent increase in the concentration of calcium and aluminium ions until a saturation point is reached. This is the point which is the equilibrium solubility of the anhydrous phases with the hydroxylated surface layers. After

* cement notation is used where C=CaO, A=Al_2O_3, H=H_2O the dissolution phase there follows an induction period during which nuclei attain a critical size and quantity.

Once this is achieved the nucleation phase is followed by a rapid and massive precipitation of the hydrates, leading to a drop in solution concentration. This is a dynamic process which continues to consume anhydrous cement. In a physical sense, it is the growth of these hydrates which interlock and bind together to provide mechanical resistance. The ambient temperature significantly modifies the hydrates[3] (table 2) that result due to the fact that their solubility changes with temperature[5].

Table 2 : Hydration scheme for monocalcium aluminate

Temperature	Hydration		Reaction
< 10°C	CA + 10H	→	CAH_{10}
10 – 27°C	2CA + 11H	→	$C2AH_8 + AH_3$
	CA + 10H	→	CAH_{10}
> 27°C	3CA + 12H	→	$C_3AH_6 + 2AH_3$
	$2CAH_{10}$	→	$C_2AH_8 + AH_3 + 9H$
	$3C_2AH_8$	→	$2C_3AH_{10} + AH_3 + 9H$

The hydration mechanisms and kinetics such as those described above can be significantly modified[6] when fine reactive powders (fillers) and additives are present as in the case of deflocculated castables. The nature of these reactions depends to a large extent upon each unique combination of CAC, fillers and additives. It is possible in these systems to distinguish short-term hydration reaction effects, related to castable placing characteristics, from longer-term hydration mechanisms that are linked to castable hardening. Figure 1 shows a general scheme linking the various reaction mechanisms that take place in deflocculated castable systems through the castable placing chain. The reactions that have been identified[7,8,9] are complex and occur at all stages through the castable processing chain. The identified chemical reactions occur via solution and are based upon ionic dissolution, complexing mechanisms and precipitation reactions. The effect of the surface must also be considered. A variety of analytical techniques have been developed to identify reactions at various stages of the castable placing chain.

The experimental techniques and examples that follow illustrate some of these underlying mechanisms and their impact upon castable placing characteristics. The early dissolution and precipitation reactions, as shown in figure 1, have been attributed[10] to the interactions of additives and the fine fillers (typically fumed silica and alumina) with the CAC. The following examples will show how these interactions have a marked impact upon castable placing characteristics.

Figure 1 : The nature of reactions within deflocculated castable systems

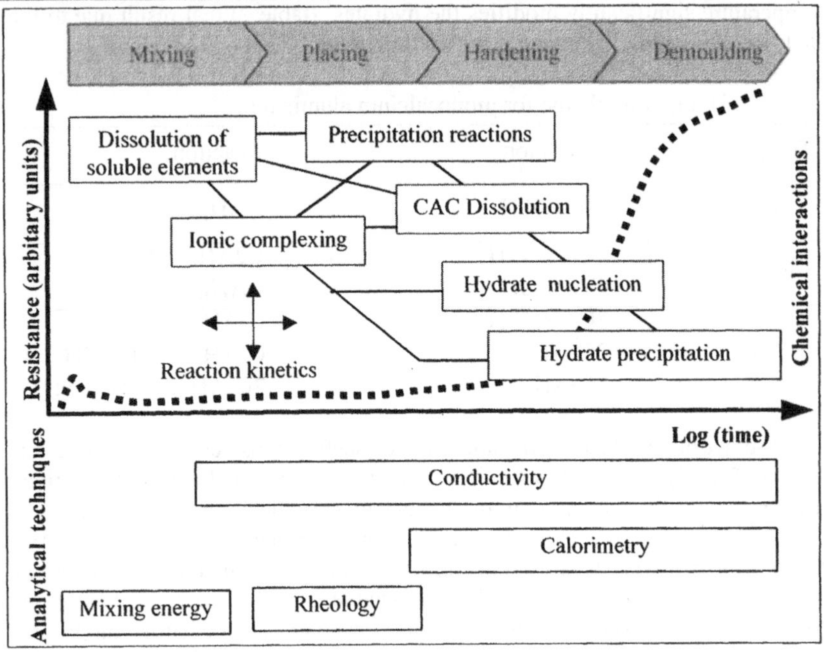

3 EXPERIMENTAL TECHNIQUES

3.1 Mixing Energy

This technique monitors the change in force measured by a piezo electric sensor attached to a modified laboratory planetary mixer. The initial phase of mixing of a castable after water additions can be observed and quantitative data gathered as to the « wet out » time of deflocculated castables as well as to the initial rheology. It is particularly useful in assessing the behaviour of self-flow castable types, which can often display unusual rheological characteristics during the mixing process.

3.1.1 Experimental procedure

A planetary type mixer (type Perrier) is modified with the addition of a piezo-electric sensor at the point of linkage between the mixing bowl and the mixer frame. An interface board is used to capture the data to a PC file. The sensor measures the changing force (tension in volts), which alternates between traction and compression as the mixing paddle rotates within the bowl. This generates a sinusoidal wave. Due to the elliptical rotation of the mixing paddle, a second sinusoidal wave will also be detected. A treatment of the raw data, captured continuously for a specific time period, transforms the data to positive values and calculates the change in the mixing energy as a function of time. The measurement generates a large number of data points (~4000-10000 per run) which makes analysis of individual data difficult. It is far more convenient to derive a single curve from each data set. The resulting data points can then be modelled, using the maximum and minimum values and the corresponding time intervals, to produce a single curve representing the change in mixing energy with time.

The collection of data is started with the mixing of the dry castable followed by the addition of a predetermined amount of water (water addition calculated to give an initial vibration flow value of 100% as measured by the ASTM method). The mixer is then left to run for a predetermined period of time, typically 250 seconds.

3.1.2 Results

Various low cement castables (LCC) have been evaluated along with classical high cement castables. Deflocculated castables can be characterised by a curve which tends to rise to a maximum, then decreases exponentially to a minimum. This is in contrast to classical castables, which tend to rise to a maximum more quickly and have no subsequent decrease. These differences are presumed to be due to the impact of additives, which when active, cause the « wet out » of LCC castables and a corresponding reduction in mixing energy. The « wet out » time is the point at which the castable appears to be uniformly mixed and the mass appears to be wet and fluid. The period preceding this « wet out » time represents the maximum mixing energy prior to the arrival of the fluidifying effect of the additives. Maximum fluidity is normally achieved when the volts tension decreases to 1,5-2 N/volts.

The same base model castable has been used in all these examples, with only the additives and the fine reactive fillers changed. The first example (figure 2) uses a low cement castable with 5% of a 70% alumina cement that has a sintered alumina aggregate system (particle size distribution based upon an Andreassen distribution modulus of q=0,25). The addition levels of the reactive alumina (Pechiney type P152SB) and the fumed silica (Elkem 971U) are kept constant at 15% wt. addition but the relative proportions are altered. The same 4 additive system (0,03% Darvan 7S, 0,02% sodium tripolyphosphate (TPP), 0,0015% sodium bicarbonate and 0,002% citric acid) is used in the three examples shown.

The water addition was a constant 5% in all cases. Depending upon the choice of reactive fillers, the mixing energy profile changes significantly. The system with the highest amount of fumed silica showed the most rapid decrease in mixing energy after the « wet out » peak. The final energy value increased as the fumed silica content decreased. The maximum mixing energy was similar in all cases irrespective of the filler type used, although the duration of this maximum peak changes with higher fumed silica levels generating a sharper (narrower) peak.

Figure 2 : The change in voltage (force) versus time for 3 different castable systems

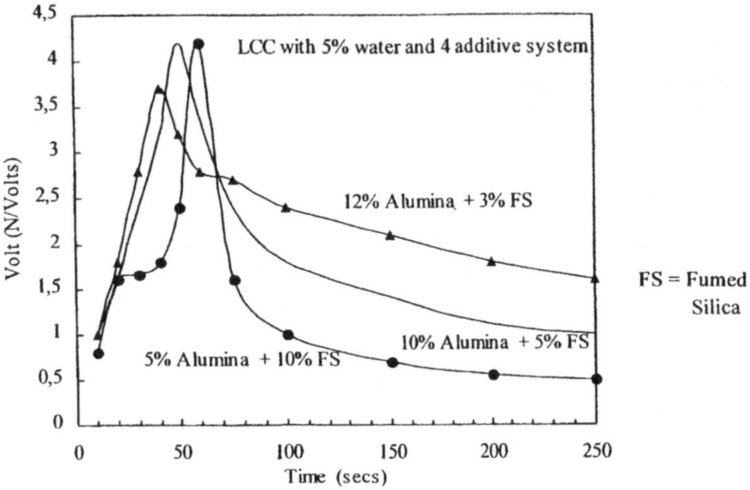

A second example of this technique is shown in figure 3. Here the base castable system is the same as the previous example, with the reactive alumina fixed at 10% and the fumed silica level at 5%. Two additive systems are investigated, the first is based upon a ternary system using poly-napthalene sulphonate (0,2%), citric acid (0,04%) and sodium carbonate (0,035%). The second additive system is the quaternary system used in the previous example. The differences between the two additive systems during mixing and "wet out" are quite marked. However the final values for both systems are similar. The quaternary system shows a lower overall mixing energy is required (lower peak and more rapid decay to the minimum level). This suggests that the ternary additive system would need more careful on site mixing to yield similar placing characteristics to that displayed by the quaternary additive system.

Figure 3: The change in voltage (force) versus time for different additive systems

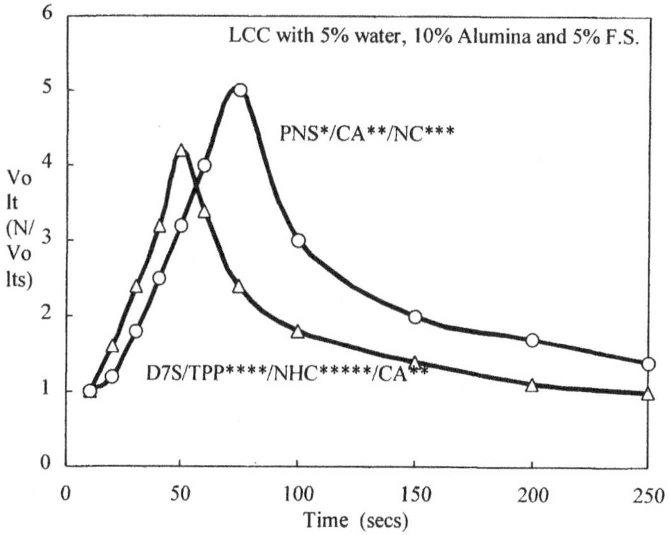

The mixing energy technique gives important information about the first interactions within deflocculated castables, namely the "wet out time". This global indicator encompasses the effects of additive solubility and time to disperse the reactive fine fillers. This technique can also be used to assess the optimum type of mixing regime necessary for a given type of castable. Furthermore, the efficiency of different combinations of additives and fine fillers upon the « wet out » time can be assessed. The impact of additive dissolution on the mixing energy profile is large. This suggests that the selection of a specific grade of additive (particle size, purity and morphology) should be made carefully. Important information can be gleaned as to the behaviour of deflocculated castables during and immediately after mixing. This can be used to design castables which are inherently easy to mix (i.e. requiring low mixing energy) and which would enhance the placing reliability of such

* PNS : polynapthalene sulfonate
** CA : citric acid
*** NC : sodium carbonate
**** TPP : sodium tripolyphosphate

systems. No information, however, can be gained as to the evolution of the castable after mixing and in particular information relating to the point at which the CAC becomes active. It is necessary to use alternative analytical techniques to provide information concerning the subsequent castable processing steps.

3.2 Conductimetry

The conductivity of a dilute cement suspension is measured as a function of time, allowing the complete hydration process to be followed. This technique has been applied equally to classical refractory systems and to deflocculated castables.[11]

3.2.1 Experimental procedure

Sixteen grams of solids are weighed into a double walled beaker which is connected to a constant temperature water bath. Eighty grams of distilled water at 20°C are then added to the beaker and the solution stirred continuously. A conductivity probe is inserted into the beaker which measures the conductivity continuously in milli-siemens/second for a period of 1440 minutes. The data is captured via a data logger to a personal computer. The results on identical samples yield reproducible results with coefficients of variation around 1-2%. A generalised curve of a 70% alumina cement (Secar 71) is shown in figure 4. The three phases of hydration can be clearly distinguished.

Figure 4 : Typical conductimetry curve for a 70% alumina cement (Secar 71)

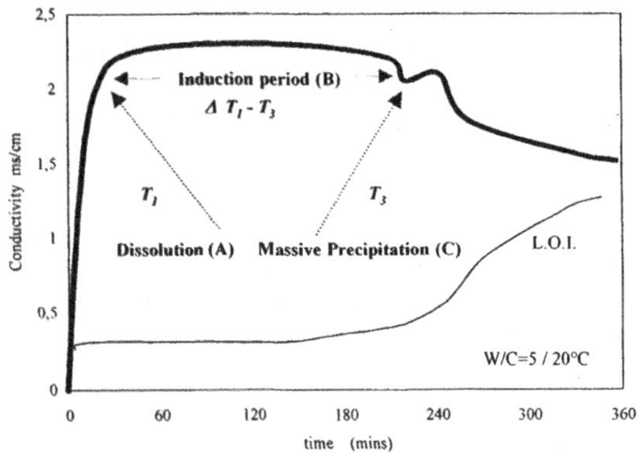

Each cement will give a unique characteristic trace which can be used to follow the hydration process. The hydration process is broken into three regions, dissolution (A from figure 4), nucleation (B) and type of massive precipitation (C). Furthermore, this same type of measurement can be performed on the matrix part of the « binder phase » of deflocculated castables.

3.2.2 Results

Figure 5.1 and 5.2 shows the effect of two classical additives, lithium carbonate (LC) and tri-sodium citrate (TSC), upon the hydration of a 70% alumina cement. These well-known additives are used[12] to accelerate hydration in the case of LC and retard hydration in the case of TSC. Conductivity analysis shows the mechanism of action for each of these additives. With increasing dosage of LC, the massive precipitation time is reduced significantly while the dissolution time remains constant. In fact, the nucleation period decreases. The inverse is true for TSC, where the nucleation time increases as the dose increases. The actual mechanism can often be deduced[13] by withdrawing samples of the suspension at specific intervals and « stopping » the reaction by washing the sample with alcohol, followed by subsequent (ICP, XRD) analysis of the solution. The shortening of the massive precipitation time in the case of LC has been found to be due to an increased nucleation rate of alumina trihydrate (AH_3).

Figure 5.1 : The effect of lithium carbonate on the hydration of a 70% alumina cement

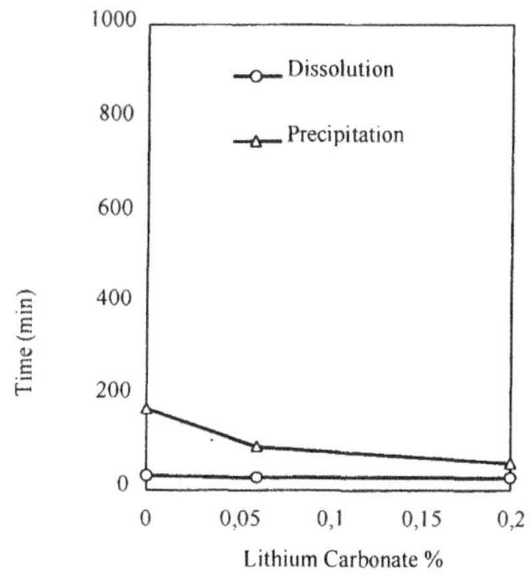

Figure 5.2 : The effect of trisodium citrate on the hydration of a 70% alumina cement

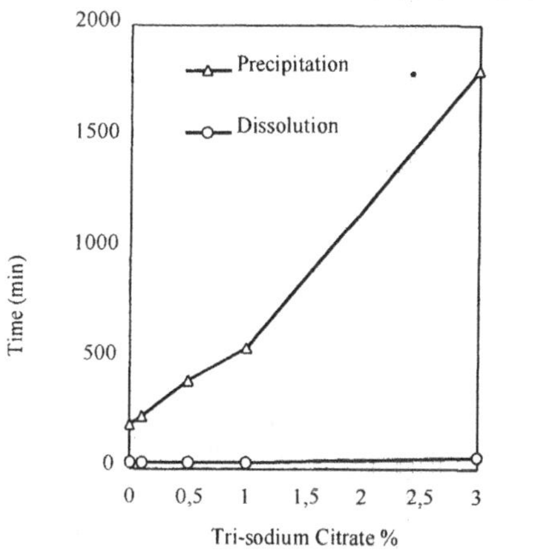

The retardation of the massive precipitation seen with TSC is due to a lengthening of the induction period. This is believed to be due, from solution

chemistry analysis, to the formation of a gel containing citrate, calcium and aluminium ions. These interesting results show that the action of these additives upon the three stages of CAC hydration is specific. This opens the possibility to envisage the use of multiple additives to optimise castable systems, each acting upon a specific phase of the hydration process. Thus, the combination of these two additives in a castable would give a castable with a long working time followed by a rapid hardening. Furthermore, conductivity analysis offers a rapid technique by which additive dosages may be optimised.

Conductivity can also be used to study deflocculated castable systems[14] In this case, the binder phase or matrix part of a refractory castable is used. Table 3 gives details of a typical binder phase for a low cement castable (LCC) based upon a 70% alumina cement, which was used for conductivity analysis. The percentages shown refer to the total castable, including aggregates and other components, on a weight basis.

Table 3 : LCC Binder phase composition

Binder material		% weight in castable
Calcined alumina	: Pechiney P153SB	10
Fumed silica	: Elkem 971U	5
Cement	: Secar 71	5
Additive	: Sodium Tri-poly phosphate	0,03 - 0,2

The hydration mechanism of a 70% alumina cement is significantly modified in LCC systems compared to classical systems, where the cement represents the total binder phase. This can be clearly seen in Figure 6, which shows the resulting conductivity curve for the LCC binder phase system. The initial dissolution is delayed compared with the cement only system. Also, the dissolution occurs at a much slower rate with the LCC system. However, the precipitation occurs during a similar time period for both the deflocculated system and the cement alone. The precipitation time corresponds well with the onset of hardening and the acquisition of mechanical resistance in castables. The initial level of conductivity is derived from the TPP additive which is added directly into solution. This also offers a possibility of using conductivity as a means to control the TPP dose in castables. The initial conductivity level has been found to be strongly correlated with the TPP dose.

Figure 6 : LCC conductimetry curves versus time for CAC with/without TPP

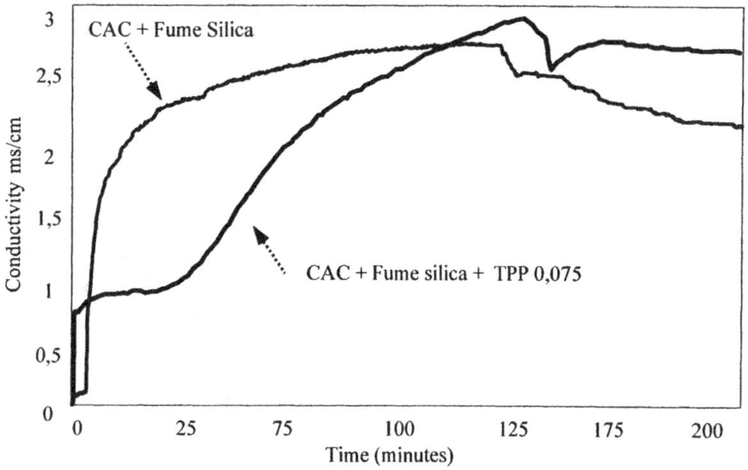

The delay in CAC dissolution has been shown[14] to be the mechanism which provides the castable with sufficient working time. The longer the delay in dissolution, the longer the working time becomes. This is believed[10] to be due to surface reactions that result in the formation of calcium phosphate precipitates, which tend to block CAC dissolution. This retarding effect on CAC dissolution is not seen when binary combinations of phosphate and CAC or fumed silica and CAC are analysed. It is presumed, therefore, that the surface of the fumed silica in the presence of phosphates also plays a key role. Maeda[15,16] proposed a mechanism whereby the fumed silica interacts with the surface of the cement grain. The addition of TPP together with fumed silica and reactive alumina, (fine fillers), clearly results in interactions that delay the dissolution of the CAC. It can be considered that the length of the working time of low cement castables results from a series of interactions between the additives and the fine fillers, together with the CAC. The working time is not simply dependent upon the cement alone as in the case of classical or conventional castables with a high cement dose. Thus, by studying the duration of this delayed dissolution through conductimetry, a rapid method exists by which castable working times can be measured and optimised more precisely than through classical methods.

The same binder phase was also used to study the effect of ambient temperature on LCC castable systems. The effect of differing TPP dose upon the dissolution and precipitation reactions within the castable was measured at 20°C, 15°C and 5°C. The time to maximum dissolution and the first precipitation minima were recorded. The results are shown in figure 7. The temperature dependence of the system is extremely marked at low ambient temperatures (5°C) with an extension of both the dissolution time and the time to massive precipitation. Nucleation times remain relatively constant. The sensitivity to low temperature becomes noticeably more marked at higher TPP doses. At doses of +0,15% TPP (weight basis – total dry weight), the dissolution and precipitation times increase by almost three fold. These results suggest that lower TPP dose rates will result in a more robust castable system at 5°C. The practical implications of these conductivity results would be a long working time and hardening time at 5°C, especially in the case of higher TPP dose rates. The results show that LCC systems can display a large variation in placing property behaviour according to the ambient temperature. This variation is due not only to the CAC, but also to the nature of interactions between the CAC the fine fillers and the phosphate.

Figure 7 :The effect of TPP dose upon CAC dissolution and precipitation

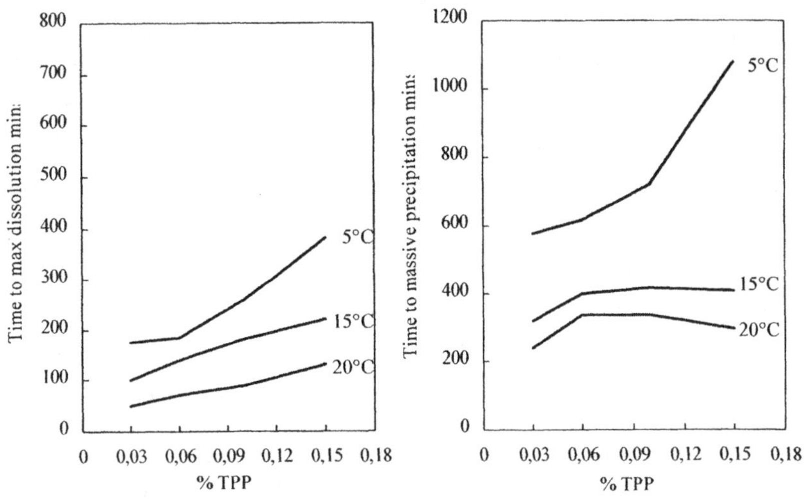

The dosage level of the additives needs to be carefully controlled to avoid extreme prolongation of castable working and hardening times. Conductivity provides a useful insight into the mechanisms that control the CAC hydration and consequent effect upon castable placing characteristics.

The temperature dependence of the system is extremely marked at low ambient temperatures (5°C), with an extension of both the dissolution time and the time to massive precipitation. The nucleation times remain relatively constant. The sensitivity to low temperature becomes noticeably more marked at higher TPP doses. At doses of +0,15% TPP (weight basis – total dry weight), the dissolution and precipitation times increase by almost three fold. These results suggest that lower TPP dose rates will result in a more robust castable system at 5°C. The practical implications of these conductivity results would be a long working time and hardening time at 5°C, especially in the case of higher TPP dose rates. Results show that LCC systems can display a large variation in placing property behaviour according to the ambient temperature. This variation is due not only to the CAC, but also to the nature of interactions between the CAC, the fine fillers and the phosphate. The dosage level of the additives needs to be carefully controlled to avoid extreme prolongation of castable working and hardening times.

Conductivity provides a useful insight into the mechanisms that control CAC hydration and its effect upon castable placing characteristics.

3.3 Calorimetry

This technique is particularly useful for the analysis of exothermic reactions which arise during the course of CAC hydration. The technique has been used by Lafarge Aluminates for many years[17,18]. It has more recently been used to follow the chemical reactions that take place within a variety of deflocculated castable systems[6,19,20]. The massive precipitation is an exothermic reaction which liberates heat. Through the measurement of this exothermic heat, an indication of the advancement of the hydration process can be obtained. In the case of a deflocculated castable system, an initial peak is often found which can be attributed to the dissolution phase of the hydration process. It is particularly useful for measuring the end point of castable working time.

3.3.1 Experimental

The technique illustrated here uses a specially developed calorimeter designed to measure the heat flow associated with chemical reactions occurring in castables as a function of time. Normally, the apparatus is used to measure the heat evolution due to wetting-dissolution that occurs when water is added to a calorimetric cell containing a dry cement. The technique has been modified in that a ready mixed paste is introduced into the calorimeter so that the heat flow associated with wetting is removed. Only the subsequent chemical reactions are measured and followed. The time to maximum heat flow is evaluated, which is not used to derive information relating to the reaction kinetics. The maximum heat flow will be dependent upon the hydrates[17] (heat of reaction decreases from CAH_{10} to C_2AH_8 to C_3AH_6) formed as well as the relative proportion of cement used. In the case of a mechanistic study of deflocculated castables, it is particularly useful to study the castable binder phase rather than the whole castable.

3.3.2 Results

Figure 8 shows a heat evolution curve for the LCC system (table 3). Two exothermic peaks can be identified. The two peaks have been generically labelled as Pi and Pm. The points Pi and Pm are taken as the points of maximum heat flow and the time taken to reach these peaks is recorded. The peak Pm is believed to correspond to the massive precipitation of the hydrates, whereas Pi is associated with the dissolution phase of the cement. In terms of castable placing characteristics, Pi is associated with the loss of workability and castable flow decay. Pm corresponds to the onset of strength development during castable hardening.

Figure 8 : Heat flow for an LCC system as analysed by micro calorimetry

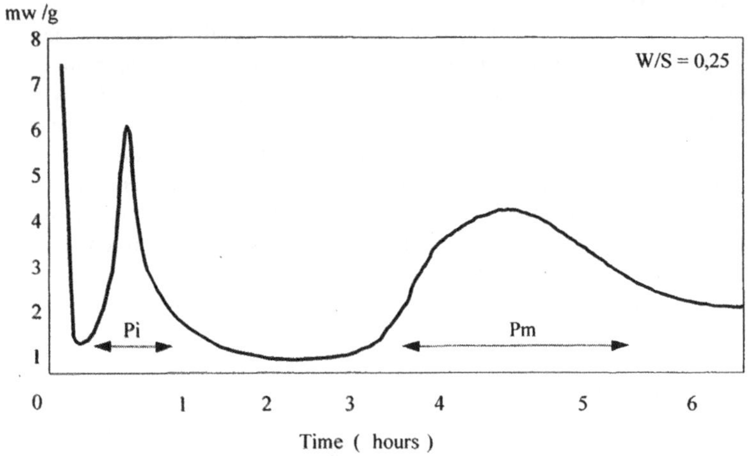

A study of numerous low cement binder phase systems (variations of both fillers and additives) has shown[6] significant variation in the time to reach Pi. There exists a relationship between working time of the full castable and the apparition of this peak, but the relationship depends upon the composition of each specific binder phase composition (fumed silica, alumina etc). Figure 9 illustrates this effect. The curves represent the peak Pi for two different systems of reactive fillers. The graph on the left shows a system containing 25% of a 70% alumina cement, 50% of a reactive alumina (CT3000SG) and

25% of fumed silica (Elkem 971U). The additive system is a binary system (sodium polymethacrylate and citric acid) added at 0,5% by weight of the total castable binder phase. The system in the right hand graph is identical except for a substitution of the fumed silica by the reactive alumina (i.e. CAC=25% and reactive alumina = 75%). The dotted line in both graphs shows the measured working time of each castable which corresponds to the apparition of Pi in both cases. The graphs provide some explanation of the well known effect that removing fumed silica from the system results in a much shorter working time of the castable.

Figure 9 : The impact of different filler systems upon the first calorimetric peak. Additive is sodium polymethecrylate and citric acid.

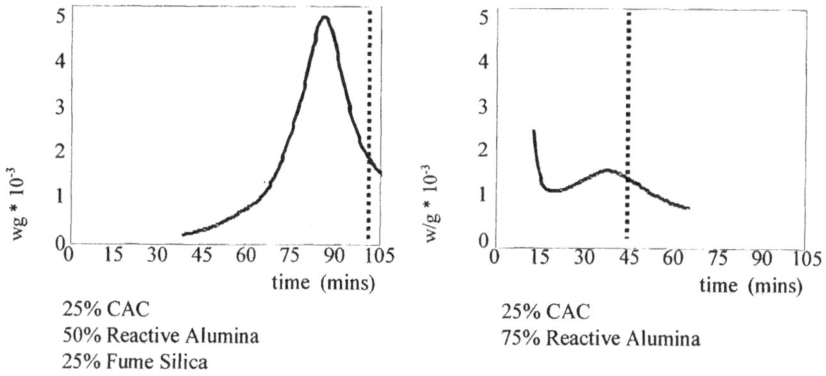

The heat flow data clearly shows that there is a chemical reaction related to the fillers and that, although they are not hydraulic binders, they act in a chemical sense within low cement castable systems. In effect, the fumed silica provides a strong retarding effect upon the cement dissolution, which gives a usable working time to the castable. By contrast, the system without fumed silica does not have this retarding effect upon cement dissolution. As a consequence, a short working time and rapid flow is expected. In the case of an actual castable, this would need to be corrected through a modification of the additive system. It can be seen that calorimetry provides useful information regarding the underlying chemical reactions, and particularly, the castable flow decay. Thus, a study of these reactions and the occurrence of Pi can provide a way to optimise the castable flow decay and working time.

A further example is exhibited in figure 10, where calorimetry is used to optimise the castable working time. The studied system uses two binder phase compositions, one based upon 100% of a 70% alumina cement and the second on a 50:50 composition of cement and fumed silica. In both cases a single additive, TPP, is used at varying doses. The time to reach Pi is plotted as a function of the TPP dose. It should be noted that the TPP dose quoted is based on the binder phase and not the total castable, as in the previously discussed analytical techniques. Once again the participation of the fumed silica in chemical reactions can be seen as it provides a strong retarding effect compared to the system without fumed silica.

Figure 10 : The impact of changing TPP additions on the first calorimetric peak

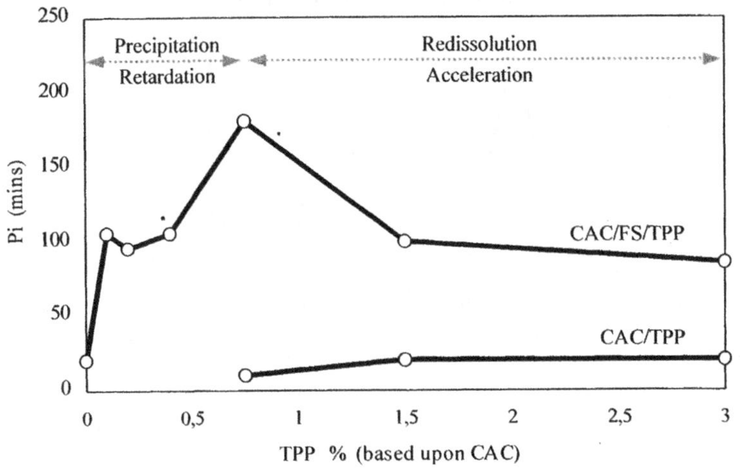

The retarding effect depends upon the TPP dose. At lower dose rates, a strong retardation of Pi is seen which transforms into a reduced retardation (or acceleration relative to the largest Pi value) at higher additions of TPP. This has been attributed[10] to the formation of a complex calcium phosphate precipitate on the surface of the cement/fumed silica. This then provides a diffusion barrier through a semi permeable layer and delays dissolution. At higher concentrations of TPP, this complex appears to re-dissolve, reducing the retarding effect.

4 DISCUSSION

The three techniques of mixing energy, conductivity and calorimetry have shown how it is possible to follow the hydration process of CAC within deflocculated castable systems. This can be related to the practical placing properties of the actual castables. The techniques provide insight into the underlying mechanisms and the subsequent optimisation of the castable system through formulation control.

The study of chemical reactions reveals the impact of the various active components upon the castable behaviour, and more specifically, the CAC. It is evident that the hydration process can be significantly modified through the use of additives and fine reactive fillers. Therefore, the optimisation of castable behaviour can be only be done by the consideration of these components as an interdependent system. This must remain as one of the key formulation challenges for the future given the unpredictable behaviour of some of the active binder phase components. For example, it has been shown that multiple combinations of additives and fillers can be used to good effect to simultaneously control castable behaviour. However, the notion that multiple additives can be used to enhance castable flow and flow decay needs to be considered within the context of each filler system. This is because the response might vary according to each specific combination.

It should be noted that this paper has only considered the chemical reactions that take place within castables. In order to develop a more complete model, effects such as the particle packing density as well as surface effects would need to be considered. The provision of an integrated model is perhaps one of the key areas for the future. It is also evident that no single technique is able to provide a complete understanding of the interactions that occur within deflocculated castables during their placing. The use of multiple techniques allows a model of some of the key interactions to be established. Future developments will need to focus on the provision of more predictive techniques, which will allow a more efficient optimisation of castable placing characteristics. Within this context, the illustrated techniques in this paper have focused upon the placing properties. The risk is that they might be optimised at the expense of other properties such as the hardening kinetics. Therefore a potential area for the future development of analytical techniques lies in the provision of a more complete understanding of the hardening process.

5 CONCLUSIONS

The application of a specific analytical technique to assess the underlying mechanisms responsible for castable placing characteristics is dependent upon the type of diagnostic required. It is clear that no single technique can provide all the necessary information as to the chemical reactions within castables and that those complimentary techniques in tandem should be considered.

The study of chemical reactions that occur during the castable placing process has shown that they result from a series of interactions between the various binder phase components (fillers, additives and calcium aluminate cement). The type of reactions that are seen (dissolution, complexing and precipitation reactions) are complicated in nature and often dependant upon each specific castable system. Therefore, an effective optimisation of castable systems can only be effected via an understanding of the underlying mechanisms. No longer can it be considered that all optimisation problems linked to castable flow and hardening are due simply to the calcium aluminate cement.

It is likely that the future generations of castables will become increasingly sophisticated as installation techniques and performance constraints become more demanding. The development of these castables will only be successful if they are derived from an understanding of the underlying mechanisms and interactions. Thus, the continued development of techniques able to study these complex interactions is forecast to be a key success factor in the future.

6 ACKNOWLEDGEMENTS

The authors would like to thank all the co-workers at Lafarge Aluminates who contributed to the preparation of this paper and who performed the experimental analysis.

7 REFERENCES

[1]Th. Bier, N.E. Bunt, C. Parr, "Calcium Aluminate Bonded Castables – Their Advantages and Applications", Alafar proceedings, Bariloche, Argentina, Vol. 2, 1996, pp. 73-84.

[2]C.M. George, "Aspects of Calcium Aluminate Cement (CAC)," Hydration *Procedings of the Refractories Symposium*, American Ceramic Society, St. Louis Section, St. Louis, 1994.

[3]K. L Scrivener and A. Capmas, Calcium Aluminate Cements"; Chapter 13 in *Lea's Chemistry of Cement and Concrete*, 4th ed. Edited by Peter C. Hewlett, John Wiley and Sons New York, 1998.

[4]D. Sorrentino, F. Sorrentino, C.M. George, "Mechanisms of Hydration of Calcium Aluminate Cements," in Scalny J.P. ed. *Materials Science of Concrete*, vol IV. Amercian Ceramic Society 1995; 41-90.

[5]A.Capmas, D. Ménétrier-Sorrentino, D. Damidot, "Effect of Temperature on Setting Time of Calcium Aluminate Cements", in *Calcium Aluminate Cements*, ed. R.J. Mangabhai, Chapman and Hall, 1990.

[6]H. Fryda, K. Scrivener, Th. Bier, B. Espinosa, "Relation Between Setting Properties of Low Cement Castables and Interactions Within the Binder System (CAC-Fillers-Additives-Water)", *Vol. 3 UNITECR 97 Proceedings*, USA, 1997, pp.1315-1323.

[7]Th. A. Bier, Ch. Parr ; "Admixtures with Calcium Aluminate Cements and CAC Based Castables" *presented at the 28th Annual SA Ceramic Society Symposium at Johannesburg*, RSA 1996.

[8]Th. A. Bier, A. Mathieu, B. Espinosa and C. Marcelon ; "Admixtures and Their Interactions with High Range Calcium Aluminate Cement." ; *UNITECR Proceedings*, Kyoto, Japan, Vol. 1,1995, pp. 357-364.

[9]C. Parr, C. Revais, D. Jones, M. Bennet, "Ageing of a Low Cement Castable", *UNITECR 97 Proceedings*, New Orleans, USA, Vol. 1, 1997, pp. 81-90.

[10]J.P. Bayoux, J.P. Letourneux, C. M. George, "Fumed Silica and Cement Interactions Parts 1-3 ", *UNITECR 89*, Anaheim, USA, 1989.

[11]Th. A. Bier, A. Mathieu, B. Espinosa and J. P. Bayoux, "The Use of Conductimetry to Characterise the Reactivity of Calcium Aluminate Cements", *UNITECR*, Sao Paulo, Brazil, 1993, pp.705-716.

[12]D. Sorrentino, J.P. Bayoux,R. Montgomery,A. Mathieu, A. Capmas, "The Effect of Sodium Gluconate and Lithium Carbonate on the Properties of Calcium Aluminate Cements", *UNITECR 91*, Germany 1999, pp. 530-533.

[13]S. Möhmel, W. Gessner, B Ködderitzsch, "A Comparison of Two Different Methods for Stopping the Hydration of Monocalcium Aluminate at 20°C", *International Cement Conference*, Goteborg, Sweden, 1997.

[14]A. Mathieu, B. Espinosa, "Electrical Conductimetry to Control LCC's", *Proceedings of Alafar Conference*, Mexico, 1994, pp. 139-156.

[15]K. Jono, E. Maeda, K. Sorimochi, "Reaction Between Alumina Cement and Phosphate," Vol 50 No. 4, *Taikatbutsu*, Japan, 1998.

[16]E. Maeda, K. Jono, "Adsorption of Silica onto the Cement Particle Surface in Castables", Vol 52 No. 3, *Taikatbutsu*, Japan, 2000.

[17]C. Fentiman, C. M. George, R. Montgomery, "The Heat Evolution Test for Setting Time of Cements and Castables ", *Advances in Ceramics*, Amercian Ceramic Society, 13 ; 1985.

[18]N. Bunt, "Advanced Techniques for Measuring the Rheology of Cement Based Refractories", *Proceedings of the Refractory Sypmposium*, St. Louis Section, American Society, 1993.

[19]Th. A. Bier C. Parr et al, "Workability of Calcium Aluminates Cement Based Castables Containing Magnesia" ; *Proceedings of Alafar Conference*, Bariloche, Argentina, Vol. 2, 1996, pp. 19-29.

[20]Th. A. Bier, Ch. Parr, C. Revais, H. Fryda ; "Chemical Interactions in Calcium Aluminate Cement Based Castables Containing Magnesia.", *UNITECR 97*, New Orleans, Vol. 1, 1997, pp. 15-21.

HIGH TEMPERATURE MECHANICAL BEHAVIOR OF MAGNESIA-GRAPHITE REFRACTORIES *

Carmen Baudín
Instituto de Cerámica y Vidrio, CSIC
Ctra. Valencia km 24.300
Arganda del Rey, Madrid, Spain 28500

ABSTRACT

Technology of magnesia-graphite refractories is symbolic of changes in refractory practice in the last quarter century –relatively expensive lining materials that last long time and function in large part by in situ changes. High temperature mechanical behavior of refractories determine their resistance to wear and thermal shock and, finite element methods used to design structures require accurate mechanical properties of the materials as data inputs. Microstructural modifications and phase formation in magnesia-graphite bricks at high temperature heavily influence their mechanical behavior. Due to the intrinsic chemical instability of these materials, experimental approaches to their mechanical characterization vary widely. In this paper, phase development and microstructural modifications in magnesia-graphite
materials at high temperatures are reviewed in relation with their mechanical behavior. Some examples of mechanical properties are shown for magnesia-graphite and magnesia-graphite-aluminum bricks.

INTRODUCTION

Magnesia-graphite refractories are an essential part of the steelmaking advancement in the last 25 years [1-3]. Their extensive use took place originally to form the working linings in electric arc furnaces. These bricks were latter applied to basic oxygen converters in which the use of magnesia-graphite materials has increased the lining lives from several hundreds to several thousands of heats per campaign.

More recently their use has expanded to other areas of the steelmaking. Magnesia-graphite bricks are employed to form the slag line and hot spots of secondary steelmaking ladles (replacing fused-grain and fused-cast chrome-magnesia), especially in plants operating ladle arc furnaces. Moreover, small quantities of magnesia-graphite materials are used in components, such as sliding nozzle plates or immersion nozzles, for the control of steel flow during continuous casting, specially for calcium-alloy treated steel [(4-5)].

Since 1994 [(6)], magnesia-graphite castables are been developed, and even adopted for routine use [(7)], to repair slag lines of ladles lined with magnesia-graphite bricks, because durability is approximately twice as much as that of alumina-magnesia castables.

Nowadays, magnesia-graphite bricks used to line about 100% of Electric Arc Furnaces (EAF) in Japan, North America and Europe, and about 100% of the converter linings in Japan and North America. But, for historical reasons on the one hand and for reasons of availability of raw materials on the other, the solutions chosen in Europe for the lining of converters and steel ladles sometimes completely deviate from those selected in North America or Japan [(8)]. In Europe, only 80% of the converters and hardly 50% of the slag lines of ladles are lined with magnesia-graphite and these mean shares are distributed very evenly. For instance, the use of graphite-enriched dolomite is generalized in Germany and Austria, and the traditional steel plants in Russia are always linked to dolomite plants. Conversely, magnesia-graphite refractories show considerable application in converters and ladles in Great Britain and Italy.

Magnesia-graphite and magnesia-graphite-antioxidant bricks are basically constituted of fine (50-500 μm) magnesia grains, magnesia aggregates (1-7 mm) and natural graphite flakes (~50-500 μm length) bonded by resins or tars. In general, magnesia proportions range from 80-93 wt.%, graphite flakes from 7-20 wt.% and antioxidant additions range from 0 to about 8 wt.%. One of the most common antioxidants is Al metal, added in levels in the range of 2-5 wt.%.

Magnesia-graphite materials present characteristics that clearly differentiate them from traditional refractories: *as fabricated* bricks lack of ceramic bonds, and they incorporate two major phases that are non-compatible thermodynamically and have highly different morphological characteristics.

Graphite flakes in their long direction can be dimensionally commensurate with the large oxide particles, whereas their widths are of the same order as the finer particles. Due to morphological and size differences between the constituents, mixing is performed in powerful shear mixers, and forming is achieved by uniaxial pressing, most often in double action presses. Carbonaceous binders are added to the original mix or infiltrated after pressing. Nowadays, the most common binders are phenolic resins [(9)], due to health problems associated with other carbonaceous compounds such as pitch or tars, and new combinations of binders to optimize the amount of residual carbon and minimize emissions are being developed [(10)]. Resin-bonded bricks are usually low temperature cured.

Perhaps the most interesting aspect of magnesia-graphite based refractories is that their primary constituents are not thermodynamically stable under the working conditions normal to steelmaking. Oxidation of secondary carbon from binders and graphite and carbo thermal reduction of magnesia are main sources of brick instability. On the other hand, different metals or metal alloys are added to the bricks as "antioxidants", following the basic idea that a constituent with greater tendency to oxidize will protect the carbon phase [(11-17)]. Most investigations report the formation of products of the reaction of these additives with the primary constituents of the brick that might be solid, liquid or gases [(17)]. In particular, solid phases such as Al_4C_3 and $MgAl_2O_4$ have extensively be reported to form in Al-containing bricks [(18-28)].

Impurities in the raw materials, specially reducible oxides such as silica which have less negative formation energy than magnesia, play also an important role in material modifications during use [(21, 27)]. Moreover, as most reactions taking place in magnesia-graphite based bricks involve gases as reactants or products, atmosphere is determinant of brick performance.

As a consequence of the special characteristics that present magnesia-graphite materials, the hot strength, that was the key issue for the performance of basic refractories, is not determined by liquid formation as in the traditional magnesia ones. Moreover, a full understanding of the relationships between the properties of specimens tested in the laboratory and the material performance in use would imply a careful analysis of the use conditions [(28)] and a strict control of testing [(23, 29-30)].

In this paper, the high temperature mechanical behavior of magnesia-graphite refractories is reviewed. In particular, the contribution of the different constituents to the mechanical behavior of the brick, and how microstructural modifications taking place in the materials influence the mechanical behavior are addressed. As atmospheric conditions have an important role in these microstructural modifications, special attention is paid to location in the specimen as well as environment. Some examples of the mechanical behavior of magnesia-graphite without antioxidant and magnesia-graphite-aluminum bricks are used to illustrate the discussion.

ROLE OF GRAPHITE

One of the most important reasons for corrosion of refractories is the penetration of slags into pores, therefore, a classical way to reduce corrosion is to reduce porosity of the materials. But magnesia bricks, as most oxide refractories, present high thermal expansion ($\alpha \approx 14 \cdot 10^{-6} C^{-1}$) and Young's

Modulus ($E \approx 40$-70 GPa) and low thermal conductivity ($k \approx 0.02\ J\cdot m^{-1}s^{-1}K^{-1}$) and, consequently, a decrease in material porosity by improving the ceramic bond is always linked to a decrease in thermal shock resistance.

One of the methods used to diminish porosity of traditional magnesia refractories has been to optimize the amount and distribution of the residual carbon from the organic binders used to conform the brick. But, even in bricks with high residual carbon contents (5-7wt.%), the carbon layer that surrounds the magnesia aggregates is thin (100-300μm), and long term performance of these bricks at high temperature is poor due to the high reactivity of residual carbon [(31)].

Graphite is one of the oldest constituent of refractories, together with clay [(32)]. Added to its high sublimation temperature and non-melting characteristics at normal pressures -graphite does not melt and reaches a vapor pressure of 1mmHg at around 3000°C- graphite has very desirable refractory properties.

Due to its crystalline structure, graphite presents good resistance to wetting by molten slags and metals, and higher oxidation resistance than residual carbon from binders. In a simple way, any reactivity must be seen as deriving from crystal edges because graphite does not have propensity to bond or react across the basal planes. Classical data, reported by T. Hayashi [(33)] show how graphite additions dramatically improve the resistance to slag penetration of magnesia refractories (e.g: decrease of the depth of slag penetration from ≈60mm to ≈5mm for graphite contents as low as 5wt.%).

Moreover, the resistance of magnesia refractories to thermal stresses is highly improved by graphite additions. First, the rather low thermal conductivity of magnesia refractories is increased (e.g: ≈50% increase for 15wt.% graphite added, [(33)]) and Young's Modulus is decreased (e.g: ≈50 decrease for 15wt.% graphite added [(33)]). Second, despite the fact that graphite has a mean thermal expansion coefficient of the same order of that of magnesia ($\alpha_{avg} \approx 13\cdot 10^{-6}\ {}^{\circ}C^{-1}$ for graphite vs $\approx 14\cdot 10^{-6}\ {}^{\circ}C^{-1}$ for MgO) and, indeed in one direction is as high as 30-$33\cdot 10^{-6}\ {}^{\circ}C^{-1}$, its contribution to the thermal expansion of composite refractories is almost negligible, due to the presence of Mrozowski cracks which introduce porosity into which the graphite moves in the expansion. The result of this effect is that many composite refractories with significant graphite levels of natural graphite flakes (up to 30wt%) have a thermal expansion in the region of half that which might be expected from the oxide phases, typically of about 3-$4\cdot 10^{-6}\ {}^{\circ}C^{-1}$ [(34)]. Last, graphite confers high work of fracture to composite refractories ($\approx 120 J/m^2$ for oxide-graphite vs 60-$90 J/m^2$ for oxide refractories) due to three processes that occur during deformation and fracture [(32, 34)]. Graphite deformation is non-linear and accompanied by strains of failure one order of magnitude larger than those of oxide refractories ($\varepsilon \approx 0.15$-0.25 for graphite vs ≈0.02-0.04 for typical oxides); the presence of graphite is responsible for crack branching phenomena during fracture, which provides a tortuous crack path; and graphite flakes confer the refractory plate-like reinforcement through pull out and plate failure.

The mentioned properties: thermal conductivity, k, Young's Modulus, E, thermal expansion coefficient, α, and Work of fracture, γ_{wof}, are involved in the modified classical thermal shock parameter, R_{st}', that describes the resistance to thermal shock of low strength materials with large initial cracks, such as refractories:

$$R_{st}' = k \cdot \gamma_{wof} / \alpha^2 \cdot E$$

Moreover, as Mrozowski cracks can be reversibly open and closed under thermal stresses, they act as thermal stress relievers [(34)].

As a consequence, magnesia-graphite refractories show high thermal shock resistance that increases with graphite content as demonstrated by data on repeated quench tests (e.g: 50% Loss of Young's Modulus after 10 cycles for 5wt.% graphite addition vs 10% Loss for 10wt.% addition [(35)]).

MECHANICAL TESTING

It is well known that mechanical properties and stress-strain relations, together with thermal properties, are needed for finite element analysis of the stress state of structures and pieces. In a first approximation, the behavior of the materials can be considered as linear and the stress state and fracture limit can be calculated from material properties -thermal properties, Young's modulus and fracture strength- at the use temperatures and the external solicitations.

The most used parameter to characterize the mechanical behavior of refractories in general is the fracture stress of parallelepiped bars in three point bending, or modulus of rupture (MOR), and most mechanical data for magnesia-graphite materials are MOR values obtained at room temperature after different thermal treatments, or at high temperature under different atmospheric conditions. Less common strength data are tensile stresses of diametrically compressed discs. Dynamic and static Young's modulus at room and high temperature have been also reported. As it will be seen in what follows, the above parameters are determined by different characteristics of the specimens which, in turn, are greatly affected by temperature and the atmospheric conditions of the test.

Even though the hot face temperature in a steelmaking refractory lining can be as high as 1700°C , the largest number of reported data on modulus of rupture of magnesia-graphite bricks has been determined at temperatures around 1400°C and, in general, good correlation is found between the obtained mechanical parameters and the performance of the materials under abrasive and corrosive solicitations at high temperatures. For instance, classical data reported by S. Takanaga [36], show a direct correlation between modulus of rupture determined at 1400°C and abrasion index by molten steel at temperatures between 1650 and 1700°C.

Several reasons are to explain these agreements because, in practice, magnesia-graphite linings are frequently exposed to temperatures around 1400°C under conditions of high oxygen availability. In converters, preheating treatments by upper lance charging of pure oxygen gas at T≈1400°C, once the brickwork is finished and before operation are reported [37]. In treatment ladles, preheating treatments are performed at 1400°C and temperature of the hot face changes from ≈1650°C to ≈1400°C during use [38].

Moreover, calculations for magnesia-graphite ladle linings indicate that hot face temperatures around 1650°C may imply temperatures at the interface between the working lining and the baking lining, where oxygen is available, around 1400°C [39].

Last, even though atmospheric conditions inside the bricks are favorable for carbo thermal reduction of magnesia and, according to Yamaguchi [40], kinetics of the magnesia-carbon reaction are high from temperatures around 1400°C, the bricks do not volatilize during use, which implies that at some point of the heating treatment the lining must become a closed system. Therefore, testing at temperatures around 1400°C would provide a good image of the capability of the brick to protect itself by in situ microstructural changes.

In simple oxide or non-oxide systems it is always possible to find a neutral atmosphere to perform testing. On the contrary, there is not neutral atmosphere for mechanical testing of magnesia-graphite based refractories, as a consequence of internal instability. For instance, even for testing in Ar flow at oxygen partial pressures for mechanical testing furnaces as low as 10^{-3}, high temperature modulus of rupture values are greatly affected by microstructural modifications in the surface region of the samples such as magnesia dense zone formation or additive-derived new phases [26-27]. Moreover, as gases are reactants and products of the reactions between the material components and the external atmosphere, atmospheric conditions during testing are variable as a function of temperature as well as the volume of material tested [23].

As a consequence, it is not always possible to compare mechanical data obtained in different laboratories and, in many cases, contradictory conclusions about questions such as the performance of one additive or another are reported (e.g: MOR data at 1400°C for MgO-C bricks containing Al, Al-Si or Al-Mg as additives [41-42]). Extreme examples are data, reported by Tammerman and Le Doussal [29], that demonstrate that just by increasing the heating rate of the specimens before testing, MOR values at 1500°C in reducing atmosphere ($Ar+5vol\%H_2$) can be multiplied by seventeen.

FABRICATION-DERIVED EFFECTS ON THE MECHANICAL PROPERTIES

Forming of magnesia-graphite bricks is achieved by uniaxial pressing of usually very large pieces (90-150x250-800x150-210 mm^3), most often in double action presses. Binders are added to the original mix or infiltrated after pressing. In current industrial practice, bricks are nor fired prior to supply to the customer; the resin-bonded bricks are usually low temperature cured.

In principle, high variability of the mechanical parameters of a series of nominally identical magnesia-graphite bricks, as a function of the brick as well as of the zone within each brick, could be expected, as it occurs in other shaped refractories. Conversely, a large number of modulus of rupture determinations at 1450°C, performed in a specially developed apparatus [43] under the optimum testing conditions previously established [23-24] and described elsewhere (flowing Ar, $P_{O2}<10^{-3}$) has demonstrated that variability of this parameter is very small. As an example, for a commercial MgO-C-Al material shaped in blocks of 800x210x150mm^3, ten samples taken from different zones of each of five bricks were tested and all modulus of rupture values could be included in the interval 9±1 MPa.

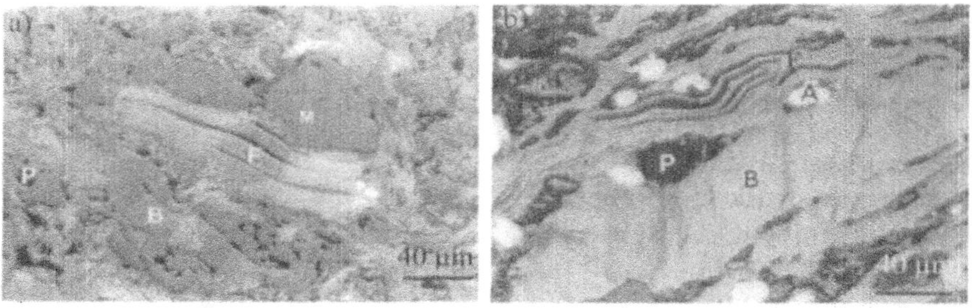

Figure 1: Optical microscope micrographs of polished surfaces of commercial magnesia-graphite refractories. Major components and microstructural features are indicated as follows: M: Magnesia; P: Porosity; B: Graphite flake showing its basal orientation; F: Graphite flake orientated showing its foliated orientation.
a): MgO-C; b): MgO-C-Al.

Due to the high aspect ratio of graphite and the forming method, graphite flakes are partially orientated during uniaxial pressing of the magnesia-graphite bricks [44-45]. Figure 1 shows how in cross surfaces of magnesia-graphite materials basal and foliated oriented graphite flakes are found. Point counting, on optical microscope micrographs of polished sections cut parallel and perpendicular to the pressing direction of commercial MgO-C and MgO-C-Al bricks, demonstrated that the larger quantities of basal graphite were found in surfaces with the surface vector parallel to the pressing direction than in surfaces which vector was perpendicular to the pressing direction (36±2 vs 26±2 surface%) and that, consequently, the opposite trend was followed by the foliated graphite (64±2 vs 74±3 surface%). Conversely, the low porosity present in these bricks (≈3.5vol%) was evenly distributed across the parallel and perpendicular surfaces.

This textural anisotropy affects the values of highly directional mechanical parameters such as dynamic Young's modulus determined by sonic transmission. Reported room temperature values [46] for MgO-C samples coked at different temperatures indicate that when the sound wave propagates perpendicularly to the pressing direction of the brick, E_{per}, values are larger than those obtained when the pressing direction and that of the propagation of the wave are parallel, E_{par}. This difference ($E_{par} \approx 0.25 \cdot E_{per}$) is found in *as received* as well as in coked (T≤1500°C) specimens.

Conversely, dynamic Young's modulus calculated from the resonance frequency of flexured bars after impact is not affected by textural anisotropy, as shown in figure 2. This parameter is a volumetric property which involves small and instantaneous deformations in the samples, therefore, values are controlled by the most rigid phase present in the material, the magnesia aggregates and, consequently, orientation effects are not observed.

Figure 2: Effect of texture on room temperature dynamic Young's modulus, E, of a series of *as received* MgO-C and MgO-C-Al bricks. E was determined from the resonance frequency of bars tested by impact in flexure.
E Perpendicular: impact perpendicular to the pressing direction of the brick.
E Parallel: impact parallel to the pressing direction of the brick.

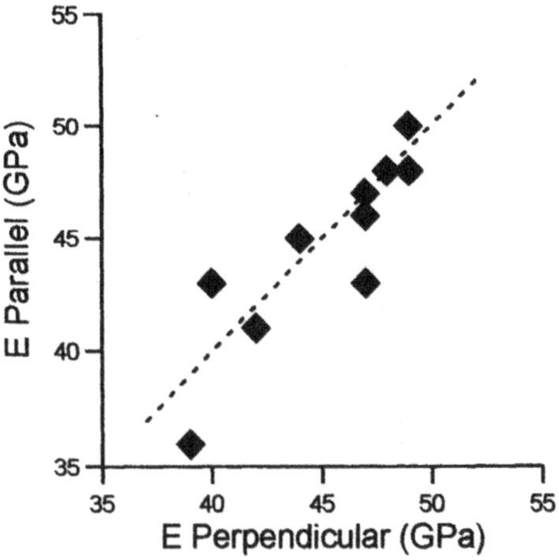

Another mechanical parameter for which no orientation effects are found is the tensile strength of as received and low temperature coked (N_2 flow, 900 °C- 3h) samples determined by the diametrical compression of discs [45]. From a mechanical stand point the material may be considered as constituted by a large discontinuous phase –the aggregates- surrounded by the matrix -graphite, fine magnesia grains, carbonaceous binder and additives-. Magnesia aggregates give texture to the brick whereas the graphite flakes are responsible for the texture of the matrix, in which the fine MgO grains are located. In *as received* and low temperature coked samples, in which no ceramic bond exists, a system of flaws which determines the fracture path is created around the aggregates because organic binders are low fracture resistance paths. Therefore, magnesia aggregates control the fracture and no orientation effects are found.

LOW TEMPERATURE MECHANICAL BEHAVIOR

The fact that from a mechanical standpoint the MgO aggregates give texture to the *as received* materials, implies that room temperature MOR values for magnesia-graphite bricks are determined by the size of the largest magnesia aggregates and the amount of graphite, as shown in figure 3.

This dependence of MOR on the size of the magnesia aggregates is maintained in the low temperature range ($T \leq 500°C$) for materials fabricated using the same organic binder, even though absolute values are affected by the characteristics of the binders and their evolution with temperature [47-49]. Values in figure correspond to materials containing phenolic resin as additive, that presents a maximum of strength at its hardening temperature (≈150-200°C) followed by a continuous drop due to the course of the pyrolisis [48-49].

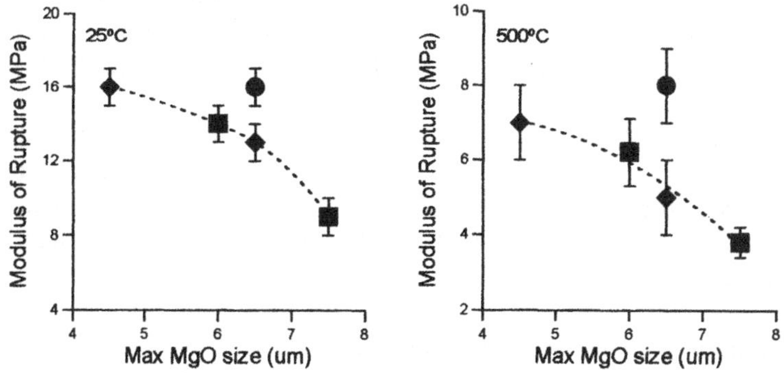

Figure 3: Modulus of rupture at the indicated temperature versus maximum size of the MgO aggregates for MgO-C-Al materials with different compositions:
■ : 2.5wt% Al, 14wt% C; ◆: 5wt% Al, 14wt% C; ● : 5wt% Al, 7wt% C

Summarizing, low temperature mechanical behavior of magnesia-graphite materials is determined by binder characteristics, magnesia aggregate size and graphite content, as a consequence of the heterogeneous microstructure as well as the lack of ceramic bond.

HIGH TEMPERATURE MECHANICAL BEHAVIOR

The dependence of modulus of rupture values of MgO-C-Al bricks on MgO aggregate size is lost when testing temperature increases and a strong dependence on
aluminum content appears. As shown in figure 4, this dependence is only maintained within materials with equal composition and in the higher temperature range (1200-1450°C). Moreover, orientation effects have been observed on room temperature tensile strengths of MgO-C and MgO-C-Al samples after thermal treatments at temperatures between 1200 and 1450°C, determined by the diametrical compression of disks [44-45]

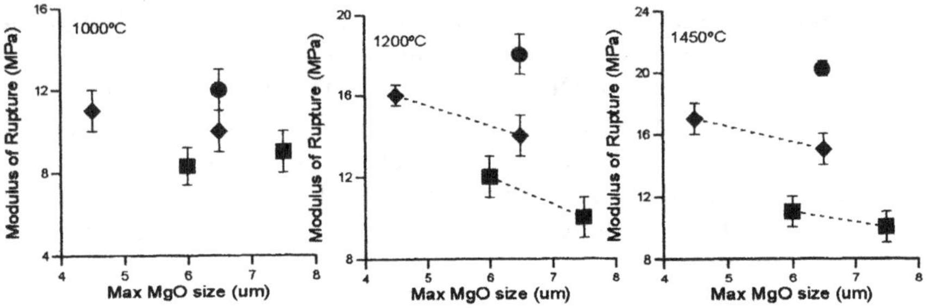

Figure 4: Modulus of rupture at the indicated temperature versus maximum size of the MgO aggregates for MgO-C-Al materials with different compositions:
■ : 2.5wt% Al, 14wt% C; ◆: 5wt% Al, 14wt% C; ● : 5wt% Al, 7wt% C

These observations show that the matrix, in which graphite controls texture, is the determining factor for fracture behavior of specimens tested at high temperature and at room temperature after high temperature heat treatments. Microstructural modifications that take place during heat treatments are responsible for this change of behavior.

Magnesia-graphite materials without antioxidants

Due to the mixed oxide-non oxide characteristics of magnesia-graphite (MgO-C) materials, as well as to the fabrication procedure, the main processes that lead to the modification of their microstructure at high temperature are the following: loss of volatile compounds from binders, reduction of oxide impurities, oxidation of residual carbon and graphite and carbo thermal reduction of magnesia. All these processes imply porosity increase in the bulk of the specimens. As discussed before, carbo thermal reduction of magnesia occurs in the bulk of the materials from 1400°C, temperature from which kinetics of the solid state reaction between magnesia and carbon are large enough [40], and leads to reprecipitation of $Mg_{(g)}$ in the form of MgO in areas accessible to oxygen.

The formation of a dense layer of MgO due to reprecipitation of $Mg_{(g)}$ in the surface region of specimens during standard mechanical testing practices at temperatures larger than 1400°C has been extensively reported [26-27, 40, 50-55]. This layer is very rarely observed in post-mortem specimens from the hot face of the linings, probably due to the abrasive effect of steel. Conversely, reprecipitated magnesia is found in the colder areas. It has been proposed that the role of the dense magnesia zone is to seal off the hot face of magnesia-carbon bricks and, consequently, to slow down magnesia reduction as well as oxidation of carbon [53, 55]. The same effect would have the reprecipitated magnesia found in the colder areas of the linings, usually exposed to large oxygen availability as discussed before.

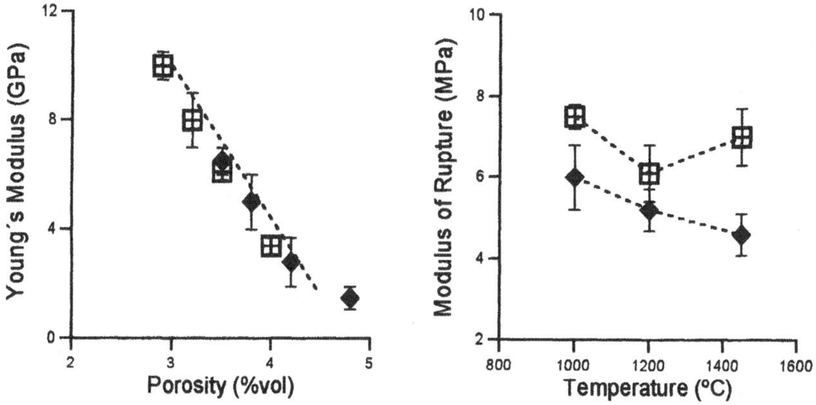

Figure 5: High temperature mechanical properties of two MgO-C materials with nominally the same composition (85wt%C, 15wt%C)

In MgO-C bricks, as in classical refractory systems, the evolution of high temperature static Young´s modulus with temperature is determined by porosity development in the bulk of the samples during testing. In figure 5, this evolution is plotted for two different ladle bricks with nominally identical composition (15wt.%C-85wt.%MgO). Increasing porosity values correspond to increasing testing temperatures (25, 1000, 1200 and 1450°C).

Characteristic evolutions of MOR of MgO-C refractories with temperature are represented by the values determined for the two bricks above described. At low temperatures (T≤1200°C), MOR values follow the same trend as Young's modulus, decreasing as testing temperature increases. Conversely MOR values are strongly affected by modifications in the surface region of the specimens when the testing temperature exceeds 1400°C. MOR values at high temperatures (≈1450°C) are larger than those determined at lower temperatures when a dense magnesia layer that seals defects is formed in the surface region of the specimens. Discontinuous layers made of independent magnesia grains (fig. 6 a), have much less reinforcing capability. The micrograph shown in figure 6 a corresponds to the material for which a decrease of MOR according to Young's modulus is plotted in figure 5.

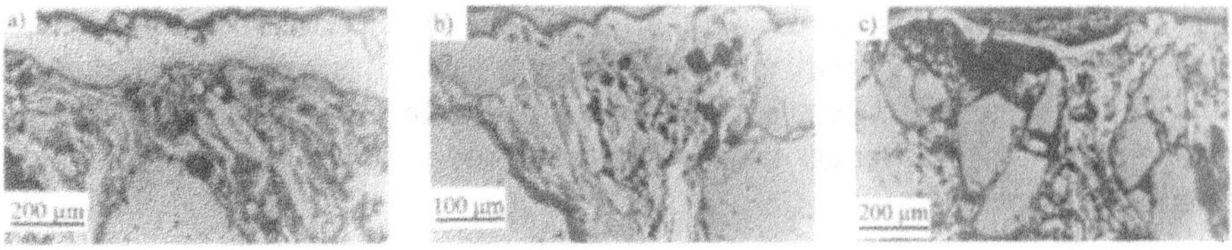

Figure 6: optical microscopy micrographs of polished cross sections of specimens tested (MOR) at 1450°C. Dense magnesia layers are observed in the tension surfaces (upper parts of the micrographs).
a) 85wt%MgO-15wt%C, discontinuous magnesia layer with independent grains; b) 86wt%MgO-10wt%C-2.9wt%Al, microstructure shown in fig. 8 a; c): 80wt%MgO-14wt%C-2.9wt%Al, microstructure shown in fig. 8 b-c.

Magnesia-graphite-Al materials

In magnesia-graphite bricks containing antioxidants, as in MgO-C, porosity increases at high temperature due to the loss of volatile compounds, oxide impurity reduction, residual carbon oxidation and carbo thermal reduction of magnesia. But, in general, materials with antioxidants experience a series of added microstructural modifications originated by reactions of the antioxidants with the components of the brick.

Aluminum-related reactions in MgO-C-Al materials lead to larger porosity increases with temperature than in the in the non-Al containing ones [(26-27)].

In laboratory, aluminum carbide has been reported to form in the bulk of magnesia-graphite-aluminum specimens isothermally treated at temperatures between 750 and 1400°C, in a wide range of atmospheric conditions [(15-16, 22-28, 30, 56)]. Stoichiometric magnesium-aluminate spinel forms in the surface region of specimens tested at temperatures between 1200°C and 1450°C and a highly non-stoichiometric spinel is found in the bulk of specimens tested at temperatures higher than 1400°C [(23-27, 30)]. Moreover, the width of the dense magnesia zone formed in the surface regions of specimens tested at 1450°C has a systematic increase with greater Al contents [26-27].

The new phases formed in isothermally treated specimens correspond to those reported for laboratory specimens tested in temperature gradient [18-19]. At the hot face (T≈1600°C), a dense magnesia zone followed by an area with individual particles of stoichiometric spinel are observed. In the bulk of the specimens, non-stochiometric spinel in the areas subjected to the largest temperatures (≈1550-1300°C) and aluminum carbide in the colder areas (≈1300-1200°C), are found.

As in the case of MgO-C bricks without antioxidants, reprecipitated magnesia is rarely observed in post-mortem Al-containing specimens from the hot face of the linings but it is found in the colder parts. Conversely, extensive formation of stoichiometric spinel is observed in the colder areas as well as in specimens from the hot face, and an aluminum-rich phase identified as spinel [18-19] or aluminum nitride [20] is detected in the bulk of the post-mortem bricks. Up to the knowledge of the author, no aluminum carbide has been reported to appear in post-mortem specimens. The extremely low equilibrium partial pressures of oxygen and of carbon monoxide for Al_4C_3 stability (e.g: 10^{-6}, for CO and 10^{-29}, for O_2, at 1200°C [26]) are locally reached in the bulk of the specimens during short term isothermal treatments, such as standard mechanical testing. Conversely, aluminum carbide is not an equilibrium phase in post-mortem specimens, that have been exposed to long term and variable high temperatures. As a matter of fact, Al_4C_3 is no longer found in MgO-C-Al laboratory specimens exposed to cyclic thermal treatments between room temperature and 1300°C [25]. In this case, aluminum carbide disappears and leads to the formation of the equilibrium phase, magnesium-aluminate spinel, after the first thermal cycle.

Schematic plots of typical evolutions of the mechanical properties of MgO-C-Al bricks are shown in figure 7. Young's modulus values at all temperatures reflect the presence of increasing contents of a low stiffness phase as graphite. The dependence of Young's modulus with temperature is easily explained by the formation of new phases in the bulk of the specimens: aluminum carbide in the low temperature range and of magnesium-aluminate spinel at higher temperatures.

Aluminum carbide is a high modulus phase formed by direct reaction between Al and the secondary carbon originated from the binders, which is more reactive than graphite [26-27]. Aluminum carbide grows in a tabular habit linking the pores associated with its formation reaction and, consequently, the overall effect on the specimen is a stiffening effect. The increase in Young's modulus in the temperature range 1000-1200°C is larger as aluminum content increases, as schematized in figure 7.

Conversely, the magnesium-aluminate spinel that appears in the bulk of the specimens is a low stiffness phase which forms by direct reaction of aluminum carbide and magnesia, having $Mg_{(g)}$ as a product, and presents a highly porous microstructure.

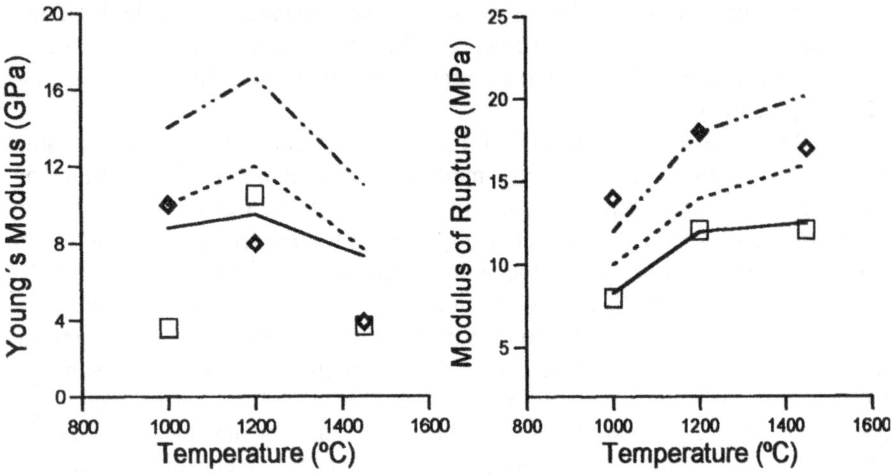

Figure 7: High temperature mechanical properties of MgO-C-Al materials with different compositions. Lines indicate general behavior, simbols represent materials with the special microstructural features shown in figure 8.

————: 2-3wt% Al, 14-15wt% C, 82-84wt% MgO.

---------: 3.5-5wt% Al, 14-15wt% C, 80-83wt% MgO.

—··—··—: 5wt% Al, 7wt% C, 88wt% MgO.

◆ : 2.9wt%Al, 10wt%C, 86wt%MgO. Microstructure shown in fig. 8 a.

□ : 2.9wt%Al, 14wt%C, 80wt%MgO. Microstructure shown in fig. 8 b-c.

Therefore, Young's modulus of MgO-C-Al bricks decreases in the temperature range in which aluminum carbide is no longer present in the specimens (T>1200°C). The same drop has been reported for Young's modulus of specimens tested at lower temperatures (≈1000°C) after one thermal cycle, in which aluminum carbide was not longer present [25].

Modulus of rupture values do not follow the non-monotonous evolution of Young's modulus with temperature. The increase of MOR in the low temperature range corresponds with that of Young's modulus and is due to the stiffening effect of aluminum carbide. The sealing effect of the dense magnesia zone is responsible for MOR to increase in the temperature range in which Young's modulus decreases, as occurred in materials without antioxidants. In specimens constituted by similar quantities of the same MgO and graphite raw materials, larger dense magnesia zones for increasing aluminum contents are formed, because the reaction of aluminum carbide with magnesia is an added source of $Mg_{(g)}$; and, in general, larger zones imply larger values of MOR (e.g: for MgO-14wt.% C, with 2.5, 5 wt.% Al; zone widths=60±10, 100±10μm; MOR=10, 14 MPa; respectively [(26)]).

The stoichiometric spinel particles located close to the surface of the specimens, which are formed from temperatures as low as 1200°C, also improve the mechanical behavior of these bricks in terms of MOR [(23, 26-27, 30)].

The effects of aluminum carbide formation, on Young's modulus and MOR, and of the development of the dense magnesia zone in the surface region of the samples, on MOR, are so strong that even highly non homogeneous Al-containing materials present improved properties. As examples, figure 8 shows two extreme microstructures, that correspond to the materials which mechanical properties are plotted as symbols in figure 7. In one of the materials aluminum grains are large (≈30-700μm) and non homogeneously distributed (fig. 8 a). The other material, present large amounts (≈3.4vol%) of non homogeneously distributed porosity (fig. 8 b-c). The

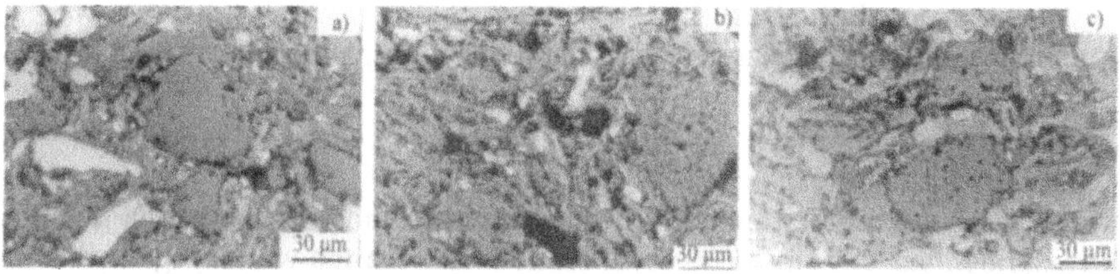

Figure 8: Optical microscope micrographs of polished surfaces of *as received* Magnesia-graphite-aluminum materials with special microstructural features:
a) 2.9wt%Al, 10wt%C, 86wt%MgO. Large aluminum grains are apparent.
b), c) 2.9wt%Al, 14wt%C, 80wt%MgO. Non homogeneously distributed porosity is shown.

specimens of these materials tested at 1450°C presented continuous dense magnesia zones, as shown in figure 6 b-c.

For both materials, Young's modulus values at the temperature of maximum Al_4C_3 formation (1200°C) and MOR values at the temperature at which the dense magnesia zone is well developed (1450°C), are of the same order as the rest of the values schematically plotted in figure 7. Conversely, Young's modulus at the temperatures at which aluminum carbide formation is incipient (1000°C) or has disappeared (1450°C) is very low due to porosity.

CONCLUDING REMARKS

There are two main subjects to be highlighted from the previous discussion: the role played by major components of the magnesia-graphite based bricks on the mechanical behavior, and the meaning of the classical mechanical parameters modulus of rupture and static Young´s modulus.

In magnesia-graphite materials each constituent has a different contribution to the overall mechanical behavior:

- Magnesia aggregates give strength and stiffness to the brick and the largest of them constitute the critical defects.
- Graphite flakes give flexibility to the brick and contribute to the increase of fracture energy.
- The low temperature mechanical behavior of the bricks is determined by the binders.
- The high temperature mechanical behavior of the bricks is determined by the matrix, which is modified at temperature by reactions between the components. These reactions are affected by impurities and atmosphere.

In order to get a meaningful mechanical characterization of the materials it is important to decide which kind of result is being searched:

- Microstructural modifications in the bulk of the specimens determine Young´s modulus values.
- Surface modifications greatly affect modulus of rupture values.

Acknowledgements

It has been possible to carry out the research line on magnesia-graphite refractories thanks to the close collaboration between the working groups from the Instituto de Cerámica y Vidrio (Madrid, Spain) - C. Baudín, C. Alvarez , A. Caballero and E. Criado- and the Ceramic Engineering Department from the U. Missoury-Rolla (Rolla, MO, USA) –R.E. Moore, J. Smith, B. Headrick and M. Karakus-.

REFERENCES

[1] C. Alvarez, E. Criado, C. Baudín, "Refractarios de Magnesia-Grafito", *Bol. Soc. Esp. Ceram. Vidr.*, **31** [5] 397-405 (1992).

[2] J. Mosser, H. Baumgarten, G. Karhut, "Global Perspectives for the Refractory Steel Industry", pp. 5-24 in *Proceedings of the XXVII ALAFAR Meeting.* Asociación Latinoamericana de Fabricantes de Refractarios, Lima, Perú, 1998.

[3].- W.E. Lee, R.E. Moore, "Evolution of in Situ Refractories in the 20th Century", *J. Am. Ceram. Soc.*, **81** [6] 1385-1400 (1998).

[4] K. Akamine, S. Nitawaki, T. Kaneko, M. Harada, "MgO-C Sliding Nozzle Plate for Casting Calcium-Alloy-Treated Steel, *Taikabutsu Overseas,* **18** [1] 22-27 (1998).

[5] Y. Rikimaru, S. Iitsuka, T. Kaneko, M. Harada, "Application of MgO-C Immersion Nozzles for Casting of Ca-Si Treated Steel", *Taikabutsu Overseas,* **17** [1] 48-52 (1997).
[6] S. Sakamoto, Y. Ono, "Graphite Containing Unshaped Refractories", *Taikabutsu Overseas,* **14** [1] 47-51 (1994).
[7] H. Teranishi, T. Kawamura, K. Yasui, I. Imai, "Application of MgO-C Castable to Ladle Furnace Slag Line", *Taikabutsu Overseas,* **18** [1] 38-42 (1998).
[8] J. Piret, "European Tendencies and New Developments of Refractory Materials for Steel Works", *Interceram,* **40** [3] 179-183 (1991).
[9] M.C. Franken, P. Vd. Ham, "Emission of Resin Bonded Refractories: Labtest Practice", in *Lining techniques, Slag Control and Refractory Wear during Steel Production, Proceedings of the XXXIXth International Colloquium on Refractories.* Aachen, Germany, September 24th-25th, 1996.
[10] H. Kato, T. Takahashi, I. Ueno, Sh. Endo, " Development of MgO-C Bricks with Pitch-Phenolic Resin Binder for BOF Lining", in ref. 9.
[11] B. Brezny, "Magnesia Brick Containing Metallic Magnesium"; in *Proceedings of The 65th Steelmaking Conference.* The Iron and Steel Society of the AIME, Pittsburgh, Pennsylvania, USA, 1982.
[12] A. Yamaguchi, "Affects of Oxygen and Nitrogen Partial Pressure on Stability of Metal, Carbide, Nitride and Oxide in Carbon-Containing Refractories", *Taikabutsu Overseas,* **7** [1] 4-13 (1987).
[13] B. Nagai, T. Matsumura, S. Uto, T. Isobe y H. Ohsaki, "Behavior of Magnesia Carbon Bricks Containing Aluminum under Vacuum at High Temperature"; pp. 925-933 in *Proceedings of The 2nd International Conference on Refractories.* Technical Association Refractories of Japan, Tokyo, Japan, 1987.
[14] T. Rymon-Lipinski, "Reaktionen von Metallzusätzen in Magnesia-Kohlenstoff-Steinen in Einem Sauerstoffkonverter"; *Stahl u. Eisen,* **108** [22] 1049-1059 (1988).
[15] A. Watanabe, T. Matsuki, H. Takahashi, M. Takahashi, "Effects of Metallic Elements Addition on the Properties of Magnesia Carbon Bricks"; pp. 125-134 in *Proceedings of The First International Conference on Refractories: Refractories for the Steel Industry.* Technical Association Refractories of Japan, Tokyo, Japan, 1983.
[16] A. Yamaguchi, "Thermochemical Analysis for Reaction Processes of Aluminium and Aluminium-Compounds in Carbon-Containing Refractories"; *Taikabutsu Overseas* **7** [2]11-16 (1987).
[17] M. Rigaud, "New Additives in Carbon-bonded Refractories"; pp: 399-413 in Advances in Science and Technology, Vol. 3 A, *Ceramics: Charting the Future.* Edited by P. Vicenzini. Techna Srl., Florence, Italy, 1995.
[18] R.E. Moore, J.D. Smith, M. Karakus, "Evaluation of Magnesia-Graphite-Metal Bricks Subjected to Thermal Gradient", pp. 124-131 in *Proceedings of the*

UNITECR'95 Congress, vol 3.Technical Association of Refractories of Japan, Kyoto, Japan, 1995.
[19] J.D. Smith, *"Reaction Chemistry and Thermochemistry of Magnesia-Graphite Systems Containing Antioxidants"*; Ph. D. Thesis, University of Missouri-Rolla, Rolla, Missouri, USA, 1993.
[20] M. Karakus, J.D. Smith, R.E. Moore, "Mineralogy of the Carbon Containing Refractories", pp. 745-753 in *Proceedings of the UNITECR'97 Congress*. Edited by M.A. Sett. The Am. Ceram. Soc., Westerville, OH, USA, 1997.
[21] R.E. Moore, J.D. Smith, D. Cramer, C.W. Ramsay, "Reaction between Magnesia-Graphite Metal Components of BOF Type Bricks; Role of Impurities"; pp. 238-241 in ref. 11.
[22] J. Mapiravana, B.B. Argent, B. Rand, "Reactions of Silicon and Aluminium in MgO-Graphite Composites: II. Reaction Products"; pp. 251-253 in *Proceedings of the UNITECR'91 Congress*. The German Refractories Association, Bonn, Germany, 1991.
[23] C. Alvarez, C. Baudín, "Caracterización mecánica en caliente de refractarios a base de grafito", pp. 812-828 in *Proceedings of the XXIII ALAFAR Congress*. Asociación Latinoamericana de Fabricantes de Refractarios, Monterrey, México, 1994.
[24] C. Baudín, C. Alvarez, "Thermal History and Mechanical Behavior of MgO-C Based Refractories", pp. 84-91 in ref. 18.
[25] S. Uchida, K. Ichikawa, K. Niihara, "High Temperature Properties of Unburned MgO-C Bricks Containing Al and Si Powders", *J. Am. Ceram. Soc.*, **81** [11] 2910-16 (1998).
[26] C. Baudín, C. Alvarez, R.E. Moore, "Influence of Chemical reactions in Magnesia-Graphite Refractories: I, Effects on Texture and High-temperature Mechanical Properties", *J. Am. Ceram. Soc.*, **82** [12] 3529-23 (1999).
[27] C. Baudín, C. Alvarez, R.E. Moore, "Influence of Chemical reactions in Magnesia-Graphite Refractories: I, Effects of Aluminum and Graphite Contents in generic Products", *J. Am. Ceram. Soc.*, **82** [12] 3539-48 (1999).
[28] A. Yamaguchi, "Considering the Evaluation and Development of Refractories", *Taikabutsu Overseas*, **17** [4] 6-12 (1997).
[29] E. Tammerman, H. Le Doussal, "Influence of Experimental Conditions on Hot Mechanical Strength of Refractory Products Used in the Iron and Steel Industry", pp. 133-135 in ref. 22.
[30] C. Baudín, B. Headrick, R.E. Moore, "Young's Modulus and Fracture Strength Testing of Magnesia-Graphite Refractories at High Temperature, pp. 413-15 in *Proceedings of the UNITECR'99 Congress*. The German refractory Association, Bonn, Germany, 1999.
[31] K.K. Kapmeyer, D.H. Hubble, "Pitch-Bearing MgO-CaO Refractories for the BOP Process", pp. 2-76 in *High Temperature Oxides*. Edited by A.M. Alper. Academic Press, New York, USA, 1970.

[32] C.F Cooper, , "Flake Graphite, Its Function in Modern Refractories", *Br. Ceram. Trans.* **84** [2] 48-53 (1985).
[33] T. Hayashi, "Recent Development of Refractories Technology in Japan", pp. 5-33 in ref. 15.
[34] C.F. Cooper , I.C. Alexander, C.J. , Hampson, C.J. , "The Role of Graphite in the Thermal Shock Resistance of Refractories", *Br. Ceram. Trans.* **84** [2] 57-62 (1985)
[35] B. Brezny, T.F. Vezza, T.A. LEITZEL, "Selected Refractory Advances in Steel-Handling Systems", *M.R.S. Bulletin*, **14** [11] 38-44 (1989).
[36] S. TAKANAGA, "Wear of Magnesia-Carbon Bricks in BOF", *Taikabutsu Overseas*, **13** [4] 8-14 (1993).
[37] Y. Kubo, J. Yagi, K. Doi, Y. Tanno, M. Satoh, A. Mori, "Improved Design for BOF Lining", pp: 712-723 in ref. 13.
[38] E.S. Chen, H. Ozturk, C. Roulston, "Selection of Ladle Lining Materials for Thermomechanical Performance", pp. 379-388 in ref. 20.
[39] D.A. Bell, F.T. Palin, "Computer Modelling of Mechanical Behaviour of Refractories in Iron and Steel-Making Applications", pp. 480-85 in *Proceedings of the UNITECR'89 Congress*, vol. 1. Edited by Trostel Jr. The Am. Ceram. Soc. Inc., Westerville, OH, USA, 1989.
[40] A.Yamaguchi, "Control of Oxidation-Reduction in MgO-C Refractories", *Taikabutsu Overseas*, **4** [1] 32-37 (1984).
[41] B.H. Baker, R.L. Shultz, "The Importance of High Temperature Properties in Refractory Material Selection", pp: 396-415 in ref. 39.
[42] P.O.R.C. Brant, A.O. Junior y G.I. Lara: "Evolution of BOF Refractory Lining in Brazil"; pp. 1-12 in ref. 22.
[43] C. Alvarez, E. Criado, C. Baudín, G. Duphia, H. Kelichaus, "Hot Modulus of Rupture Automatic Testing Machine"; pp. 435-441 in *Proceedings of the UNITECR'93 Congress*. Asociación Latinoamericana de Fabricantes de Refractarios, Sao Paulo, Brasil, 1993.
[44] C. Alvarez, C. Baudín, "Influence of Graphite-Flake Orientation on Textural Modifications of Graphite-Based Refractories at High Temperature", pp. 108-115 in ref. 18.
[45] C. Alvarez, C. Baudín, E. Criado, "Influence of Heat Treatments on Room Temperature Tensile Strength of MgO-C Refractories Determined by the Diametral Compression Method", pp. 249-257 in *Third Euroceramics*, vol. 3. Edited by P. Durán, J.F. Fernandez. Faenza Editrice Iberica, Castellón de la Plana, Spain, 1993.
[46] S. Hayashi, "Techniques for Evaluating Carbon-Containing refractories", *Taikabutsu Overseas*, **17** [4] 23-30 (1997).
[47] A.M. Fitchett, B. Wilshire, "Mechanical Properties of Carbon-Bearing Magnesia-II. Pitch Bonded Magnesia-Graphite", *Br. Ceram. Soc. Trans. J.*, **83** [2] 59-62 (1984).

[48] A.M. Fitchett, B. Wilshire, "Mechanical Properties of Carbon-Bearing Magnesia-III. Resin Bonded Magnesia and Magnesia-Graphite", *Br. Ceram. Soc. Trans. J.*, **83** [3] 73-75 (1984).
[49] G. Kloβ, W. Schulle, W. Zednicek"Contribution to Microstucture Evaluation of Differing Carbon_Containing Model Compositions", *Radex-Rundschau.*, [1/2] 419-432 (1991).
[50] B. Brezny, I. Peretz, "Reactivity of Periclase with Carbon in Magnesia-Graphite Refractories"; pp. 369-395, vol. 1 in ref. 39.
[51] C.D. Pickering, J.D. Batchelor, "Carbon-MgO Reactions in BOF Refractories"; *Amer. Ceram. Soc. Bull.*, **50** [7] 611-614 (1971).
[52] R.J. Leonard, R.H. Herron, "Significance of Oxidation-Reduction Reactions Within BOF Refractories"; *J. Amer. Ceram. Soc.*, **55** [1] 1-6 (1972).
[53] S.C. Carniglia, "Limitations on Internal Oxidation-Reduction Reactions in BOF Refractories"; *Amer. Ceram. Soc. Bull.* **52** [2] 160-165 (1973).
[54] B.H. Baker, B. Brezny, R.L. Shultz, "Role of Carbon in Steel Plant Refractories", *Amer. Ceram. Soc. Bull.*, **55** [7] 649-654 (1976).
[55] B.H. Baker, B. Brezny, "Dense Zone Formation in Magnesia-Graphite Refractories"; pp. 242-247 in ref. 22.
[56] H. Toritani, T. Kawakami, H. Takahashi, I. Tsuchiya, H. Ishii, "Effect of Metallic Additives on the Oxidation-Reduction Reaction of Magnesia-Carbon Brick"; *Taikabutsu Overseas*, **5** [1] 21-27 (1985).

NEEDED FUNDAMENTAL THERMOMECHANICAL MATERIAL PROPERTIES FOR THERMOMECHANICAL FINITE ELEMENT ANALYSIS OF REFRACTORY STRUCTURES

Charles Arthur Schacht, Ph.D.,P.E.
Consulting Engineer
Schacht Consulting Services
12 holland Road
Pittsburgh, PA 15235

ABSTRACT

The use of computer methods such as finite element analysis (FEA), continues to gain popularity in evaluating the complicated thermomechanical behavior of refractory structures (refractory lining systems). Refractory structures are exposed primarily to thermal type loading. The thermal loads are a result of the thermal expansion of the refractory material and the restraint of this expansion by the support structure.

The thermomechanical analysis of refractory structures is divided into two primary parts, the thermal analysis and the thermal stress analysis. The thermal material properties are often known and commonly provided by the refractory manufacturer. All of the mechanical material properties are typically not provided by the refractory manufacturer, particularly static compressive stress/strain (SCSS) data. There are a limited number of test laboratories that can provide SCSS data. However each laboratory has their own technique for developing SCSS data. In order to provide uniform SCSS data for the engineering community an ASTM standard should be developed.

INTRODUCTION

Computer methods of structural analysis, such as the finite element analysis (FEA) method, continue to gain popularity in the engineering community in solving a multitude of complex structural problems. Because of the complex thermomechanical behavior of most refractory structures (refractory lining systems), FEA is well suited investigation of refractory structural behavior. The thermomechanical behavior of refractory structures is highly complex for lining systems exposed to operating temperatures in excess of about 450 to 550°C. Above these temperatures most of the refractory undergoes plastic deformations, a material nonlinear behavior. In addition brick type joints will support only compressive loads which can be defined as a nonlinear geometric behavior. FEA thermomechanical analysis is required to determine these nonlinear behaviors of the lining.

Often FEA investigative analysis is required to determine if expansion allowance is required to eliminate or minimize mechanical deterioration of the refractory lining. In some cases the interaction refractory lining system with the support structure must be understood to eliminate or minimize distortions in the support structure. To conduct these types of investigations, a reliable temperature dependent stress/strain relationship of the refractory material is required. Static compressive stress/strain (SCSS) data evaluated for the temperature range of interest provides this necessary stress/strain relationship.

REVIEW OF MATERIAL PROPERTY REQUIREMENTS

Both the thermal material properties and mechanical material properties are often sensitive to temperature. Therefore the evaluation of these material properties should take into account the effects of temperature.

The thermal material properties required for thermal analysis are thermal conductivity, specific heat and density. If only steady-state thermal analysis is required, then the later two thermal material properties are not required.

The mechanical material properties required are the coefficient of thermal expansion, the density (if gravity loads are also required) and the stress/strain behavior. In most refractory structures the thermal expansion stresses are often considerably greater than gravity load stresses. Therefore the density is of less importance in the mechanical analysis.

The stress/strain properties are obtained from the SCSS data. An example of SCSS data is shown in Figure 1[1)] for a fired 70% alumina brick. SCSS data curves were developed for temperatures of 22, 800,1100 and 1200°C (70, 1500, 2000 and 2200°F). Note that the stress/strain relationship becomes nonlinear at low values of stress and for temperatures of 800°C and higher. Each SCSS data curve is estimated for the FEA investigation by a bilinear representation also shown on the Figure 1. The bilinear representation curve, for the purpose of establishing definitions, is made up of two primary slopes. The first slope of the first portion, starting at the origin, is the modulus of elasticity (MOE) while the second slope portion is defined as the modulus of plasticity (MOP). The intersection point of these two slopes is defined as the yield point (S_Y). For the SCSS data curves below 800°C the yield point is set at the crushing strength and the MOP is set at a very low value. The bilinear curves provide a reasonable representation of the SCSS data curves. More accurate representation of the SCSS data curve can be made using a series of multiple sloped curves rather than the bilinear curves. If should be noted however that the use of

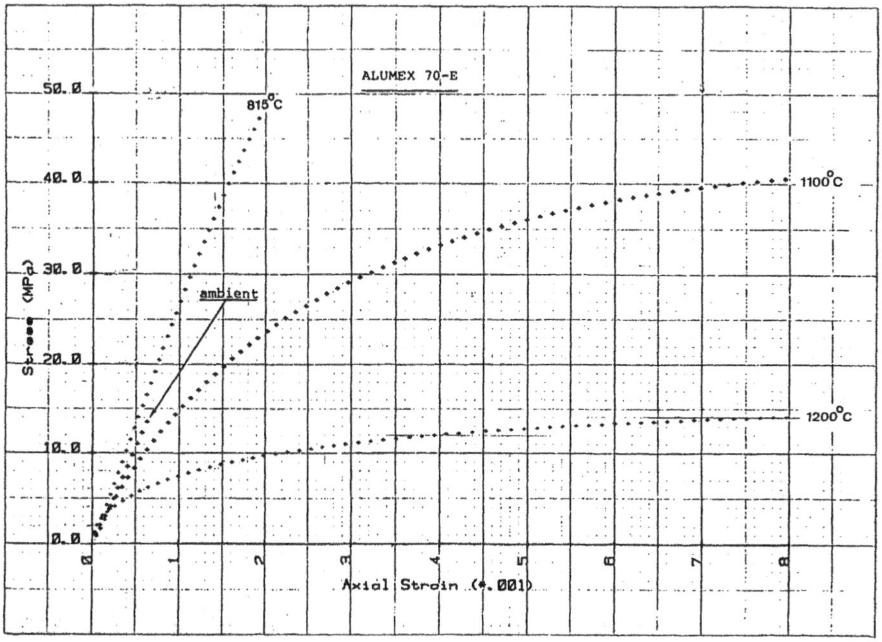

Figure 1 Static Compressive Stress/Strain Data of 70% Fired Alumina Brick

bilinear or multilinear curves is determined by the elastic/plastic analysis capability of the FEA program. If only linear elastic analysis can be conducted by the FEA computer program then the nonlinear SCSS stress/strain data cannot be used.

Figure 2 is a plot of the temperature dependent MOE, for the temperature range of 1100 to about 1400°C, of the fired 70% alumina brick used in the following

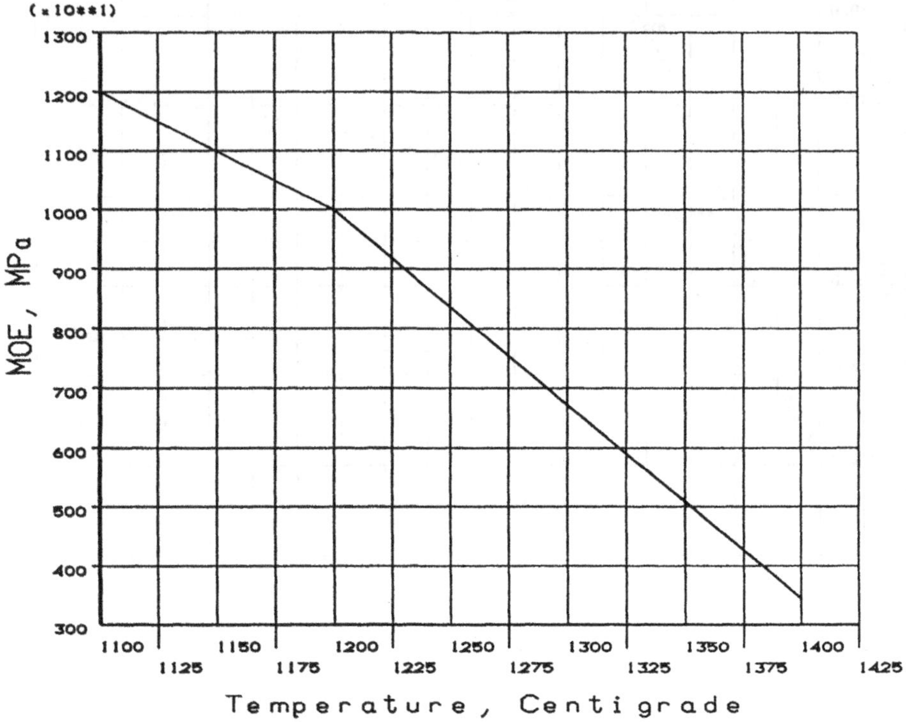

Figure 2 Modulus of Elasticity (MOE) of Fired 70% Alumina Brick

example FEA investigation. The MOE at 1400°C was estimated based on the trend between 1100 and 1200°C. Figure 3 shows the bilinear representation of the SCSS data. One bilinear curve is used for the temperature range of 20 to 800°C.

A comparison of sonic and SCSS MOE data on similar fired 70% alumina brick is shown in Figure 4. Note that the sonic method predicts an unrealistically higher MOE across the temperature range of interest. The second disadvantage of the sonic method is that no information is provided on the nonlinear plastic strain response of the refractory brick especially at the higher temperature range.

CURRENT SCSS TESTING PROCEDURES

Three prominent laboratories have the capabilities to obtain the SCSS data for refractory materials. However each of these laboratories conduct the SCSS tests in different ways. The following information provides a brief outline of the SCSS test

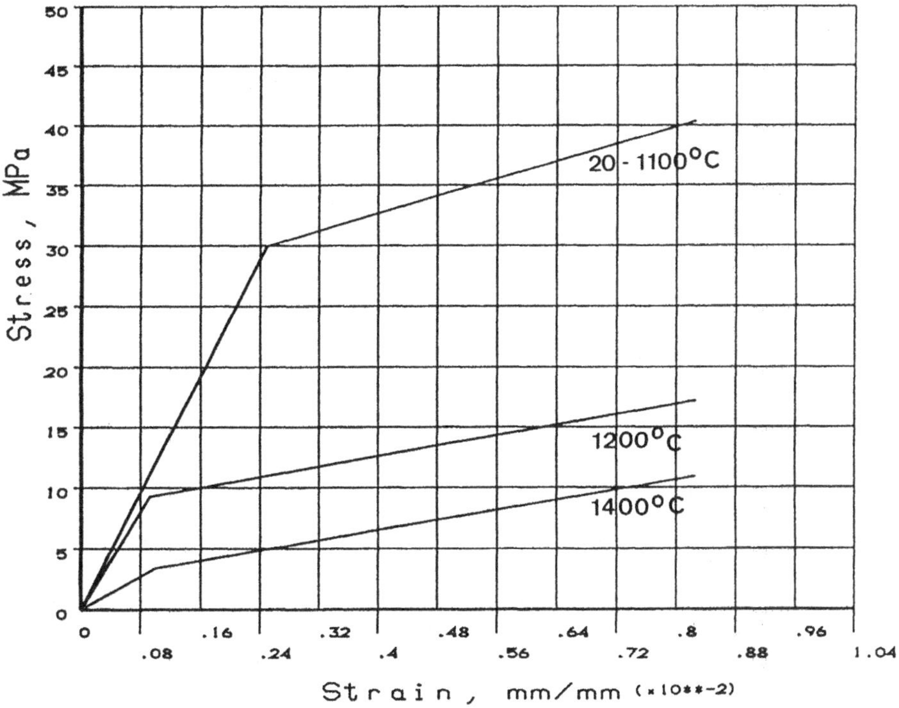

Figure 3 Bilinear representation of Compressive Stress/Strain Data

procedures for these laboratories. The order of the listing does not reflect any preference in the laboratories qualifications to conduct the SCSS tests.

Laboratory No. 1

Sample Size - 63.5 mm (2.5 in.) diameter x 100 mm (4 in.) long.
Strain Measured Over Central 50 mm (2 in.).
Strain Rate at 0.2% per minute.
Loaded to 10 to 50 Mpa but not to crushing strength .

Laboratory No. 2

Sample Size - 25 mm (1 in.) diameter x 50 mm (2 in.) long.
Strain measured over full sample length.
Strain Rate at 0.5% per minute.
Loaded to crushing.

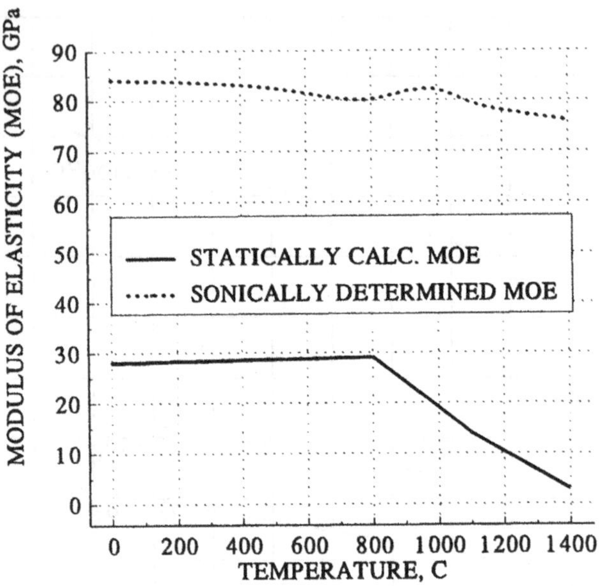

Figure 4 Comparison of SCSS MOE and Sonic MOE

Laboratory No. 3

Sample Size - 35 x 35 mm (1-3/8 x 1-3/8 in.) square x 40 mm (1.6 in.) long.
Strain Measured over full sample length in hole drilled in 40 mm length.
Strain Rate at 0.5% per hour.

The three different laboratory test procedures for developing SCSS data will most likely produce three different SCSS data results. There is a need to establish an ASTM test procedure for determining consistant SCSS data for refractory materials.

EXAMPLE FEA INVESTIGATION

The following FEA example is used to demonstrate the complexity of the thermomechanical behavior of refractory structures. The example describes the complex nonlinear thermomechanical of a refractory sprung arch. The arch had an internal radius of 3658 mm (144 in.) and a span of 3658 mm. The arch 70% alumina brick had a thickness of 343 mm (13.5 in.). A total thickness of 203 mm (8 in.) of light weight insulating material was used on the cold face of the arch. The vertical buckstays supporting the skews are pinned at their base. The top ends of the skews are connected by a tie rod with threaded ends and nuts for adjustment. The hot gas operating temperature was about 1375°C (2500°F). The SCSS data in this example was developed at laboratory No. 1.

The FEA analysis of the refractory sprung arch was evaluated for the steady-state operating temperature of the arch. This refractory arch forms the roof of a tunnel kiln. Each arch skews (refractory brick shape at each end of arch) are supported by the tunnel kiln refractory brick walls (not shown).

Figure 5 shows the temperature contours in the 70% alumina arch brick for the steady-state operating conditions. The through-thickness temperature is nearly linear. As expected at the skew region the arch temperature distribution is greatly influenced by the skew.

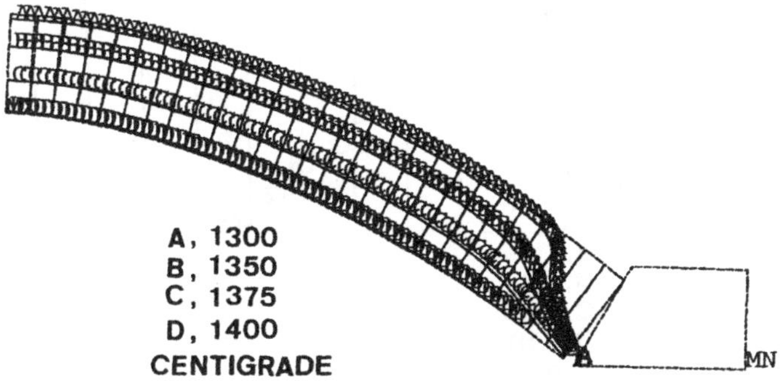

Figure 5 Steady-State Temperatures in Arch Brick

Figure 6 shows the initial thermal expansion displacements just after heat-up of the arch and supporting buckstays. The resulting arch displacement is due to the outward

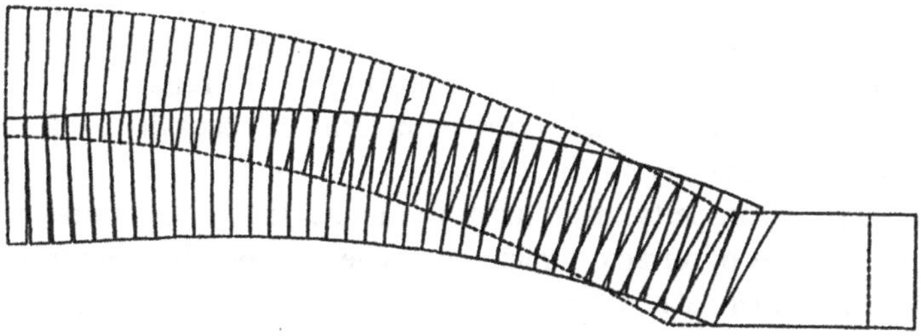

Figure 6 Arch Displacement Due to Thermal Expansion Combined with 125 mm Outward Skew Movement

movement of the skews due to the adjustments made on the threaded tie rod nuts. Figure 7 is an amplification of Figure 6 in the crown region of the arch. The circumferential stress contours are also shown. Peak compressive stresses in the range of 0.70 to 1.4 MPa (100 to 200 psi) in the cold face side of the crown. Likewise similar compressive stresses occur on the arch hot face brick in the region of the skews.

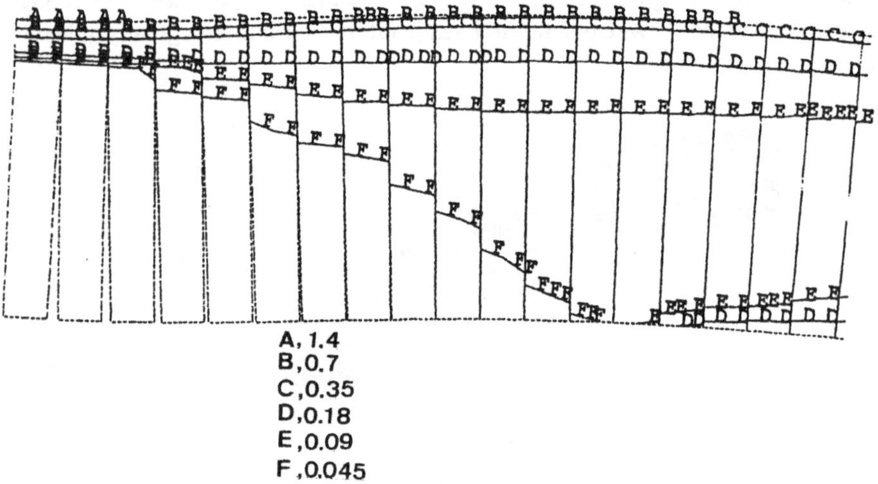

Figure 7 Compressive Stresses in Crown Region of Arch

SUMMARY

The use of computerized structural analysis methods such as FEA is a powerful tool to better understand the complex nonlinear thermomechanical of refractory structures. The example used in this paper of a refractory sprung arch serves to demonstrate the complex thermomechanical behavior of refractory structures and the versatility in the FEA method. SCSS data is an important mechanical material property data used to define the temperature dependent stress/strain properties of refractory materials. The example also demonstrates that the thermal loading (thermal restraint loading) is the most significant load imposed on refractory structures. However there is no ASTM standard for determining the temperature dependent nonlinear stress/strain behavior of refractory material. There are ASTM standards for determining the thermal material properties of refractory materials. To better unify thermomechanical investigations and because thermal restraint loadings are typically the most significant loading on refractory structures existing ASTM standards should be selected and new ASTM standards be developed for thermomechanical investigations of refractory structures.

RECOMMENDATIONS

Based on modern needs to better understand the thermomechanical behavior of refractory structures to improve the life and thermomechanical behavior of refractory structures, the following recommendation is made:

An ASTM standard should be defined specifically for determining the thermomechanical material properties of refractories. The intended use of these thermomechanical refractory material properties would be for formal thermomechanical investigations (such as finite element investigations) of refractory structures. This standard would specify the procedure for determining both the thermal material properties and the mechanical material properties of refractory materials. Existing ASTM standards would be required for establishing thermal material properties to be used in thermomechanical investigations of refractory structures. Existing ASTM standards would be required for establishing the coefficient of thermal expansion of refractory materials to be used in thermomechanical investigations of refractory structures. New ASTM standards would be established for determining the temperature dependent SCSS data for refractory materials. All of the subsets of ASTM standards would be included under one general ASTM standard defined specifically for determining the thermomechanical material properties of refractories intended for FEA investigations.

REFERENCES

[1] G. C. Angell, National Refractories & Minerals Corporation, Refractories Division Research Laboratory, Compressive Stress/Strain Data of Fired 70% Alumina Brick, Circa1980.

POROUS CERAMIC SIMULATION OF RESERVOIR ROCKS DETERMINATION OF POROSITY BY ELECTRIC PERMITTIVITY MEASUREMENTS

Joaquín Lira-Olivares, Diana Marcano and Cyril Lavelle
Surface Engineering Center, Department of Materials Science
Simón Bolívar University
Caracas VENEZUELA

Francisco García Sánchez
Solid State Electronics Laboratory
Simón Bolívar University
Caracas VENEZUELA

ABSTRACT

Control of porosity in ceramics is a mayor task for scientists and engineers. Pores usually have deleterious effects in mechanical properties of ceramic materials, however, they are desired features in fine ceramics, as those used for filters, bone implants, and membranes. A good knowledge of pore measurement and detection is useful not only for the optimization of processes in ceramics, but also for the study of porous rocks. In porous rocks, different physical properties such as density are measured in order to determine quantitatively their porosity. However, a quantitative method for determining porosity by means of electrical properties has not been used yet. The present article sumarizes some of the important characteristics of pores, and some results of porous materials with controlled porosity. Also the effect of porosity at room temperature and atmospheric pressure on the conductivity and on the complex permittivity of porous ceramics that simulates reservoir rocks is reported. When there is a system made of several components labeled "mixture", the permittivity of the mixture can be estimated, from the permittivities and concentrations of each one of the components. Using as primordial materials silica, kaolin, shale and feldspar, ceramics which simulate reservoir rocks were sintered with porosities less than 5% and between 20-30%. These ceramics were characterized by DRX, MEB, picnometry and porosity measurements using classic methods. Also conductance and capacitance measurements were performed in order to determine the ceramic dielectric properties. The measurements were made from 10 Hz to 1 MHz at room temperature. By comparing with the Archimedes method, the porosity could not be estimated from conventional "mixture laws". However, it was possible to calculate the porosity of the system using a fractal description of the medium. The permittivity method and fractal approach seems to detect smaller pores than those detected by Archimedes method.

INTRODUCTION

Pores can be deleterious when accidental or could be desired "defects". A heat insulator, for instance, performs better if there are pores in it. A rock bearing oil would permeate better its treasured fuel, if it has a high density of interconnected pores and a self lubricating bearing will also perform better with high porosity, filters could not exist without pores. On the contrary, negligible porosity is needed in most the utilitarian and decorative ceramics, as well as bearing ceramics, metals or plastic pieces.

While ceramic processing is usually aimed at removing pores to give best mechanical behavior, some porous ceramics or refractories with high surface areas, are used commercially as thermal insulators due to the pores` low thermal conductivity, and as catalysts (transition aluminas), catalyst supports (cordierite) in car exhaust convertes and in molten metal filters. Also, porous ceramics are used for bone implants. Often they are made with a foam like structure by, for instance, green processing them with polystyrene balls which are later burned out. Sponges are also used to ensure interconnected pores and Hot Isostatic Pressure (HIP) has been lately used for oriented porosity. The presence of porosity also has significant effects on optical properties of ceramics. The 5% residual porosity in sintered ceramics will, if the pore size is greater than the wavelength of light, scatter light and render the component opaque.

Porous ceramics based on alumina, silicon carbide, mullite, cordierite and other compositions have been used as filters and honeycomb elements for applications in aeration, nuclear generation plants, dust collectors, acoustic absorbers, catalyst carriers for chemical plants and automobiles, engine components, and heat exchangers. Even electronic materials are being used: a method has been patented for selective filtering of a fluid using porous piezoelectrics (lead zirconate titanate or PZT). Porous ceramics with foamlike structures are used in some of these applications, as well as for thermal insulation, electrodes, light structural laminates, and as burner materials.

The biotechnology and biomedical industries are also using porous ceramics. MgO-$Si0_2$ and diatomite ceramics are being used as ceramic carriers, with pores ranging between 0.02 and 0.01μm, for bioreactors used in fermentation processes, reaction times can be significantly reduced [1]. For bone replacement, porous hydroxyapatite combined with partially stabilized zirconia has been successfully used. Though there are many other applications, this article will focus on the research on porous ceramics towards understanding thermal barrier coatings and bone implants and also porous rocks. In this work the dependence of electrical conductivity and dielectric permittivity on porosity is studied, using porous ceramics that simulate reservoir rocks.

The oil industry has a special interest in studying the porosity of reservoir rocks. The sintering of ceramics which simulate them is a common methodology used to study their physical properties. Previous authors have applied resistivity and dielectric permittivity measurements to detect fluids within porous rocks [2,3,4]. However the porosity determined by other methods [2] still has not been calculated by electrical methods [5]. The porosity of ceramic materials, being these natural or man made, affects their mechanical and electrical properties.

The effect of porosity on the conductivity and on the permittivity of porous ceramics is a subject of current interest [6,7,8,9,10,11,12]. If there is a mathematical model of a system composed of several components labeled "mixture", it is possible to estimate the physical property (permittivity) of the mixture knowing the permittivities and concentrations of each one of the components.

In this work we evaluate the porosity from dielectric permittivity (real part) measurements, in a two component system (matrix + pores). To accomplish this we used porous ceramics which simulate reservoir rocks and we performed conductance and capacitance measurements in order to determine their dielectric properties. The measurements were made in the frequency range from 10 Hz to 1 MHz at room temperature.

Usually, the permittivity in these systems is estimated from the well accepted "mixture laws" [13,14,15,16,17] knowing in advance the permittivities and concentrations of the components. We also study some of the "mixture laws" commonly found in the literature in order to calculate the porosity of the ceramics knowing the dielectric permittivities of the components and determining experimentally the permittivity of the mixtures. However, this formula can not be extrapolated to all situations, due to the fact that they are obtained empirically based on the description of the pore system in terms of simple geometries (spheres, circles, needles, etc).

Some authors using computerized analysis of sedimentary rocks micrographs taken with SEM (Scanning Electron Microscope), have shown that the porous medium [18,19,20,21] can be described using fractal geometry. In this work we propose a "mixture law" considering the natural geometry of the system: irregular (fractal) instead of considering simple geometries. This is achieved taking into account that the dielectric relaxation process develops over a fractal structure, i.e., performing a fractal description of the ceramic (including the pores) structure. The fractal dimension is then linked to the dielectric relaxation process [22].

Also, the porosity can be calculated from the fractal dimension [23,24]. In order to establish comparisons, the porosity of the system is calculated by the Archimedes method.

DEFINITION OF PORE

Cavities, bubbles, cracks and pores, are often called pores. All of them have a common quality; they are volumes within a solid, devoid of solid matter. They also have similar, but not identical consequences on the mechanical properties, as well as the electrical, magnetic and thermal properties of the solids. Their size, morphology, geometrical position within the solid, density and devoid volume content (gas, liquid or vacuum), change greatly their influence on their hostage solid piece.

Usually we refer to cavity, as an intragranular volumetric defect, empty of gas, while bubble is a cavity filled by a gas. Cavities in metals, for instance, are the product of the coalescence of atomic vacancies, thus only stable far from the grain boundaries and after growing few hundreds of microns, while bubbles are stabilized by the gas pressure inside them, even if small. These can often survive near grain borders and surfaces, growing to macroscopic sizes. Cracks preclude the formation, after solidification, of two new surfaces, that if joined, reproduce the perfect solid.

Cracks are the product of bond braking. Pores, on the other hand, are voids produced during the solidification or sintering of the solid, usually by trapped gas, and they are mostly between the particles or grains, not intragranular as the cavities or bubbles. However, pores could appear within a particle.

The sizes of pores vary, but are usually smaller than the surrounding particles. They are formed internally, but may appear as dimples or superficial porosity. If they are large enough to be seen by the naked eye, we usually talk about holes. The usual locus of large pores is at the union of three or more particles and small pores are usually trapped between the particle walls.

For most purposes, it is important to differentiate interconnected pores from isolated pores. The interconnection of large pores is commonly accomplished through smaller, sometimes capillary pores. The pores surface, that is, its wall, is many times covered by a different material than that of the original particles themselves, in the case of metals, maybe its oxide and in the case of ceramics, a ceramic material of different stoichiometry. If the material is immersed or contains a fluid (liquid or gas) it will most often occupy the pores. It could permeate through the solid, like oil in rock reservoirs, if the pores are interconnected, that is, if the material (rock) has open porosity. Sometimes we talk about effective open porosity. That is, porosity that is open for our purposes. Maybe there are other open pores, but for the physical condition we work under, they did not effectively participate in the permeation of the gas or liquid in a detectable manner.

In fused metals and polymers, the void volume left after solidification gives the porosity. The porosity ϕ, in sintered materials, is expressed with the following relation:

$$\varphi = 1 - P_f \tag{1}$$

P_F is the packing factor. It is impossible to fill the space with particles of irregular shape. Therefore, any particle arrangement to form a solid, implies an apparent porosity or a packing factor less than unity. The packing factor depends on: (1) the shape of the particles, (2) the arrangement of the particles and (3) their size distribution. Small particles might fill the interstices between large particles, thus diminishing porosity. A simple relation such as:

$$\Delta P_F = k\, d / D \tag{2}$$

relates the reduction of the packing factor to the grain size of small particles (d) and that of large particles (D). Agglomerants and/or a liquid phase could further collaborate on diminishing porosity in sintering. Surface diffusion contributes to erase small pores, while bulk diffusion can bring the center of the grains closer together, decreasing in that manner the large voids or pores. Diffusion, as it is known, depends on temperature and surface diffusion is activated at lower temperatures than bulk. Pressure can also help on diminishing porosity. The higher the pressure, the more compact the solid. Thus a temperature and pressure combination allows us to control porosity in most cases.

Pore Measurement

The open or apparent porosity can be measured by absorption. Usually that involves measuring the mass of certain values of dry and saturated materials. If water is used to fill the voids then the following expressions can be utilized

$$(\%\phi) / 100 = (S-D)/V_a \tag{3}$$

where S is the saturated weight, D is the dry weight and V_a is the apparent volume. When others liquids are used to avoid dissolving or hydrating (that is disintegration) we have to make the adequate correction for the specific mass. When there is an alloy or the volume of the particles is of irregular shape, the porosity could be measured by immersion using the following expression:

$$(\%\phi)/ 100=(S- D) /(S - M) \tag{4}$$

where M is the suspended weight. Three types of densities and also of specific mass are found. The real density that is the mass divided by the real volume; the apparent total density or bulk density which is the mass per the total volume, which include all the porous and the occupied space. The apparent density, which is the mass per the apparent volume includes the compact material and the close pores. The zone of the compact volume and the apparent porous space is equal to the apparent volume.

Permeability

If the distribution of grain sizes and other factors are equal, a change in the mean grain sizes of the particles would not affect the porosity of the aggregate, however it would affect its permeability. The speed V with which a gas or liquid will traverse a porous aggregate, is given by:

$$V = K^* D^2_P(-\Delta P)F_P / \eta L \quad (5)$$

where K is a proportional constant, D_P is a characteristic dimension of the particle as its diameter or granulometric classification, η is the viscosity of the fluid, and $-\Delta P/L$ is the pressure gradient. F_P is a packing coefficient that changes with the apparent porosity and the spheroidicity of the particle. If the porosity and the spheroidicity diminish, F_P diminishes [25]. Then we can notice that the effect of particle size, as a result of its second power, is significantly high. From the previous equation K, D^2_P, and F_P are sometimes combined as the coefficient of permeability, K′, giving:

$$V=K'(-\Delta P)/\eta L \quad (6)$$

As permeability depends on porosity, as seen in equation (1), it could be used to measure porosity.

Rule of Mixtures

We can see that ceramics are best not considered as single phase materials since they usually contain a continuous second phase at grain boundaries and also an extensive porosity. Usually this grain boundary phase is a silicate glass and it is common for the properties of, for example, an alumina ceramic to lie between those of pure alumina and those of glassy silicate ceramics such as porcelains. For many physical properties the rule of mixtures can be applied to correlate behavior with the oxide/silicate composition so we had better review this rule. Properties of composite materials are commonly analyzed in terms of properties of contributing materials. The mixture rules can be used to determine the expected properties as a function of the distribution and content of each material.

The simplest rules are for scalar properties such as density. The density of a mixture, ρ_m, is:

$$\rho_m = V_1\rho_1 + V_2\rho_2 + \ldots \quad (7)$$

or simply:

$$\rho_m = \Sigma\, V_i\, \rho_i \quad (8)$$

Where V_i is the volume fraction of each contributor.

Thermal conductivity on the other hand is directional so that the geometry of the mixture must enter the calculation. Three cases can be taken into account. Consider the coefficient of thermal conductivity k. When the conduction is parallel to the plane of the slabs and dominated by the better conductor the equation:

$$k_{//} = V_1\, k_1 + V_2\, k_2 + \ldots \quad (9)$$

applies. When the layers are in series, thermal conduction is perpendicular to them and dominated by the lower conductivity ceramic leading to a reciprocal mixture rule

$$1/k_{perp} = V_1/k_1 + V_2/k_2 + \ldots \quad (10)$$

More often microstructures involve a dispersion of two or more phases. To a first approximation, thermal conductivity in such cases may be considered to vary linearly on a volume fraction basis. However, the equations must be modified if one of the two phases is continuous, c, and the other dispersed, d, as for many ceramics [26]

With $k_d << k_c$

$$K_m = \text{approx } k_c\,(1+2\,V_d)/(1\,-V_d) \quad (11)$$

With $k_d >> k_c$

$$K_m = \text{approx } k_c(1-V_d)\,/\,(1+V_d/2) \quad (12)$$

where V_c, and V_d, are the volume fractions of the continuous and disperse phases, respectively. The dominant role of the continuous phase is clear, even though it may be minor in amount. If the continuous phase has high conductivity it forms a continuous path for heat flow. If it is an insulator then transport is limited even if large amounts of isolated high conducting phase are present.

While many ceramic properties are additive and so amenable to these mixture rules many are interactive such as the deformation of a two phase microstructure. In this case the total ceramic response depends on the interaction between the various phases. For example, if the load is placed on a CMC (ceramic-metal-ceramic) the elastic behavior of each grain or phase is influenced by the load carried by the adjacent material. The absence of fracture with the higher Young's modulus will have a proportionately higher tensile stress, with subsequent shear stresses along grain and phase boundaries. These distributions are further complicated by differences in Poisson's ratio. Pores are flaws and stress raisers in a ceramic component and also have a direct deleterious effect on mechanical properties through the Griffith equation:

$$\sigma_f = Yk_{ic}/c^{1/2} \qquad (13)$$

where σ_f is the tensile strength, Y is a constant determined by flaw size and specimen geometry, c is the critical flaw size and K_{ic} is the critical stress intensity factor or toughness of the material. Assuming spherical pores the largest pore is the most critical although the non-spherical pores may have stress raising corners.

High open porosity and high mechanical strength are required for many applications. It is difficult, however, to obtain high strength porous materials with high open porosity, because these properties are contradictory. Although there are different methods of manufacturing porous materials, the properties of porous materials produced by these methods are not satisfactory. For instance, in sintering at lower temperature than a normal sintering, bridging parts between powder particles become weak. A new method to obtain porous materials by capsule-free Hot Isostatic Pressure (HIP) was reported [27].

Porous materials produced by a capsule-free HIP process pertain excellent properties such as high open porosity, high mechanical strength, high Young's modulus and narrow pore size distribution. In the capsule free HIP process to produce porous materials, high-pressure gas delays shrinkage of open pores during sintering. By HIP process, it is possible to sinter at higher temperatures than by conventional sintering, while keeping the same porosity for both cases. As a result, the HIPed porous materials have a well-grown bridging between particles, even though they have high open porosity. Apparently, bridges or necks, grow faster with some gases than with others. This might be due to the molecular (atomic) weight of the gases that affect thermal conductivity by convection. Neck engrossment, as is well known, depends on surface diffusion, that is thermaly activated first than bulk diffusion. The following sections are results of pore control application studies.

FRACTAL GEOMETRY AND POROSITY

A fractal is a mathematical object with a fractional or fractal dimension D_f [28]. D_f supplies a way to measure the irregularity of systems or processes. Normally we consider that lines have dimension $D_E=1$, surfaces have dimension $D_E=2$ and solids, dimension $D_E=3$. However, a rough curve is an object with dimension between 2 and 3. The dimension of a fractal object is a number that characterizes the way in which the measured longitude grows, meanwhile the scale diminishes. While the topological dimension D_E of a line is always 1, the surface's dimension always is 2, the fractal dimension of an object is any real number between 1 and 3. The fractal dimension is defined as:

$$D_f = \frac{\log(L_2/L_1)}{\log(s_1/s_2)} = \frac{\ln(L_2/L_1)}{\ln(s_1/s_2)} \tag{14}$$

where, L_1 and L_2 are two measured lengths of the same curve, and s_1, s_2 are the sizes of the yardsticks used in the measurements. When measuring the length of a coast line, due to the scale invariance present, the length of the coast line grows as the yardstick diminishes, in agreement with (14). This equation is also called a power law relationship. Many other phenomena are scale invariant. As examples, we can mention the frequency-size distributions of rock fragments [29], earthquakes [30,31], topographical characterization of rough surfaces [32,33], fractured surfaces of materials [34], pores and pore space of sedimentary rocks [18,19,21,23], the Brownian movement of particles [28] and the conductivity-dielectric permittivity of disordered materials [35,22].

It is important to emphasize that real fractals are an idealization. No curve or surface in the real world is a fractal. A fractal is a mathematical object. Real objects are produced by processes that set over a finite range of scales.
Therefore estimates of D_f can vary with scale. This variation may serve to characterize the relative importance of different processes at different scales.

Porous real materials show fractal structures over some scales. The interface solid-pore may also be a fractal. Assuming that the pore space volume is determined by the fractal structure, it is possible to find a relationship between porosity and fractal dimension [36]. Therefore for a volume element with macroscopic linear size H, the pore space volume V_P is given in units of characteristic pore size h as

$$Vp = G\,(H/h)^{D_f} \tag{15}$$

Where D_f is the fractal dimension, H and h agree with the upper and lower limits of similarity respectively. The constant G is a geometrical factor. The volume of the whole sample scales in h units as:

$$V = G(H / h)^{D_E} \tag{16}$$

Where D_E is the Euclidean dimension. Therefore, porosity can be written as [36]

$$\phi = \left(h/H\right)^{3-D_f} \tag{17}$$

where ϕ is the porosity, h and H are the lower and upper limits of similarity and D_f is the fractal dimension of the object.

Using this model implies knowing the similarity interval of the porous medium, and also that the measurement technique coincides with this interval. In addition the fractal dimension must be constant during this interval. In this case the parameter ϕ in equation (17) has the meaning of a macroscopic porosity of the material. Equation (17) has been used to obtain the porosity in sedimentary rocks from SEM measurements [21]. Using the technique of the determination of lengths with different yardsticks [18] or with the counting box method [19], the fractal dimension of the rocks were assessed. In this work we want to calculate the porosity from dielectric permittivity measurements. We show how the fractal dimension of the ceramic samples can be determined from the dielectric permittivity measurements, and then the porosity is calculated by equation (17). We assume that the porous medium of a ceramic can be described as a self-affine fractal [37] due to the fact that in different directions the irregularities can scale in different ways, i.e., in this way we can take into account the irregularities present at the structure and the irregularities present at the solid-pore interface.

POLARIZATION AND DIELECTRIC PERMITTIVITY

The term dielectric is used for any system capable of polarizing itself under the effect of an electric field [38]. When an electric field is applied to a material, in addition to a current of free charges, there is a local redistribution of bound charges to new equilibrium positions. This phenomenon of charge redistribution is called polarization. When the electric field varies sufficiently slow, bound charges can keep up with the changing field and be in equilibrium with the instantaneous value of the field. But if the field oscillates too rapidly, certain charge redistribution processes will not be able to keep up with the field and fully contribute to the polarization. The relationship linking the applied electric field (E) and the polarization (P) is given by

$$P = (\varepsilon - \varepsilon_0)E \tag{18}$$

where ε_0 dielectric permittivity of free space (ε_0 = 8,85x10'^{2} F/m) and ε is the dielectric permittivity of the material. In ceramic materials the following polarization processes can take place [39] electronic polarization, ionic polarization, dipolar orientation polarization and Maxwell-Wagner-Sillars (MWS) [40] polarization. The last one occurs in heterogeneous materials where the polarization is consequence of solid interface effects. These effects, strongly dependent on microstructure and constituents materials, are usually important at frequencies of 100 Hz and lower. In the ceramic samples used in this work, as they do not have permanent dipoles, the possible polarization processes are the ionic, electronic and MWS. However, due to the frequency range and the temperature used, the relevant polarization phenomena observed is MWS.

In general the parameter ε can be a tensor and also a complex quantity. The imaginary part of this quantity represent energy dissipation during charge redistribution. So in this case

$$\varepsilon^* \equiv \varepsilon' - i\varepsilon'' \tag{19}$$

Where ε^* the complex permittivity. The current resulting from the application of a sinusoidal voltage V, with angular frequency ω to a capacitor with capacitance C, made up of two flat electrodes with parallel faces of area S and separation d between which there is a dielectric material, is $I=\omega iCV$. If the capacitor is not perfect, that is to say, if the material undergoes dielectric losses, the resulting current is then written $I=i\omega C^*V$, where $C^*=C' - iC''$ is the capacitance of this non-ideal capacitor. By measuring C' then the real part of the dielectric permittivity can be calculated by means of

$$C' = \frac{k'\varepsilon_0'(\omega)S}{d} \tag{20}$$

Where k' is the relative dielectric permittivity. The imaginary part of the dielectric permittivity can be obtained [41] from

$$\varepsilon'' = \frac{\sigma(\omega) - \sigma_f}{\omega\varepsilon_0} \tag{21}$$

DIELECTRIC PERMITTIVITY OF A MIXTURE

In order to understand the behavior of ceramic materials [42] which are polycrystalline and polyphasic, the analysis of the dielectric properties must include the effects produced by porosity and existing phases. This requires the consideration of the dielectric properties of the mixture and the consideration of

MWS polarization that may arise in mixtures of components with different dielectric behaviors.

A dielectric mixing law aims to express the dielectric permittivity of a heterogeneous mixture in terms of the permittivities of the components. Knowing the permittivities and volumetric fractions of the components, the permittivity of a mixture can be determined. For simplicity in notation we consider only the real parts of the permittivities, so k_m is the relative dielectric permittivity (real part) of the mixture, K_1 and k_2 are the relative dielectric permittivities (real parts) of constituents 1 and 2, and v_1 y v_2 corresponds to the volumetric fractions of each constituent, i.e., $v_1 + v_2 = 1$. Also recall that the relative dielectric permittivity of a material is $k = \varepsilon/\varepsilon_0$.

Mixtures in ideal dielectrics [42] can be considered as simple models if the different constituents are in layers in order to simulate a system in series or in parallel to the applied electric field. When the layers are parallel to the capacitor faces, the structure corresponds to capacitive elements in series, then for a two component system (matrix + pores), the corresponding mixture law is:

$$\frac{1}{k_m} = \frac{\phi}{k_a} + \frac{(1-\phi)}{k_{cer}} \qquad (22)$$

Where k_a is the relative permittivity of k_{cer} is the relative dielectric of the matrix (ceramic) and ϕ is the air volumetric fraction or porosity. On the contrary, when the elements are normal to the capacitor faces, the applied electric field is similar to each element, whence the capacitance's are additive and

$$k_m = \phi k_a + (1-\phi) k_{cer} \qquad (23)$$

Equations (22) and (23) are special cases of the general empirical relation

$$k_m^n = \sum_i v_i k_i^n \qquad (24)$$

where n is a constant which is related with the system geometry, v_j is the volumetric fraction of phase i, and when v □ □ k $_v$ □ 1 + vλvk, equation (24) can be written a

$$\ln k_m = \sum_i v_i \ln k_i \qquad (25)$$

therefore for our two component system, the previous equation is:

$$\ln k_m = \phi \ln k_a + (1-\phi)\ln k_{cer} \tag{26}$$

Assuming that it is possible to consider an averaged electric field in a volume which is big compared to the distances at which the heterogeneous character of the medium can be appreciated, the mixture is an homogeneous and isotropic medium and it can be characterized by effective permittivity values designated by ε_m. In this context Landau [43] derived a semi empirical mixture equation

$$k_m^{1/3} = (1-\phi)k_{cer}^{1/3} + \phi k_a^{1/3} \tag{27}$$

where ϕ is the porosity, k_m, k_a and k_{cer} are the relative dielectric permittivities of the mixture, air and ceramic respectively. However this equation does not give any information neither about pore geometry nor possible interactions within pores (open porosity).

In the refractive complex index method (CRIM) [16], the mixture law represents an average temporal equation, i.e., it was derived taking into account that the travel time of an electromagnetic wave through the mixture is equal to the travel time of the wave trough the pore plus the travel time of the wave through the matrix. It is an empirical equation and it is defined by

$$k_m = \phi^2 k_a + (1-\phi)^2 k_{cer} + 2\phi(1-\phi)\sqrt{k_a k_{cer}}\,\frac{\cos(\delta/2)}{\sqrt{\cos\,\delta}} \tag{28}$$

where δ is the loss angle. In deriving this equation, it was not taken into account the pore geometry, nor how they are distributed inside the matrix.

The most commonly used mixing laws are derived under the effective medium approximation (EMA) [15]. This theory is based on the polarization induced by an externally applied, uniform electric field on isolated spherical inclusions located within a host material. The depolarization field due to the inclusions [43] is then calculated. If spherical pores are assumed, then following expression is obtained [15]:

$$k = k_0\left[1 + 2\sum_i v_i \frac{k_i - k_0}{k_i + 2k_0}\right]\left[1 - \sum_i v_i \frac{k_i - k_0}{k_i + 2k_0}\right]^{-1} \tag{29}$$

where k is the effective relative dielectric permittivity (permittivity of the mixture), k_i are the permittivities of the inclusions, k_0 is the permittivity of the lost material and v_i are the volumetric fractions of the inclusions. In a two component system (matrix + pores) where a volumetric fraction ϕ (ϕ>>1) of material with dielectric constant k_a is immersed in a material of dielectric constant k_{cer}, equation (29) can be written as

$$\frac{k_m - k_a}{k_m + 2k_a} = (1-\phi)\frac{k_{cer} - k_a}{k_{cer} + 2k_a} \tag{30}$$

where the approximation k_0=k_{cer} has been used. This equation is known as Clausius- Mossotti (CM) or Maxwell-Garnet [14] equation. If, instead, in equation (29) the approximation k_0=k is used, the mixture law equation for our two component system can be written as

$$\phi\frac{k_a - k_m}{k_a + 2k_m} + (1-\phi)\frac{k_{cer} - k_m}{k_{cer} + 2k_m} = 0 \tag{31}$$

This approximation treats all components in the same way (symmetrically) [14]. This equation is known as symmetrical effective medium equation (SEM). The symmetric term means that if the components are interchanged, interchanging also their respective volumetric fractions, the resulting mixture will have the same effective dielectric constant when compared to the original mixture [44,45].

When non spherical inclusions are considered for deriving the "mixture law" equations, the depolarization factors can also be calculated [43] and the following equations are obtained

Needles

$$k_m = \frac{9k_{cer}^2(k_a + k_{cer}) + 2\phi k_{cer}(k_a - k_{cer})(k_a + 5k_{cer})}{9k_{cer}(k_a + k_{cer}) - \phi(k_a - k_{cer})(k_a + 5k_{cer})} \tag{32}$$

Disks

$$k_m = \frac{9k_{cer}^2 k_a + 2\phi\phi_{cer}(k_a - k_{cer})(2k_a + k_{cer})}{[9k_{cer}k_a - \phi(k_a - k_{cer})(2k_a + k_{cer})]} \tag{33}$$

Many other formulas have been previously obtained [13], however, they all have empirical or semiempirical character and were derived for particular situations. In this work, we use equations (22), (23), (26), (27), (28), (30), (31), (32) and (33) to evaluate the porosity of ceramic samples, knowing in advance, the permittivity of the components and finding experimentally the permittivity of the mixture. However, as will be shown later, this "mixture laws" do not provide satisfactory results.

Further, it is shown how to obtain a general "mixture law", based on a mathematical model that describes better the pore geometry of the system: fractal geometry.

Kohlrausch - Williams - Watts (KWW) function and fractal model of dielectric relaxation

The frequency dependency of dielectric permittivity, for semiconductor and insulator materials, have often been described using empirical dielectric relaxation models such as the Debye model [46], the Cole-Cole model [41], and the Davidson-Cole [47] model. However, it has been observed that these materials show a "universal" behavior identified as a power-law behavior [48], i.e., the relaxation phenomena is not described by an exponential, instead it is described by a potential function. This function represents a non-exponential decay function or an "anomalous" decay, which is slower than the exponential. An empirical function which has been studied [49,50,51] to describe this decay is the KWW function or also called "stretched exponential" function. This function is given by

$$\psi(t) = \exp(-(t/\Gamma))^{\beta} \qquad (34)$$

where $<\tau> = (\tau/\beta)\Gamma(1/\beta)$ is the average relaxation time, Γ is gamma function [52] and slow relaxation is represented by $\beta<1$. For physical systems $0<\beta$□1. When $\beta=1$ the relaxation decay is exponential and the behavior of the system is reduced to a Debye behavior.

Phenomenologically, equation (34) describes the retarded decay of the polarization of a dielectric after the removal of the electric field. For dynamic electric fields the dielectric constants $\varepsilon(\omega)$ is introduced as

$$\frac{\varepsilon(\omega)-\varepsilon(\infty)}{\varepsilon(0)-\varepsilon(\infty)} = -\int_0^{\infty} e^{-i\omega t}\left[\frac{d\psi(t)}{dt}\right]dt = \varepsilon'(\omega)-\varepsilon''(\omega) \qquad (35)$$

i.e., the experimentally observed quantity is related to the function $\psi(t)$ by a Fourier transformation, as a consequence of the Wiener-Khintchine [53,54] theorem. In the classical Debye relaxation theory $\psi(t)$ drops exponentially with $\beta=1$, then using equation (33) we obtain the well known results;

$$\varepsilon'(\omega) = \frac{1}{1+\omega^2\tau^2} \; ; \varepsilon''(\omega) = \frac{\omega\tau}{1+\omega^2\tau^2} \tag{36}$$

Equations (34) and (35) are related to the following general equation by a Fourier transform:

$$\varepsilon'(\omega) = A\omega^{-\beta} \tag{37}$$

where A is a constant and β apparently depends on the inclusion geometry, as will be seen further. Both equations (34) and (37) represent a common characteristic of the dielectric properties in ceramic, insulator and semiconductor materials [35]. In disordered materials (heterogeneous) [22] this behavior is also observed. The existence of this relationship denominated a "power-law" behavior, suggests the similarity property as a characteristic of the physical process and its relationship with the dielectric relaxation process which has been investigated as an explanation to the observed permittivity-frequency behavior [55,56,57]. The application of fractal geometry is performed through considering Brownian movement of the charge carriers in the material, which is a process that has its own fractal dimension [35,58,59], also, is realized trough describing the medium as a fractal structure [10,48,21]. If the random movement of charge carriers (Brownian movement) takes place over the fractal structure, the length L(t) covered by the movement of the charge carriers scales with time trough the dimension D_w [35]:

$$L(t) \approx t^{1/D_w} \approx \omega^{1/D_w} \tag{38}$$

where D_w is the fractal dimension of the movement of charge carriers, besides, $1<D_w<2$. Also, the conductivity of these charge carriers is related to the Brownian movement through the Einstein diffusion coefficient [35]:

$$\sigma(\omega) = \mathrm{en}(\mathrm{eM}(\omega)/\mathrm{KT}) \tag{39}$$

where e is the electron charge, n is a charge density, $M(\omega)$ is the diffusion coefficient in the frequency domain, K is Boltzmann's constant and T is temperature. Also $M(\omega)$ can be written as

$$M(\omega) \approx \frac{L^2(t)}{t} \approx \omega\, L^2(\omega) \approx \omega^{1-(2/Dw)} \tag{40}$$

where we had used equation (37). The AC conductivity that takes place in the fractal structure obeys then the following law [35]:

$$\sigma(\omega) \approx (L(\omega))^{Df-2} M(\omega) \tag{41}$$

where D_f is the fractal dimension of the structure and satisfies $2<D_f<3$. Then, substituting equation (40) in equation (41) we obtain:

$$\sigma(\omega) \approx \omega^{1-Df/Dw} \tag{42}$$

According to Niklasson [35], if the charge transport takes place as a diffusive process in the fractal interface (which can be considered as a bidimensional discontinuity), we can take D_w □ 2 and the previous equation can be written as

$$\sigma \approx \omega^{n} \tag{43}$$

where $n=(D_f-l)/2$ and D_f is the fractal dimension of the structure. In materials which exhibit a negligible DC conductivity at low frequencies and the previous behavior is recognized, the following empirical relations can be established describing the behavior of dielectric permittivity [22]:

$$\varepsilon'(\omega) = \varepsilon'(0)\,,\ \varepsilon''(\omega) = \sigma_f / \omega\varepsilon_0 \qquad \omega<\omega_c \tag{44}$$

$$\varepsilon'(\omega) \approx \varepsilon''(\omega) \approx \omega^{n-1} \qquad \omega>\omega_c$$

where $\varepsilon(\omega) = \varepsilon'(\omega) + \iota\varepsilon''(\omega)$ is the complex dielectric permittivity, σ_f is the conductivity due to free charges, ε_0 is the free space permittivity, and ω_c is a "crossover frequency" where the "power-law" behavior begins to be observed, and n is related to the fractal dimension through $n=(D_f-l)/2$, also $0<n$□1. In addition, ω_c, identifies a region of fractal behavior which crosses over low frequencies to a non fractal behavior, i.e., the fractal structure persists until some "correlation longitude", in which the structure is more regular; ω_c then corresponds to the correlation length.

In experimental situations it is important to distinguish over which range of frequency prevails the fractal or "power-law" behavior. It is important because using this fractal description of the dielectric relaxation process it is possible to relate the fractal dimension of the structure to the porosity (equation 17), where A= h/H defines the similarity range (fractal) in the system, since h and H are the lower and upper limits of the similarity. Therefore a "mixture formula" based on the fractal description of the relaxation phenomena can be defined.

SAMPLE PREPARATION AND DIELECTRIC PROPERTIES MEASUREMENTS

Two groups of samples (A and B) were used. Both groups of samples were prepared with the same constituents (20% Silica, 30% Kaolin, 25% Shale, 25% Feldspar). The chemical composition of these ceramics as well as their porosity simulate reservoir rocks.

The first group (A) were pressed in order to make samples with low connected porosity to form cylinders of the same dimensions as those from the second group. These samples were prepared starting with powders with different particle sizes. Subgroup A_1 with particle sizes less than 38 μm and pressed under a 15, 20 and 25 ton load, and subgroup A_2 with particle sizes less than 48 μm were pressed under a 10 ton load. Then both subgroups (A_1 and A_2) were sintered at 1250 ^{0}C for 1.5 hours. The second group (B) from a previous work [60], prepared by casting in order to obtain a connected porous system and sintered at 1200^{0}C for 2 hours. The sintering temperature used for A group samples, was selected so that in both groups of samples the same phases were present. Besides the particle size in A group samples was selected with the intention that these samples had total porosity less than 5%. Samples of group A, were manufactured because "mixture law" formulas require the κ_{cer} value, i.e., the dielectric permittivity of the matrix value with $\phi = 0\%$. These samples were cut in order to obtain cylinders of 25mm. diameter and 4mm. height. Afterwards porosity calculations were performed over all samples using Archimedes method [61] and picnometry.

DRX analysis were made to both group of samples in order to determine the present phases. This was done to guarantee that the only different variable in the samples was the porosity so that comparisons of the dielectric properties between the samples could be performed. In order to study the samples microstructure, micrographs taken with SEM were analyzed. Also it was possible to determine pore sizes from the images. In addition pore sizes were determined by BET analysis.

All samples were exposed to an ultrasonic bath for 15 min. in a water-acetone solution in order to clean the samples and then allowed to dry in a furnace at 100 ^{0}C for 24 hours. Immediately the conductivities and the complex permittivities measurements from 10Hz to 1MHz at room temperature were determined using impedance spectrometry with parallel electrodes technique. In this method an Impedance Analyzer equipment (model 4192A, Hewlett-Packard) was used. The quantities to measure were the conductance σ[S] and the capacitance C'[F], as a function of frequency. Thus using equations (20) and (21) the dielectric permittivity can be determined. By measuring the permittivity of samples A_1 the κ_{cer} value can be obtained by extrapolation. By measuring the dielectric permittivities of samples A_2 and B the dielectric permittivity value κ_m can be determined.

Porosity determination by dielectric properties measurements.

Determination of porosity using the "mixture law" formulas, involves the values κ_m, κ_a and κ_{cer} which is the value of κ_m, for ϕ=0%. Having these values it is possible to estimate the total porosity ϕ of the samples using equations (22), (23), (26), (27), (28), (30), (31), (32) and (33).

By determining the permittivity-frequency response $\varepsilon'(\omega)$ of the samples, this can be fitted (determining β) to the equation $\varepsilon'(\omega) \sim \omega^{\beta}$ where β=n-1 is a parameter related to the fractal dimension through equation (43). Finally the fractal dimension can be associated to the total porosity through equation (17). In advance, the parameter A = h/H has to be determined, where h and H coincides with the lower and upper limits of similarity, i.e., they correspond to the frequency limits given by "the correlation length" ω_c and ω, in which the system experiences the fractal behavior. If the experiment were to determine the porosity from micrographs, h represents the pore size and H is given by the sample size [21].

EXPERIMENTAL RESULTS

In order to confirm the phases percent in the samples, XRD analysis were performed to one sample of each group A_1, A_2 and B (figure 1). The phases found in all samples were quartz (SiO_2) and mullite ($Al_6Si_2O_{13}$). Consequently, it was possible to study the dielectric properties of the same material, that so far is only dependent on the porosity.

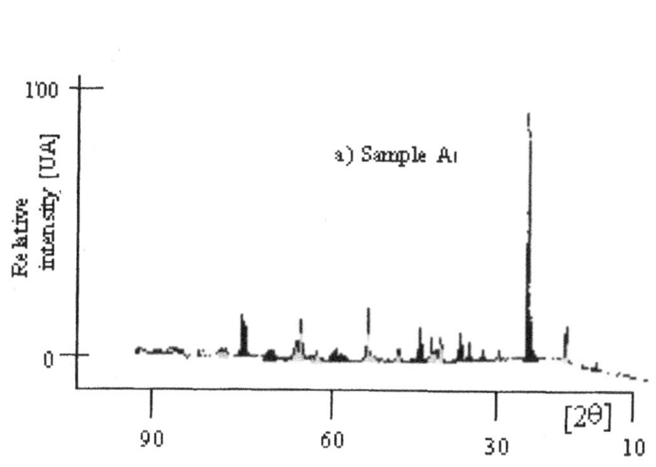

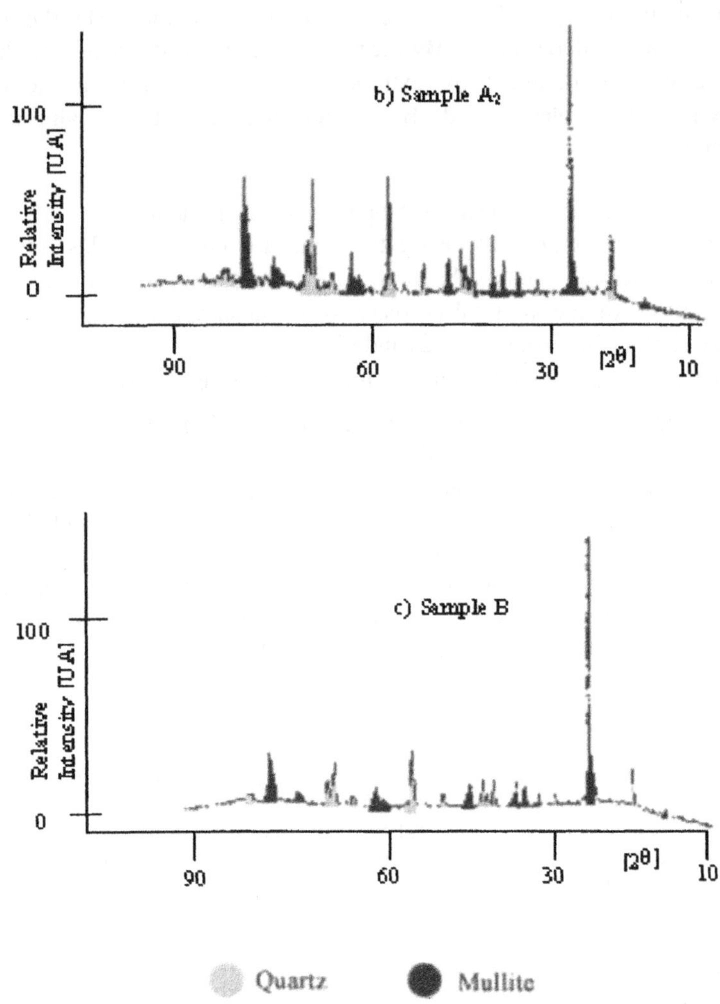

Figure 1. XRD of ceramic samples a) A_1, sintered at 1250°C for 1.5 h b) A_2, sintered at 1250°C for 1.5 h c) B, sintered at 1200°C for 2h. The same phases (mullite and quartz) were found in all samples.

Also with SEM analysis, the ceramics microstructure was analyzed (figures 2, 3 and 4). As can be seen in figure 2, which corresponds to a sample belonging to group A_1, sintered at 1250^{0}C for 1.5 h, dark zones, evidence of pores. Notice that the pores are irregular and isolated (close porosity). Also it can be observed a uniform surface, due to an almost total sintering occurring during the prolonged firing time (1.5 h), in which the vitreous phase flowed enough to fill the pores. Figure 3 corresponds to a sample belonging to group A_2, sintered also at 1250^{0}C for 1.5 h. It can be observed the presence of dark zones, a pore evidence. Notice that the pores are irregular in shape and apparently isolated (close porosity). Besides due to the fact that these samples had a particle size bigger than the used in samples A_2 and were pressed with a lower load (10 ton) than the load used to press samples A_1 there is evidence of a higher porosity than in sample A_1 (figure 2). Figure 4, corresponds to one sample of group B, sintered at 1200 ^{0}C for 2 h. The presence of dark zones, can be observed as evidence of pores. Also note that pores are irregular in shape and are interconnected. There are only a few closed pores. The large pore interconnectivity is due to the fact that these samples were prepared by casting.

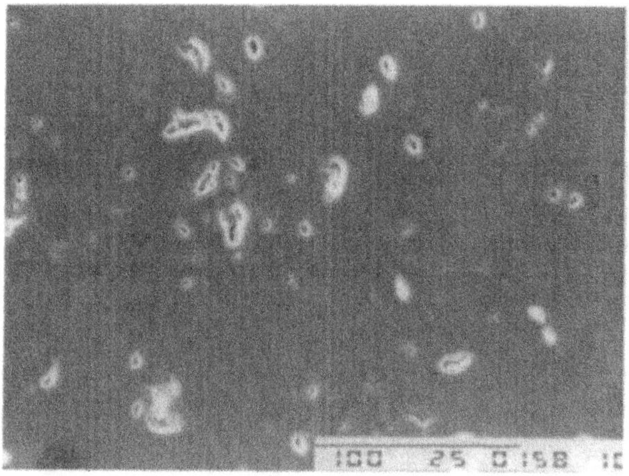

Figure 2. Photomicrography of ceramic sample A_1 sintered at 1250°C for 1.5h. magnified 350X. The darkest zones surrounded by clearer zones, are pores. Notice the irregular shape of pores. No open porosity was observed in these samples. This was corroborated by calculating the porosity by Archimedes method.

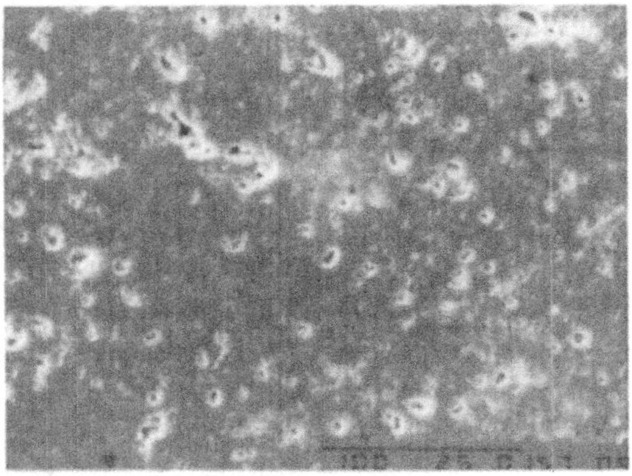

Figure 3. Photomicrography of ceramic sample A_2 sintered at 1250°C for 1.5 h. magnified 350X. The darkest zones surrounded by clearer zones, are pores. Also, notice the presence of a greater number of pores than in sample A_1, due to the lower load used to press these samples. However, it can be observed that almost all the porosity is closed. This was corroborated by Archimedes method. Also note the irregular shape of pores.

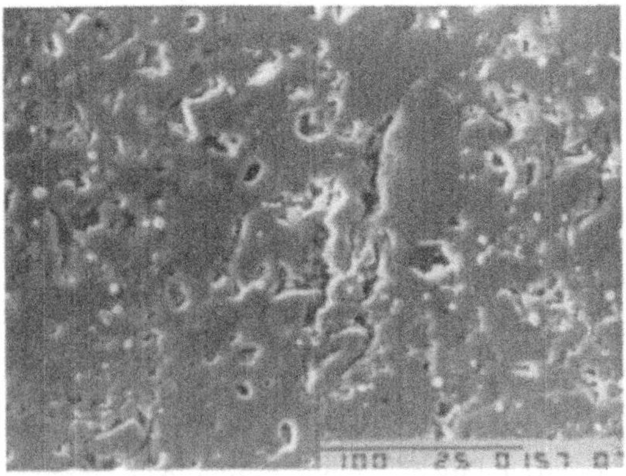

Figure 4. Photomicrography of ceramic sample B sintered at 1200°C for 2 h. magnified 350X. There is a greater number of pores than in samples Notice also that there is an open or interconnected porosity, and less closed porosity. This was corroborated by Archimedes method. Also notice the irregular shape of pores.

In order to determine the ceramic dielectric properties and study its behavior as a function of porosity, capacitance C[F](ω) and conductance σ[S] (ω) measurements, from 10 Hz to 1MHz at room temperature were taken for each one of the samples A_1, A_2 and B. The behavior of conductance as a function of frequency is similar for samples belonging to a same group, i.e., they have the same behavior, for which, in figure 5 is shown the relationship of conductance with frequency for a sample of each group. Figure 5 shows that the conductance of the samples is almost zero al low frequencies. This is the behavior expected for ceramic materials. The conductivity then grows, from a constant value σ_f (due to the free charge carriers) to a value of conductivity al high frequencies given by $\sigma(\omega) = \sigma_f + \sigma_d(\omega)$ where σ_f is the conductivity due to bound charge carries (displacement current). The difference in conductivity behavior may be explained by the difference of porosity in the samples.

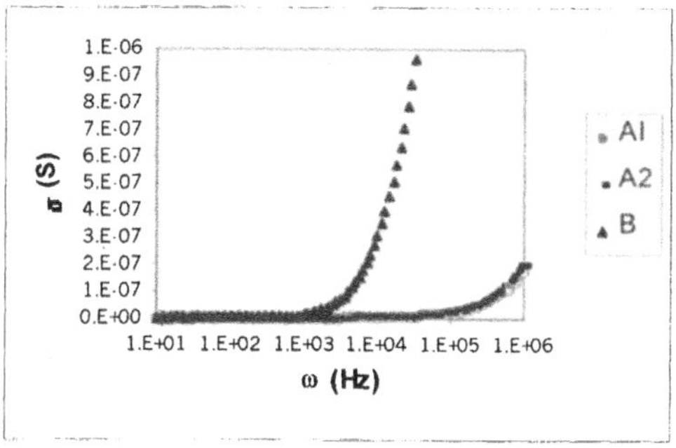

Figure 5. Conductance-frequency behavior for the ceramic samples. Note that at lower frequencies the conductance in samples B is higher than in samples A. Samples B have a higher porosity. A_1, A_2 and B are three different groups of samples.

Subsequently, with the capacitance measurements, the real part of the dielectric permittivity ε' was calculated with equation (20). Figure 6 shows the behavior of the relative dielectric permittivity as a function of frequency for 3 samples of group A_1, namely, for samples pressed with 15, 20 and 25 load. The other samples of this group have a similar permittivity-frequency behavior. Also in figures 7 and 8 the behavior of the relative dielectric permittivity as a function of frequency for some samples belonging to groups A_2 and B shown. The other samples belonging to these groups have a similar permittivity-frequency behavior. Notice that all the samples have the same permittivity-frequency behavior. The behavior of $\kappa'(\omega)$ is common with other ceramic materials and with disordered materials [56]. This behavior is called a power law behavior. Apparently, behaviors of this nature are associated with processes which take place over fractal structures [56]. Also at frequencies between 10 Hz and 100 Hz, is observed the so called anomalous dispersion due to the bound water in the system (Maxwell -Wagner - Sillars polarization) [6], which makes the relative dielectric permittivity values (real part) to be higher to those expected at these frequencies. In any case, the dielectric permittivity reaches the maximum value at low frequencies, for higher values of frequencies the relative dielectric permittivity tends to diminish, this is due to the raise in the conductance with a higher contribution of the displacement current.

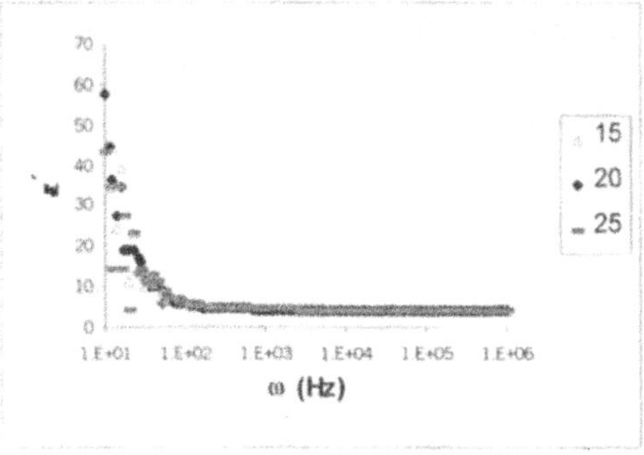

Figure 6. Relative dielectric permittivity (real part) for some samples belonging to group A. Notice that all samples have a similar permittivity-frequency behavior, in this frequency range. Also it can be observed the so called "anomalous dispersion" between 10 Hz and 100 Hz due to bound water in the system.

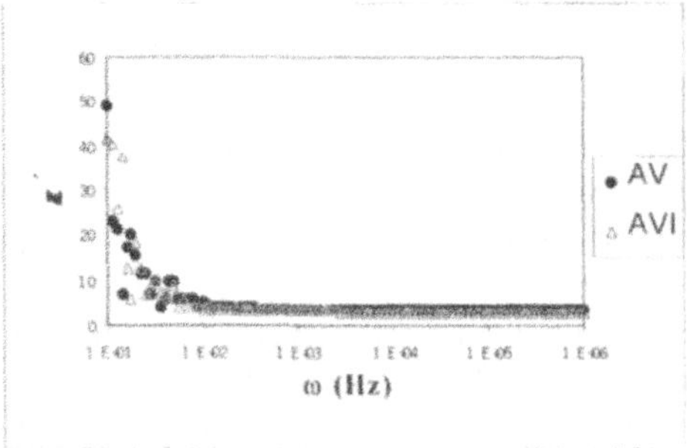

Figure 7. Relative dielectric permittivity (real part) for some samples belonging to group A_2. The other samples belonging to this group have a similar permittivity-frequency behavior. The difference can be found in the curve slopes, associated with the average relaxation time of the system. Also it can be observed the so called "anomalous dispersion" between 10 Hz and 100 Hz due to the bound water in the system. AV and AVI are samples of group A_2.

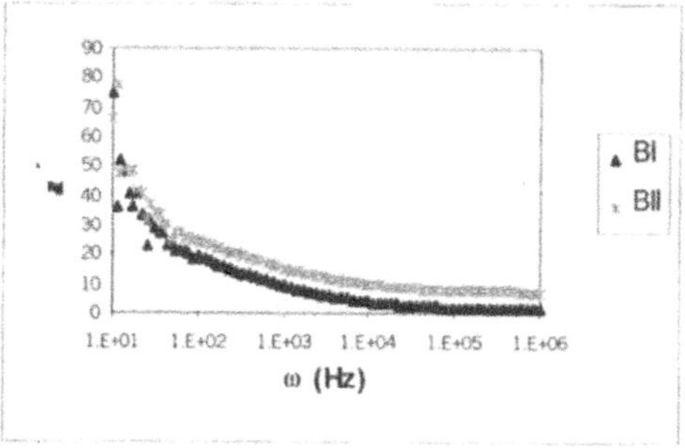

Figure 8. Relative dielectric permittivity (real part) for B group samples. The other samples belonging to group B, show a similar permittivity-frequency behavior. The difference in the curve slopes can be explained by the difference in porosity of the samples. Also it is observed the "anomalous dispersion" between 10 Hz and 100 Hz due to bound water in the system.

Table I below, shows the relative dielectric permittivity (real part) values as a function of porosity (determined by Archimedes method), for different frequency values (excluding the values at 10 Hz due to the anomalous behavior), for samples of group A_I. Also, by plotting κ as a function of porosity (ϕ), then by extrapolation of these results, κ_{cer}. values for $\phi = 0\%$ are obtained.

Table I. Permittivity – Porosity relationship for different samples of groups A_I

Sample	ϕ(total)	$\kappa'(10^2 Hz)$	$\kappa'(10^3 Hz)$	$\kappa'(10^4 Hz)$	$\kappa'(10^5 Hz)$	$\kappa'(10^6 Hz)$
AI	1.3	5.76	4.61	4.42	4.31	4.26
AII	1.5	5.76	4.32	4.09	3.91	3.89
AIII	1.8	5.76	4.61	4.33	4.19	4.11
AIV	1.2	5.76	4.75	4.42	4.26	4.19

The corresponding 10Hz values are not shown since they are bigger than the expected values due to the anomalous dispersion. Notice that at 100 Hz the permittivity values are the same for all samples. As the frequency grows, the permittivity-porosity dependence also grows.

Table II shows the κ_{cer} values that were found for different frequencies. The κ_{cer} values were determined with an error $\Delta\kappa_{cer}$ of approximately 2%. Afterwards these values were used in the "mixture law" formulas. Before calculating the porosities of the samples using the "mixture law" it is necessary to determine which of these formulas describe best the system. For this κ_{cal} is calculated (using the values of total porosity determined by Archimedes method) and is compared with κ_m where κ_{cal} is the calculated mixture permittivity and κ_m the mixture permittivity experimentally measured.

Table II. κ_{cer} values at different frequencies

ω(Hz)	κ_{cer}	$\Delta\kappa_{cer}$
10^2	5.76	$\pm$0.16
10^3	4.56	$\pm$0.16
10^4	4.47	$\pm$0.16
10^5	4.38	$\pm$0.16
10^6	4.25	$\pm$0.16

For the determination of κ_{cal} the equations (22), (23), (26), (27), (28), (30), (31), (32) and (33) were used. In table III, these values are shown (for a sample A_2 with ϕ= 24%) with their respective errors for some frequency values. The errors were calculated by the partial derivate method [62]. From table III it can be observed the difference in the values of the permittivities of the mixture determined experimentally and calculated. The difference is less at frequencies of 1 MHz, due to the fact that dielectric dissipation is less at this frequency. Of all of the "mixture law" formulas those ones that describe better the experimental permittivity-frequency behavior are the formulas corresponding to the model of spherical inclusions (MG), Landau (lan), symmetric effective medium (SEM) and logarithmic (log). However, when the porosity is calculated (table IV) using these models for 1 MHz (since κ' shows a bigger variation with porosity at this frequency) the porosity difference obtained is between 5% and 33%.

Table IIIa. Permittivity (mixture) values vs. frequency for ϕ=24%. It is shown the experimental values, and those predicted by theory for the models: series(s), parallel(p), logarithmic (1), Landau (lan) and the refractive complex index method (crim).

ω(Hz)	κ'_{exp}	κ'_a	κ'_p
10^2	5.04$\pm$0.10	4.61$\pm$0.09	2.68$\pm$0.02
10^3	3.68$\pm$0.07	3.76$\pm$0.07	2.47$\pm$0.02
10^4	3.49$\pm$0.07	3.60$\pm$0.07	2.42$\pm$0.02
10^5	3.38$\pm$0.07	3.51$\pm$0.07	2.39$\pm$0.02
10^6	3.32$\pm$0.07	3.47$\pm$0.07	2.28$\pm$0.02

Table IIIb. Permittivity (mixture) values vs. frequency for ϕ=24%. It is shown the experimental values, and those predicted by the theory for the model: spherical inclusions (cm), disk inclusions (d), needle inclusions (a), and the symmetric medium formula (sem).

ω(Hz)	κ'_{exp}	κ'_{cm}	κ'_{d}	κ'_{a}	κ'_{sem}
10^2	5.04±0.10	3.59±0.46	3.19±0.31	4.20±0.20	4.20±0.18
10^3	3.68±0.07	3.12±0.35	2.91±0.24	3.47±0.16	3.47±0.14
10^4	3.49±0.07	3.02±0.36	2.84±0.22	3.33±0.15	3.33±0.13
10^5	3.38±0.07	2.97±0.35	2.80±0.21	3.26±0.14	3.26±0.12
10^6	3.32±0.07	2.95±0.35	2.79±0.21	3.23±0.14	3.23±0.12

Note that the difference between experimentally determined and calculated values is less for ω=1 MHz, possibly because the dielectric dissipation is less at this frequency.

Table IV. Porosity values obtained the mixture formula that best describe the system, for ω=1 MHz.

ϕ (Archimedes)	ϕ (mg)	$\Delta\phi$(%) (mg)	ϕ (lan)	$\Delta\phi$(%) (lan)	ϕ (sem)	$\Delta\phi$(%) (sem)	ϕ (log)	$\Delta\phi$(%) (log)
0.245	0.16	33.13	0.21	14.39	0.22	9.04	0.17	29.94
0.266	0.25	8.86	0.30	12.52	0.32	17.07	0.26	5.13

Note that the difference in porosity between Archimedes and the dielectric permittivity method using the mixture formulas is between 5% - 33%.

The results in table III and IV show that none of these "mixture law" formulas describes adequately the behavior $\kappa'(\omega)$ and/or $\kappa'(\phi)$. This is due to:

1. These equations were formulated empirically for particular situations.
2. Were formulated for simple geometries, and also for isolated inclusions
3. It is clear that from the permittivity-porosity relationship for a particular frequency, and from the permittivity frequency relationship for a given porosity that an adequate "mixture law" formula must be a function of the frequency and the porosity simultaneously, i.e., $\kappa'(\omega,\phi)$

In order to obtain, a "mixture law" formula which takes into account an irregular non isolated pore structure and the behavior $\kappa'(\omega,\phi)$ a fractal model of dielectric permittivity was used. From the permittivity-frequency response a data fit to the equation $\varepsilon(\omega)\sim\omega^{\beta}$ was made, determining the parameter β. The β values are shown in table V. Also are shown the parameter n, the fractal dimension D_f, the parameter A and the porosity ϕ (%) values, according with equation (17). In the porosity calculations using the fractal model, it is necessary to know in advance, the h (lower limit of similarity) and H (upper limit of similarity). In this work the lower limit of similarity is taken as the crossover frequency ω_c(equation (43)) and as the upper limit as ω= 1 MHz, where it is assumed that the fractal behavior of the system is in this frequency range. These limits were obtained from the graphics of relative dielectric permittivity as a function of frequency for each sample, and the following criteria was used: the similarity range chosen was that corresponding frequency range, where the power law behavior was observed, i.e., the frequency range where the data fit made to the experimental data had the best correlation with a potential function.

Table V. Exponent (β) values, fractal dimension (D_f) and porosity (ϕ), determined by dielectric permittivity measurements.

Sample	β	n	D_f	A(h/H)	ϕ(%) (Fr)	ϕ(%) (Ar)	$\Delta\phi$ (%)
AI	-0.4681	0.5319	2.0638	0.01	1.34	1.3	3
AII	-0.4465	0.5535	2.1070	0.01	1.64	1.5	8
AIII	-0.4285	0.5715	2.1430	0.01	1.93	1.8	7
AIV	-0.4761	0.5239	2.0478	0.01	1.25	1.2	4
AV	-0.1448	0.8552	2.7104	0.01	26.35	26.1	1
AVI	-0.1432	0.8568	2.7136	0.01	26.75	25.3	5
AVII	-0.1512	0.8488	2.6976	0.01	24.85	23.8	4
AVIII	-0.1337	0.8663	2.7327	0.01	29.20	27.6	6
BI	-0.3138	0.6862	2.3725	0.1	23.58	20.0	15
BII	-0.0994	0.9006	2.8012	0.001	25.33	24.5	3
BII	-0.1021	0.8976	2.7957	0.001	24.39	24.3	0.37
BIV	-0.099	0.9067	2.8135	0.001	27.57	25.8	6

The fractal dimension was calculated using D_f =2n+1. where β=n-1. Notice that obtained values of D_f from dielectric permittivity measurements are greater than 2, i.e., they are bigger than the dimension of a surface and lower than a volume dimension (3). Besides, the difference in porosity between Archimedes (Ar) method and fractal (Fr) model is between 0.37%-15%. AI to AVIII are different samples of groups A, same for B.

Notice that the porosity determined by fractal method (Fr) tends to be higher than that determined by Archimedes (Ar). That indicates that the new method includes close porosity and possible smaller pores not detected by Archimedes method.

CONCLUSIONS

Porosity could be considered as a feature more than a defect in porous rocks. The conventional mixture laws could not describe porosity as expected. Also, Archimedes method, fail to measure small pores in rocks and ceramics.

Dielectric permittivity measurements can be used to determine the total porosity of ceramic materials. This can be done effectively by describing the porous medium as a fractal structure instead of estimating the porosity with the common "mixture law" formulas. It was observed that electric conductance grows with a raise in porosity and also with a raise in the fractal dimension of the structure. Besides, the electric conductivity and dielectric permittivity follow a "powerlaw" relationship.

The fractal method allows a higher precision and thus the detection of smaller pores. Further studies in human bones show the possibility of detecting smaller pores than those detected by classical methods (x-rays, ultrasound, etc) . A non intrusive method using eddy currents is being developed and will be presented in a future paper.

ACKNOWLEDGEMENTS

We would like to express our thanks to, Ms. Patricia Schotborgh and Ms. Dennise Urdaneta for their collaboration in the elaboration of the ceramic samples, and the dean of Research and Development of Simón Bolívar University and Department of Materials Science for their financial support. Also to Mr. César Mendoza, undergraduate student, for his help in the final version of this paper, our sincere thanks.

REFERENCES

[1].L.E. Sheppard, "Porous ceramics: processing and applications" in Ceramic Transactions, Vol. 31, pp. 3-4, 1993.

[2].AAPG. Treatise of Petroleum Geology, Reprint Series, No. 17, Formation Evaluation II, pp.2-425, 1990.

[3].Y. Guéguen, V. Palciauskas, "Introduction lo the Physics of Rocks", Cap VIII-IX. Princeton University Press, 1994.

[4]. J. S. Mendoza Sanz. "Introducción a la Física de Rocas"," Introduction to the Physics of Rocks". Equinoccio, Ediciones de la USB, Cap. 5, 1998.

[5]. P. N. Sen, "The Dielectric and Conductivity response of sedimentary rocks", J. Society of Petroleum Engineers, 9379, pp. 2-10, 1980.

[6].G. Pernier, A. Bergeret, "Maxwell-Wagner-Sillars relaxations in polystyrene-glass-bead composites" J. Appl. Phys.", 77(6), 1995.

[7]. A. Gutina, E. Axelrod, A. Puzenko, et. al. "Dielectric relaxation of porous glasses" Journal of Non-Crystalline Solids, 235-237, 302-307, 1998.

[8]. R. Hilfer, B. Nost, E. Haslund, Th. Kautzsch, B. Vingin, B. D. Hansen, "Local porosity theory for the frequency dependent dielectric function of porous rocks and polymer blends", Physica A, 207, pp. 19-27, Elsevier, 1994.

[9]. K. Wakino, "A new equation for predicting the dielectric constant of a mixture" J. Am. Ceram. Soc. 76(10), 2588-94,1993.

[10]. J.P. Calame, A. Birmau, Y. Carmel et al. "A dielectric mixing law for porous ceramics based on fractal boundaries" J. Appl. Phys." 80(7), pp. 3992-4000, 1996.

[11]. A. Puzenko, N. Kozolovlch, A. Gulina, Y. Feldman, "Determination of pore fractal dimensions and porosity of silica glasses from the dielectric response at percolation" Physical Review B, 60(20), pp. 14348-143 59, 1999.

[12]. J. Mizusaki, 5. Tsuchlya, K. Waragai, H. Tagawa, Y. Arai, Y. Kuwayama, "J.Am Am. Ceram. Soc.", 79 [1], pp. 109-13, 1996.

[13].W. R. Tinga, W.A.G. Voss, D.F. Blossey, "Generalized approach lo multiphase dielectric mixture theory" J. Appl. Phys., 44(9), pp. 3897-3902, 1973.

[14]. J. Benyman, "Rock Physics and Phase Relations" AGU Reference 3, pp. 205-228 (1995).

[15].A.Norris, P. Sheng, A.J. Callegari, "Effective medium technics for two phase dielectric media" J. Appl. Phys., 57 (6), 1985.

[16].A. Rodríguez, R. Abreu, "A mixing law to model the dielectric properties of porous media" J. Society of Petroleum Engineers, 21096, pp. 1-7, 1990.

[17].D. Bergman, "The dielectric constant of a composite material — a problem in classical physics" Physics Reports, 43, #9, 377-407, 1978.

[18].A.J. Katz, A. H. Thompson, "Fractal sandstone pores: implication for conductivity and pore formation" Physical Review Letters, 54 (12), pp. 1325-1328, 1985.

[19].J.P. Hansen, A. T. Skjeltorp, "Fractal pore space and rock permeability implications" Physical Review B, 38 (4), pp. 2635-2638, 1988.

[20].E. Guyon, C. D. Mitescu, J. P. Hulin, S. Roux, "Fractals and percolation in porous media and flows" Physica D, 38, pp. 172-178, 1989.

[21].C.E. Krohn, A.H. Thompson, "Fractal sandstone pores: Automated measurements using scanning electron microscope images" Physical Review B, 33 (9), pp. 6366-6374, 1986.

[22]. G. A. Niklasson, "Fractals and the conductivity of disordered materials" Physica D, 38, pp. 260-265, 1989.

[23]. R. R. Nigmatullin, "The generalized fractals and statistical properties of the pore space of the sedimentary rocks" Phys. Stat. Sol. B, 153, pp. 49-57, 1989.

[24]. D. L. Turcotte, "Fractals and Chaos in Geology and Geophysics", Cambridge University Press, pp. 146, 1997.

[25]. L. Van Vlack, "Propiedades dos Materiais Cerâmicos", "Ceramic Materials Properties", Edgard Blücher LTDA, Brazil, pp. 218-219, 1973.

[26]. W. E. Leeand, W. M. Ainforth, "General Influence of Microstructure on Ceramic Properties" in Ceramic Microstructunes, Chapman Hall, pp.69 — 71, 1992.

[27]. M. Nanko, K. Ishizaki, and A. Takata, "Sintering of porous materials by capsule-free HLP process", in Ceramic Transactions, Vol. 31, pp. 117-126, 1993.

[28]. B. B. Mandelbrot, "The Fractal Geometry of Nature", W. H. Freeman and Company, 1999.

[29]. D. L. Turcotte, "Fractals and fragmentation" Journal of Geophysical Research, 91 (B2), pp. 1921-1926, 1986.

[30].P. Okubo, K.Aki, "Fractal geometry in the San Andreas Fault system" Journal of Geophysical Research, 92 (B1), pp. 345-355, 1987.

[31].C.A.Aviles, C.H.Scholz, "Fractal analysis applied to characteristic segments of the San Andreas fault" Journal of Geophysical Research, 92 (B1), pp. 331-344, 1987.

[32].A. Majumdar, B. Bhusian, "Fractal Model of Elastic-Plastic Contact Between Rough Surfaces" Journal of Tribology, 113, pp. 1-11, 1991.

[33].A Majumdar, B. Bhusban, "Role of Fractal Geometry in Roughness Characterization and Contact Mechanics of Surfaces" Journal of Tribology, 112, pp. 205-2 16, 1990.

[34]. C. W. Lung, 5. Z. Zhang, "Fractal Dimension of the Fractured Surface of Materials" Physica D, 38, pp. 242-245, 1989.

[35].G. A. Niklasson, "A fractal description of the dielectric response of disordered materials" 3. Phys. Condena. Mallen, 5, pp. 4233-4242, 1993.

[36]. Idem [10], pp. 14350.

[37]. Idem [19], pp.349

[38]. D. Halliday, R. Resnick, "Física", "Physics". Vol.2 pp. 104, Compañía Editorial Continental SA., (1982).

[39]. W.D. Kingery, H.K. Bowen, D.R. Uhlmann, "Introduction to Ceramics", John Wiley and Sons, 1976. pp. 921-923.

[40]. Idem [6].

[41]. J. Bourne, "Critical Reviews in Biomedical Engineering", Vol. 24, Issues 4-6, p.p. 262, 1996.

[42]. Idem [39], pp. 947.

[43].L.D. Landau, E.M. Lifsbitz, "Electrodinámica de los Medios Continuos", "Electrodynamics of Continuous Media". Cap. II, Editorial Reverté, 1975.

[44]. F.G. Sin, Y.Y. Young, W.I.Tsui "On symmetrical dielectric mixture formulas" Journal of materials science letters, 9, pp. 948- 950, 1990.

[45]. F.G. Sun, Y.Y. Young, W.I.Tsui "Symmetrization of dielectric binary mixture formulae" Journal of materials science letters, 9, pp. 1002- 1004, 1990.

[46]. H. Ibach, H. Luth, "Solid-State Physics. An Introduction to Materials Science", Springer, 1996, pp. 287-326.

[47]. C.P. Lindsey, G.D. Patterson, "Detailed comparison of the Williams-Watts and Cole-Davidson functions." 3. Chem. Phys., 73 (7), pp. 3348-3357, 1980.

[48]. S.Thevanayagam, "Dielectric dispersion of porous media as a fractal phenomenon" 3. Appl. Phys. 82 (5), pp.2538-2547, 1997.

[49]. 5. H. Chung, 3 .J.R. Stevens, " Time-dependent correlation and the evaluation of the stretched exponential or Kolhrausch-Williams-Watts function" Am. 3. Phys, 59, pp. 1024-1030, 1991

[50]. J.R. Macdonald, R. L. Hurt, "Analysis of dielectric on conductive system frequency response data using the Williams-Watts function" 3. Chem. Phys., 84(1), pp. 496-502, 1986.

[51]. G. Weiss, J.T. Bendler, M. Dishon, "Analysis of dielectric loss data using the Williams-Watts function" 3. Chem. Phys., 83 (3), pp. 1424-1427, 1985.

[52]. W. Kaplan "Advanced Calculus", Addison-Wesley Publishing Company, second edition, pp.452, 1972.

[53]. R.G. Wilson, "Fourier Series and optical Transform Techniques in Contemporary Optics", John Wiley & Sons, INC., 1995.

[54]. L. D. Landau, E. M. Lifschitz, "Statistical Physics", Addison-Wesley Publishing Company, Cap. 12, 1969.

[55]. G.A. Nilklasson, "Comparison of dielectric response functions for conducting materials" 3. Appl. Phys., 66 (9), pp. 4350-4359, 1989.

[56]. G.A. Niklasson, "Optical properties of gas-evaporated metal particles: Effects of a fractal structure" 3. Appl. Phys., 62 (1), pp. 258-265, 1987.

[57]. L.A. Dissado, R.M. Hill, "The fractal nature of the cluster model dielectric response functions" J. Appl. Phys., 66 (6), pp.2Sll-2524, 1989.

[58]. B. Mandelbrot, "Self-Affine fractals and fractal dimension" Physica Scripta, 32, pp. 257-260, 1985.

[59]. Idem [19], pp.350-356.

[60].P.C.Schotborgh."Caracterización y estudio de permeabilidad de ceramics porosas que simulan rocas reservorio"," Characterization and study of porous ceramics simulating reservoir rocks" Thesis. Simón Bolívar University, Venezuela, pp.61-78

[61].Idem above,pp.40,41.

CORROSION OF INDUSTRIAL REFRACTORIES

M. Rigaud, Professor, École Polytechnique
CIREP-CRIQ Campus, Montréal, Québec, Canada H2M 2N9

ABSTRACT

Basic principles of penetration, oxido-reduction, dissolution, erosion and spalling are reviewed in order to appreciate the main mechanisms of corrosion-degradation of refractories. Thermodynamic, kinetic and pragmatic approaches are discussed, leading to the conclusion that corrosion resistance is not a characteristic having an intrinsic value. Corrosion testing methods are discussed. Ways to improve corrosion resistance are illustrated.

INTRODUCTION

In reference to refractories, equivalent wordings for corrosion are degradation, deterioration, decomposition or wear of refractories. This simply means that corrosion involves a combination of different mechanisms, such as dissolution and invasive penetration, where oxido-reduction reactions, adsorption, adsorption and mass transport phenomena come all into play, typically under pressure and temperature gradients.

This review will be centered mainly on the general case of liquids-solids interactions at high temperatures. "High temperatures" means at a temperature higher than the melting temperature of at least one constituent. Corrosion due to gases and dusts instead of liquids will be scarcely discussed. Considerations will be limited to chemical aspects only, hence none of the electrochemical phenomenon considered when dealing with corrosion of refractories in glasses or the low-temperature degradation of refractories due to hydration and/or dehydration will be considered.

Industrial refractories are heterogeneous, multiphase constituents. They are made of crystals, glassy phases and pores, each of different sizes. Crystalline constituents have varying degrees of purity, depending upon their specific preparation and thermal history prior to use (whether they are natural or synthetic). They have different grain morphology (spheroidal, flat, elongated) and may have highly anisotropic behavior (most of non-cubic crystal structures).

Grain size distribution is very broad, ranging from 1 to 10,000 µm, with inherent high porosity between grains ranging from 10 % to 20-25 %.

A simple, all-encompassing general theory of refractory corrosion does not and most likely will never exist. Corrosion resistance of a given refractory material is not an intrinsic property of such a product, but at best an indication of a process intensity, where many parameters need to be defined to appreciate the significance of such a characteristic. A non-exhaustive list of such parameters are illustrated in Figure 1. and include the chemical and physical aspects and the intrinsic and extrinsic properties of both the slag and the refractory. Once more, it is important to remember that the interplay between dissolution, penetration and texture is not sufficient to take into account all possible interactions. However, there are some basic principles or rules available to predict some of the most obvious practical results in terms of corrosion or degradation. After a review of these basic principles, comments on corrosion testing methods will be presented and practical considerations on corrosion resistance enhancement will be provided.

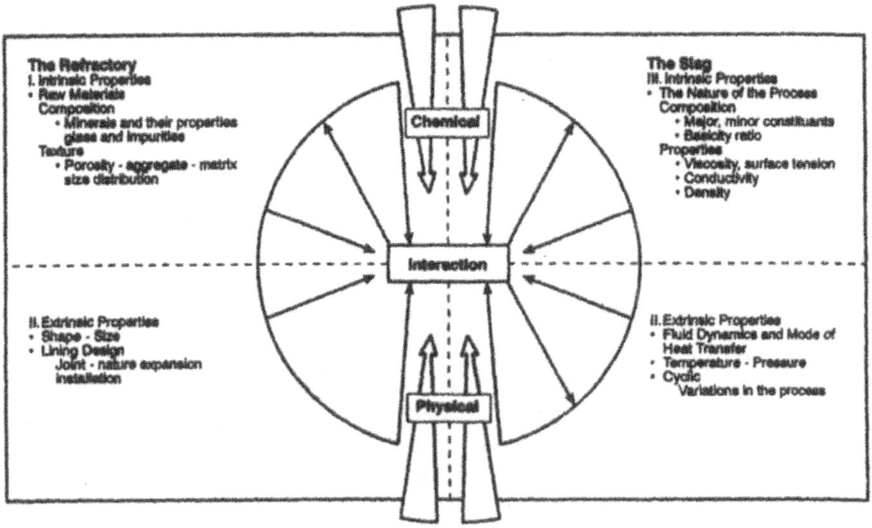

Figure 1. Illustration of the many parameters coming into play in typical slag-refractory reactions.

BASIC PRINCIPLES

To anticipate the corrosion of refractories in a given environment, it is always worthwhile to consider first the concept of acidity-basicity, secondly to estimate the driving force for corrosion using the themodynamic laws. This should be done in two steps: first to verify the available thermodynamic data for the thermal stability of each constituent, then to make the appropriate thermodynamic calculations to estimate the free enthalpy variations, $(\Delta G)_r$, for all possible reduction or oxidation (redox) reactions that may occur between constituents and between the constituents and the environment (gases and/or liquids). To understand and select the proper refractories for the proper application, it is necessary to rely on kinetic data. The principles of penetration, dissolution and spalling will be presented in order to appreciate the particularities of the liquid and of hot gases and dusts corrosion of industrial refractories.

1. The Acid-Base Effects

At first glance, one has to consider the chemical nature of the reactants (S=Solid and L=Liquid or S and G=Gas), since materials of dissimilar chemical nature, when in contact, will react. This is especially true at high temperatures.

The chemical nature of the reactants is best described by the notion of acidity and basicity. This familiar concept is of limited value since a single precise notion of high temperature acidity and basicity of all compounds (pure and in solution) is lacking. Although a full discussion of acid and gases falls outside the scope of this section, the extreme cases are indisputable.

Silica (SiO_2) is the best example of a solid acidic oxide and is to be used in applications where the destructive materials (L or G) are chemically acidic. These applications include coal gasifiers ashes, ironmaking slags, and in N_2O_5 or SO_3-SO_2 atmospheres, the most acidic gases. Magnesia and doloma are basic in nature and are to be used in application where slags or gases are generally basic such as in steelmaking slags or liquid clinker melting rotary cement kiln.

This is low ever a first approximation (a rule of thumb) which is insufficient in many cases. In many industrial processes, the corrosive environment changes from an acidic to a basic type during process operation. Nevertheless, as a first rule, it is always worthwhile to remember to make the acid/basic character of the refractory constituents similar to the corrosive fluid (liquids and/or gases) to increase its corrosion resistance.

2. Thermodynamic Calculations

The second approach to anticipate the corrosion resistance of industrial refractories is to make the appropriate thermodynamic calculations, first to verify each constituant thermal stability then to consider the melting and dissolving behaviors and finally their redox potential among each other.

2.1 Thermal Stability Prior to Melting : Several polymorphic transitions may occur which will change the microstructure integrity of the solids as they are being heated. Some are reversible and some irreversible. Well known disruptive transformations include those for silica or zirconia. Other transitions are the decomposition of mixed oxides (e.g. $ZrSiO_4$ 6 $ZrO_2 + SiO_2$), the devitrification of glassy phases, and the crystallization of high temperature phases for amorphous or ill-crystallized ones (e.g. the graphitization of carbons). For such modifications, large volume variations may occur, creating large compressive, tensile or shear forces that cause microcracks, and hence porosity, a key word to explain the corrosion of refractories.

Other volume changes or debonding may occur due to thermal mismatch when solids are heated. All non cubic lattice refractory compounds are susceptible to disruptive intercrystalline debonding since they exhibit thermal expansion anisotropy as with Cr_2O_3 (hex.), Fe_2O_3 (trig.) of ZrO_2 (monocl.). Another cause of debonding is the juxtaposition of two constituents of different nature, having different thermal expansion coefficients. Examples are alumina-mullite, mullite-silica and magnesia-chromite. In most refractories, debonding or phase boundary microcracking can be anticipated.

2.2 Melting Behavior and the Use of Phase Diagrams : It is still often the case that refractory constituants are subjected to temperatures in service higher than those attained during their fabrication history. Many times refractories cannot be regarded as having been brought to equilibrium. It is always worthwhile to calculate the amount of liquid that they may contain at a given temperature. This is achieved using the thermodynamic principles (minimization of energy techniques for multicomponent systems like the SOL/GAS mix protocole) and all data available on the thermodynamics of solutions, as well as on pure compounds under stable and metastable conditions.

Historically, compositions of refractories were limited to individual minerals as mined. With time, it was recognized that nature did not proportion the oxide contents of the minerals in the most suitable ratio for optimum refractory performance. As pure oxide became available and affordable in terms of costs, (first the alumina then the magnesia) improvements of the performances, and in particular of their corrosion resistance, were made possible. The use of phase diagrams was then recognized as a very good research tool to accelerate such improvements.

Those diagrams were and are still used to design refractories and also to understand the role of slags (of the same nature as the liquid phase formed into the refractories). The early book by Muan and Osborn (1) is then a must to read for new researchers in this field; further reviews by Kraner (2), on the use of phase diagrams in the use of fired refractories as well as by Alper (3), on fusion-cast refractories, are complementary. Phase equilibria in systems containing a gaseous component, in particular the effects of oxygen partial pressure on phase relations in oxide systems, have also been examined by many authors. The relative importance of these issues has greatly increased with the advent of the so-called carbon bonded refractories, magnesia-doloma and alumina-based system with carbon and graphite. Also, the roles of oxycarbides, oxynitrides and sialous, as well as metals and alloys in refractories, have been documented. This will be treated in some details in the next sub-section.

In essence, it is accepted that the corrosion resistance of a refractory will be high if the formation of low-melting eutectics and the formation of large amount of liquid can be avoided. Although this is just one aspect of corrosion resistance, it is sufficiently important to always try to predict it through the use of phase diagrams. Of course, as in any anticipation, the better the system is defined, the easier it is to achieve. There are nevertheless many limitations: phase diagrams are readily available for no more than 3 constituents, while in practical cases for magnesia or doloma basic refractories, as an example, one needs to elucidate the $CaO-MgO-FeO-Fe_2O_3-Al_2O_3-Cr_2O_3-SiO_2$ system. Rait (4) and White (5) have offered the most comprehensive treatment of the phase assemblages in such a system, a formidable task. A particularly important feature is the nature of the CaO/SiO_2 ratio, and its incidence on the amount of low temperature phases to be expected. It is of no surprise that this ratio is indeed related to the acid/base effect briefly described earlier.

To paraphrase Carniglia and Barna (6), Athe importance in understanding the progressive thermal softening, weakening, and ultimate destruction of refractories cannot be overstressed. It should be clear that the maximum feasible service temperature of a refractory has to be confirmed by: i) its lowest germane eutectic and the melting temperature thereof; ii) how much liquid is produced at that temperature; iii) and how rapidly the amount of liquid increases with increasing temperature above this.≅ All these characteristics are linked to the appropriate phase diagrams and the application of the lever rule.

2.3 The Oxidation, Reduction Behavior : As for melting, or S+L reactions, the thermodynamic calculations of gas-solid reactions are a powerful tool to describe the stability of refractory materials, in particular for the carbon-containing refractories, so widely used nowadays. The G+S reactions are of importance not only to deal with the direct oxidation of carbons in air, but also to consider the reduction of the aggregates (MgO, Al_2O_3, SiO_2, etc.) by carbons (indirect oxidation of carbon) and the role of the anti-oxidants, being more reducing than carbon itself in most cases. By far the most powerful use of thermodynamics, in assessing oxido-reduction behavior of refractories and ceramics, entails equilibrium calculations. Nowadays, it is possible to consider multicomponant systems, using elaborate computer programs such as Chem Sage or Solgasmix (7) (8). Graphical displays such as the Ellingham diagrams, the volatility and the predominance-diagrams are often used to describe the simplest systems, of the type $S_1 + G_1 \rightarrow S_2 + G_2$ (9) (10). For the Ellingham diagrams, all the condensed phases of the reactions are assumed to be pure phases and therefore at unit activity. Deviations from unit activity however are very common, and corrections must be applied. Predominance diagrams and volatility diagrams are graphical representations where the gaseous products are considered at various non-standard conditions. They are also used for systems involving several gases in complex environments (metal-oxygen-carbon-nitrogen for example) under a wide range of partial pressures. The predominance diagrams are equivalent to the Pourbaix diagrams used to study corrosion of a solid in an aqueous media. They predict the direction of reactions and the phases present.

From those thermodynamic calculations it is then possible to explain why and how the redox cycling is harmful to any refractory containing iron oxides as impurities; how the reduction of magnesia to magnesium gas and the reoxidation to MgO again can be used with great advantages for the MgO-carbon bricks (a destruction-reconstruction mechanism which leads to the dense zone magnesia formation theory); how non-oxide refractories (and non-oxide structural ceramics) without exception are subject to high-temperature oxidation ; and finally how, in the presence of alkalies, refractory chromites (Cr^{+3}) can be oxidized in part to toxic chromates (Cr^{+6}). The alkali chromites ($Na_2Cr_2O_7$ or Na_2CrO_4) are water-leachable and present therefore an environmental contamination risk on disposal.

3. Kinetic Considerations

Penetration, dissolution and spalling are the most important phenomena which control the kinetics of the corrosion of industrial refractories involving either a solid(S) + liquid(L) type of reactions or a solid(S) + gas(G) reactions.

L can be either slags or fluxes, molten salts or molten metals, each presenting their own peculiarities. Slags are characterized by their basicity/acidity ratio (CaO/SiO_2 ratio either bigger or smaller than 2:1) and fluidity or viscosity (very much a function of temperature and overheating , measured by ΔT = Tservice - Tmelting of the lowest melting eutectic compound to be expected in the system). Molten salts are known for their low melting temperatures, high fluidity and high fugacity (high volatility of alkalies). Molten metals are less reactive towards refractories, but are not inert.

S, the solid refractory material, has already been defined as a multiphase material with a texture, with intricated surface properties. It is most often used under a thermal gradient (a given never to be forgotten) and usually contains some inherent amount of liquid (ℓ) at the hot face, where ℓ may be equivalent to L, in terms of affinity.

G represent various hot gases, reactive or non reactive, with or without dusts. The reactivity of G can be determined starting with the acid-base series, noticing that the strongest acids and bases are volatiles. The reactivity may also be determined using the compiled thermodynamic data on predominance and volatility diagrams.

3.1 Penetration : For clarity it is good to distinguish between physical penetration and chemical invasion. Physical penetration, without dissolution, is when a strictly non-wetting liquid is forced into the pores of a solid by gravity or external forces. Chemical invasion is when dissolution and penetration are tied together. Both physical and chemical penetrations are favored by effective liquid-solid wetting and by a low viscosity liquid. In that respect, it is good to remember that silicates are usually viscous (in particular silicate glasses), that simple oxidic compounds and basic slags are less so, and finally that halides and elemental molten metals are in general the most fluid liquids).

Penetration is the result of an interplay between capillary forces (surface tensions), hydrostatic pressure, viscosity and gravity. Mercury penetration in a capillary glass tube is the best example of physical penetration. The rate of penetration is a horizontal pore dl/dt, with a pore radius r,of length, ℓ, is given by the following expression :

$$\frac{d\ell}{dt} = \frac{r\,\gamma_{\ell g}\cos\theta}{4\eta\ell} \tag{1}$$

where γ_{lg} is the surface tension of mercury in air, θ the wetting angle of mercury on the glass wall, η the viscosity of the penetrating liquid, Hg, over a length, ℓ, at time, t. This expression is only valid at constant temperature, while γ_{lg} and η are liquid properties, greatly influenced by temperature. Such an equation has often been used to describe the penetration of liquids in refractories without distinguishing between physical and chemical invasion. However, in the case of chemical invasion, the penetration-dissolution brings along changes in compositions, which in turn affect the values of γ_{lg} and η, and changes in pososity geometry .

When the pore size distribution is narrow (pores of the same size), penetration and filling of the porosity by capillarity forces produce a relatively uniform front moving gradually from the hot face and remaining parallel to it. When the pore size distribution is wide (very big and very small pores) or when open joints, cracks or gaps between bricks in a refractory wall are accessible at the hot face, rapid and irregular liquid intrusion occurs. The penetration paths are many in a refractory and the texture of the material is of primary importance. In these instances; it is important to distinguish between interconnected versus isolated porosity, between open and total porosity between pore sizes, and between unbounded boundaries between grains (aggregates and/or matrix) due to thermal mismatches during heating.

It must also be remembered that for a given temperature gradient, the pertinent eutectic temperature of the penetrating liquid determines its maximum liquid penetration depth.

3.2 Dissolution : The simplest case of pure dissolution is to consider the following reaction : $S_1 + L_1 \rightarrow L_2$ where L_2 is a solution $L_1 + S_1$. S_1 is a solid with a continuous surface.

A general equation useful to describe such a dissolution process is :

$$j=\left\{K\left(1+K\frac{\delta}{D}\right)^{-1}\right\}(C_{sat}-C_{\infty}) \tag{2}$$

where j is the rate of dissolution per unit area, at a given temperature T, K is the surface reaction rate constant, δ is the thickness of the boundary layer in the liquid phase; C_{sat} is the concentration of the dissolving solid in the liquid at the interface; C_4 is the concentration of the dissolving phase in the bulk of the liquid, and D is the effective diffusion coefficient in the solution for the exchange of solute and solvent.

The significance of the parameter δ is illustrated in Figure 2 and further defined by the expression :

$$\delta=\frac{C_{sat}-C_{\infty}}{(dc/dy)} \tag{3}$$

where (dc/dy) is the concentration at the interface.

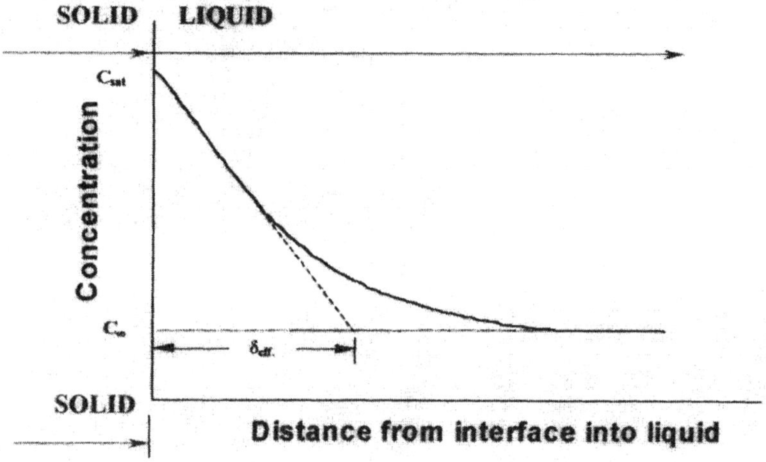

Figure 2. Evolution of the concentration of the dissolving liquid near the solid-liquid interface.

The dissolution rate, j, may be visualized and has the ratio of a potential difference (C_{sat} - C_4) divided by a resistance term :

$$K^{-1}\left(1+\frac{K\delta}{D}\right)$$

Three different cases will be briefly treated :

i) When K >>> D/δ, that is, when the chemical reaction takes place so rapidly at the solid/solvent interface that the solution is quickly saturated and remains so during the dissolution process. In this case, the dissolution rate, j, is controlled by mass transport. Equation (2) reduces to (4) :

$$j=\frac{D}{\delta}(C_{sat}-C_\infty) \tag{4}$$

The transport process is enhanced by convection due to density differences between bulk and saturated solution (natural convection) and/or by the hydrodynamic of the system, under forced convection. Expressions for the boundary layer, δ, have been derived from first principles, for both natural or free and for forced convection, and for a variety of simple geometry. Equation (4) has been validated many times, and is often called direct dissolution.

ii) When $K <<< D/\delta$, the dissolution rate is phase boundary controlled as opposed to mass-transport control. Equation (2) simply reduces to (5) :

$$j = K\,(C_{sat} - C_{\infty}) \tag{5}$$

The phase-boundary reaction rate is then fixed by the movement of ions across the interface, and hence is governed by molecular diffusion. The effective diffusion length over which mass is transported is proportional to $(Dt)^{1/2}$. The change in thickness of the specimen which is proportionnal to the mass dissolved varies with $t^{1/2}$, and is often called indirect dissolution.

iii) When $K \cong D/\delta$, both phase boundary and mass-transport are controlling dissolution. In this case of mixed control, the potential difference (C_{sat} - C_{∞}) can be seen as divided into two parts, shown in Figure 3. (C_{sat} - C^*) is the part which drives the phase boundary condition and (C^*-C_0) is the part which drives the transport process, keeping the dissolution rate for each process equal.

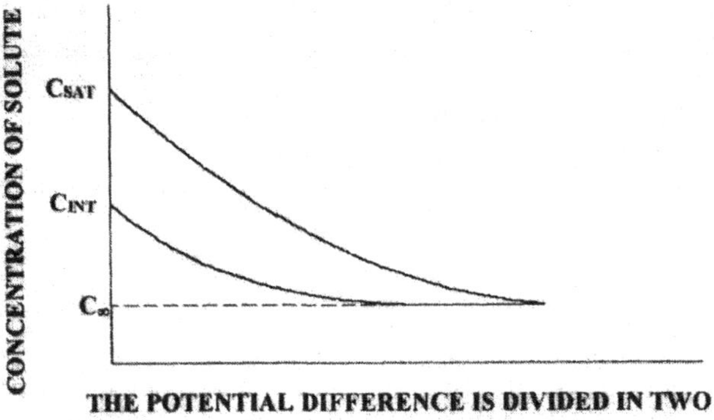

Figure 3. Schematic representation in the case of mixed control of the evolution of the concentration of the dissolving liquid near the solid-liquid interface.

The dissolution rate is extremely temperature sensitive, largely determined by the exponential temperature dependence of diffusion, and may be expressed by the following equation :

$$J_T = A \exp\left[-B\left(\frac{1}{T}\,\frac{1}{T_l}\right)\right] \tag{6}$$

where A is the dissolution rate at temperature T_1 and B is a model constant.

In most practical dissolution problems the reaction of solid into a solvent leads to a multicomponent system. No longer is the chemical composition defined by a single concentration nor is there a single saturation composition at a given temperature. Instead of pure dissolution the reaction is of the type: $S_1 + L_1 \rightarrow L_2 + S_2$. The formation of S_2 prompts the interface composition to change.

For porous solids (like refractories) with a lot of open porosity and with matrix materials being fine and highly reactive, dissolution occurs as well as penetration. Hence the majority of slagging situation involve a chemical attack of the matrix or involve low melting constituant phases, which disrupt the structure and allow the coarse grain aggregates to be carried away by the slag movement. When penetration is more important than dissolution, another mechanism of degradation needs to be considered: structural spalling.

3.3 Structural or Chemical Spalling : While spalling is a general term for the cracking or fracture caused by stresses produced inside a refractory, chemical or structural spalling is a direct consequence of a corrosion penetration phenomenon. It is not to be confused with pure thermal or pure mechanical spalling. The structural spalling is a net result of a change in the texture of the refractory, leading to cracking at a plane of mismatch at the interface of an altered structure and the unaffected material.

When slag penetration is not causing a direct loss of materials, the slag partially or completely encase a volume of refractories, reducing its apparent porosity (by sometimes more than one half). This causes differential expansion with the associated development of stresses. As a result, there is a degradation in material strength and stiffness, the appearance of microcracks and eventually total disruptive cracks parallel to the hot face of the lining. This phenomena is called structural spalling.

Quantification of the mass of spalled materials from the hot face has been attempted by Chen and Buyakozturk (11) combining the slag dissolution and spalling effects into one expression in terms of the residual thickness, X, and the rate of thickness decrease. They have proposed expressions as a function of the hot face temperature (T_H) and the maximum depth of slag penetration (D_P) and referred to the location of the hot face at time, t_i, during the $(i-1)^{th}$ and i^{th} spalling.

The same authors have developed a very interesting notion to approximate the value of D_p, using a critical temperature criterion. They have postulated that for a given system, slag penetration occurs only when temperature reaches a critical value T_c. T_c is the temperature above which a certain percentage of liquid phase may still exist in the refractory, which is always under a thermal gradient.

3.4 Gases and Dusts Versus Liquids Corrosion : The same qualitative relationships are applicable for gas and liquid penetration into a refractory, but wetting considerations are absent with a gas. In gas corrosion the driving force is a pressure gradient. Gas penetration is less rapid, in mass terms, than that of wetting liquids. Reactions do not necessarily commence at the hot face and the depth of penetration of gases can be much greater than that for liquids.

Moving down a thermal gradient, gases may condense to form a liquid. This liquid may then initiate a dissolution-corrosion process and migrate still further into the solid, or the gas may react chemically to form new compounds (liquid or solid).

Once condensation has occurred, the basic criteria for liquid corrosion apply. The condensed liquid fills the porosity and engages in debonding the refractory material, creating thermal-expansion mismatches, weakening, softening, swelling and slabbling.

Sulfur dioxide, usually from combustion or smelting operations, provides many good examples of the condensation-corrosion patterns experienced in practice. The condensation leads to the formation of sulfate in magnesia at 1100°C, in lime at 1400°C, and in alumina at 600°C. When alkali vapors (Na_2O) are present with SO_2, condensation of Na_2SO_4 overlaps with SO_2 attack. Volatile chlorides produce similar types of reactions.

When dusts are carried out in hot gases, they may cause deposition or abrasion of refractory method. Deposition leads to scaling or caking, often with beneficial or lifeshortening consequences. Dusts entrained in gases which condense within refractories may facilitate the gradual accretion of solid particles on the working surface. The resulting build up can be either a hard scale or lightly sintered cake that can act as a protective barrier. In less obvious cases, dust and condensing gases all fuse together to create an invasive liquid. In other instances, scaling and caking of refractory walls may not be acceptable to a particular process, regarding de-scaling operations that lead to refractories degradation.

When no deposition occurs, refractory working lining bombarded by dusts are subjective abrasive wear. Abrasive wear depends on the impinging particle size, shape, hardness, mass and velocity, angle of impact, the fluid dynamic of the system, the viscosity of the eluant gas, and other parameters. Abrasion is an aggravator of corrosion.

Fundamental exposure of a refractory to either a liquid or to its saturated vapor is thermodynamically the same. The differences lie in the transport mechanism. In approximate order of decreasing thermodynamic power, the oxidizing gases are NO_X, Cl_2, O_2, HCl, CO_2, H_2O and SO_2 and the reducing gases are the active metal vapors, Al, Mg, NH_3, H_2, C_xH_y, CO, common metal vapors, H_2S.

TESTING METHODS

Lord Kelvin said "When you cannot measure what you are speaking about or cannot express it in numbers, your knowledge is of a meager or unsatisfactory kind". A lot of efforts have been devoted to developing tests to measure the corrosion resistance of refractories under slagging conditions and non-oxide structural ceramics under hot gas corrosion and oxidation at elevated temperatures. Numerous methods have been tried and some reasonable correlations have been obtained for very specific conditions. Very few methods, however, have reached the status of a standard operating practice and none have yet been accepted for universal use.

The main reason is that corrosion resistance results obtained in a laboratory environment very rarely simulate the conditions which prevail in service. Sample size and geometry, state of stresses in the lining, thermal gradient and thermal cycling, as well as time, are very difficult to be scaled down to fit acceptable laboratory test conditions. It is always a good idea to remember that accelerated test, specially those done using very severe conditions, can lead to erroneous predictions.

As compared to laboratory testing, field-trial testing is much more costly, and can be unsafe. In some instances it is worthwhile to use small panel test instead of full-size tests. In panel tests, the larger the installation, the more confidence one will have in the selection of the proper material to use.

Post-mortem examination of in-service trials provides very useful insights to understanding and to determining the mechanisms causing the degradation of ceramics. Detailed investigations involved the use of all characterization methods with chemical analysis, X-Ray diffraction analysis, mineralogical analysis, and examination by scanning electron microscopy/energy dispersive spectroscopy being powerful analytical tools. Selecting samples and their preparation are often very challenging, with observations of the uncorroded part yielding frequently providing clues as to what may have happened in the corroded part.

Crescent & Rigaud (12) have reported some 106 different experimental set-ups, all falling into one of the 12 categories and schematically represented in Figure 4. Among the non-standard tests, the most commonly used are still the crucible test, the rotary disc, and the finger test.

The Refractory Committee C8, of the American Society for Testing and Materials (13) has defined two standard practices dealing with slags (C-768 and C-874). Two other tests are designed to measure the corrosion resistance of refractories to molten glass, one under isothermal conditions (method C-621) and the other standard practice using a basin furnace to maintain a thermal gradient through the refractory. These two methods could be applicable to liquid other than molten glass ; finally, there are 3 other ASTM methods that evaluate corrosion : the disintegration of certain refractories in carbon monoxide (C-288), alkali attack on carbon refractories (C-454), and alkali vapor attack on glass-furnace refractories (C-863). Only the two standard practices, C-874 and C-863, will be briefly described hereafter.

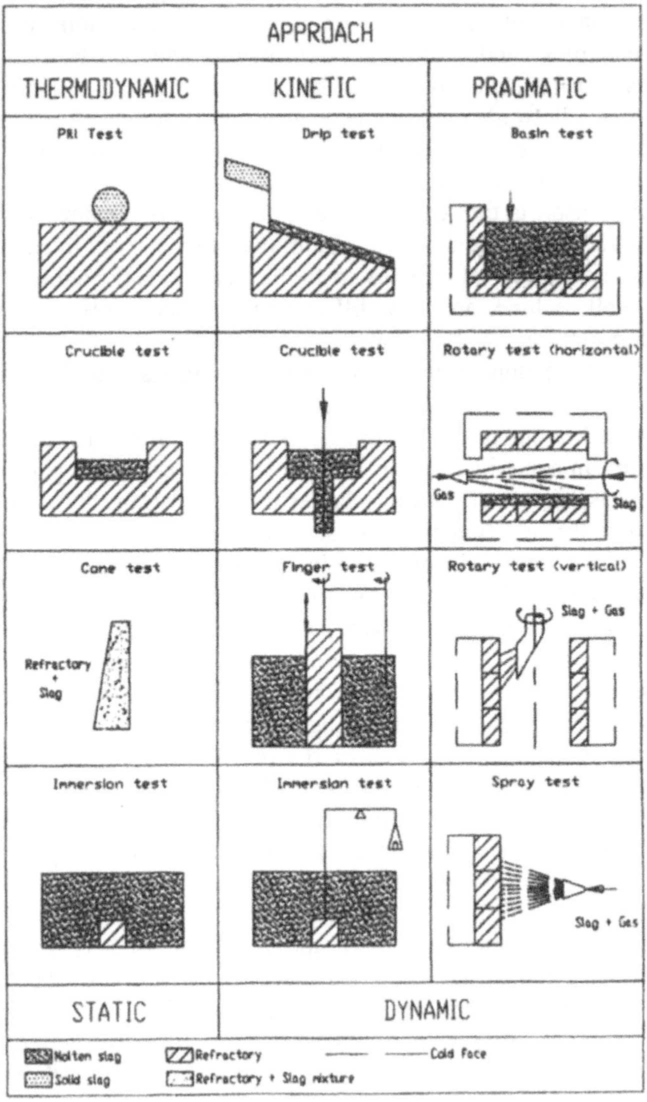

Figure 4. Slag corrosion test methods according to different approaches.

1. Rotary-Kiln Slag Testing (ASTM C-874)

This standard practice evaluates the relative resistance of refractory materials to slag erosion. The furnace is a short kiln, a cylindrical steel shell mounted on rollers and motor driven with adjustable rotation speed and tilt angle. It is heated with a gas-oxygen torch to test temperatures up to 1750° C. The kiln is lined internally with the test specimens. Normally, six test specimens, 228 mm long, and two plugs to fill the ends of the shell constitute a test lining. In principle, the furnace is tilted at a 3° angle with a rotation of speed of 2 to 3 rpm. An initial amount of slag is first fed into the kiln, which coats the lining and provides a starting bath. Extra amounts of slag are then added at regular time intervals of about 1 kg/hr for at least 5 hours. The test atmosphere is usually oxidizing. Neutral or slightly reducing conditions may be obtained using a reducing flame or by using carbon black additives to the slag mixture. The tilt angle is adjusted axially toward the burner end so that the molten slag washes the lining and drips regularly from the lower end after each addition of new slag at the higher end of the kiln. One method of assessing erosion is to measure the profile of the refractory thickness along a cut in the middle of the exposed surface of the specimen using a planimeter. Results can be reported as % area eroded from the original specimen area. It is to be noted that for this practice, a thermal gradient exists through the test specimens, which is controlled by the thermal conductivity of the specimen and the back-up material used at the cold face of the lining. The slag is constantly renewed so that a high rate of corrosion is maintained. The flow of slag can cause mechanical erosion of materials. Care must be taken to prevent oxidation of carbon-containing refractories and oxidation prone structural ceramics during heat-up. A reference refractory specimen should be used for comparison in each consecutive test run. Caution should be exercised in interpreting results when materials of vastly different types are included in a single run. A full discussion on the advantages and shortcomings of the method has been provided in reference 12.

2. Oxidation Resistance at Elevated Temperatures (C-863)

This practice covers the evaluation of the oxidation resistance of silicon carbide refractories in an atmosphere of steam. Steam is used to accelerate oxidation. The oxidation resistance measured is the ability of SiC to resist conversion to SiO_2. The volume changes of cubes (64 mm side) evaluated after exposures to at least 3 temperatures (in the temperature range from 800 to 1200°C) are recommended. The duration of the test at each temperature is normally 500 hours. In addition to the average volume change (based upon the original volume of at least 3 specimens), any weight, density, or linear changes are included as supplementary informations.

PRACTICAL CONSIDERATIONS

Detailed compilations of data on the corrosion resistance of industrial refractories is lacking because there are too many different products and too many variables to consider. Corrosion is also very seldom the only criteria to take into account when it comes to selecting a material for a given application. Other properties such as thermal shock resistance or mechanical strength have to be maximized at the same time and can be in conflict with the best corrosion resistance criteria.

Nevertheless, specific strategies have been used to enhance the corrosion resistance and improve the durability of several classes of refractories. The three sets of considerations to be optimized simultaneously are :

1. Material selection, considering both intrinsic and extrinsic characteristics, such as the design and configuration of the component.
2. Installation methods and maintenance procedures, considering that each material is a part of a bigger identity.
3. Process control to minimize variability and extreme values.

If in this section, only the first set of problems will be considered, the other two sets of problems should never be overlooked. To select the most appropriate material, the first rule is to make the acid-basic character of the refractory constituents similar to that of the corrosive fluids (liquids and/or hot gases), and then to control the penetration-dissolution mechanisms to improve the corrosion resistance. The single most important factors in terms of material properties is porosity, and, in a broader sense, the texture of the material..

1. Porosity and Texture

Improvements in the corrosion resistance of refractories have been obtained through texture control, resulting from the evolution in manufacturing processes : using much larger particle size distribution, better mixing, better pressing, when it applies, and purer raw materials when available and economical. All in all, total open porosity can be decreased to value close to 12%, but to maintain good thermal shock characteristics and acceptable insulating properties, porosity has to be maintained at such a level. At constant porosity, improvements can come from the pore size distribution, the smaller the bigger pores it is, or from the permeability of the refractory (same porosity but more closed porosity). The decrease in permeability can be accomplished, for example, through the use of fused grains instead of sintered grains. Other considerations will be discussed to minimize slag penetration in refractories.

2. Factors Governing Penetration

Once porosity has been adjusted to an optimal value, the other possibilities to minimize penetration are: i) adjust the composition of the penetrant fluid; ii) adjust the wetting-non-wetting characteristics of the same fluid; iii) adjust the thermal gradient in the material (if feasible) and iv) use of a glazing or coating on the working face, or in some specific cases, on the back or cold face, when oxidation of non-oxide constituant is important.

2.1 Composition Adjustment of the Penetrant Fluid : This happens by itself because most of the time some solid dissolution occurs as penetration continued. If the melting temperature of the liquid penetrant or its viscosity increases, the depth of penetration can be minimized. The way to do this in practice is to carefully select the matrix additives (ex : use limited amount of micro-silica in a cement-free alumina-spinel castable for certain steelmaking applications).

2.2 Changing the Wetting Characteristics : In steelmaking applications, this has been successfully achieved by using non-oxide materials in a traditional oxide composition. This can be achieved by going from fired to non-fired products. For example, by using carbon-bonded refractories and graphite. The insertion of graphite in magnesia-based materials has been shown to limit slag penetration and improve the corrosion resistance by a factor of 10. Graphite is non-wet by CaO-SiO_2 slags and acts as a barrier to its penetration. Of course, graphite can oxidize, so this solution is not valid in every case. Impregnation of pitches and tars which modify the nature of the bonding phases in refractories, also affects the penetration phenomena and explains some of the performance in some situations. Numerous other examples are available in the technical litterature, like the use of specific additives ($BaSO_4$ or CaF_2), in alumina-silica based castables containing liquid aluminum metal and its alloys.

2.3 Adjusting the Thermal Gradient : Whenever penetration is important, maintaining the freezing of the penetrant liquid as close to the hot face as possible is important to minimize corrosion. This can be achieved in adjusting the thermal gradient as steep as possible when the hot face temperature is fixed. To accomplish this, one must lower the temperature at the cold face by vigorous cooling (water cooling of shells), use of a highly conductive material where possible, or increasing the thermal conductivity of the lining by use of a conductive back up material.

2.4 Use of Glazing or Coating : Again, in specific cases where slag penetration is a key factor, the use of a sacrificial coating may become a viable solution. For black refractories (carbon containing refractories) the protection of the back face to avoid cold face oxidation has been identified as a positive solution in specific cases. Glazing, by definition low melting compound, at the hot face, has to be analyzed very carefully to prevent hot face oxidation of carbon penetration may be less where a coating is used on the hot face, but dissolution could become a new issue. Preoxidation treatment to form a protective oxide layer on Si_3N_4 or SiC materials may also be considered. The use of multiple layers on new materials such as ASialons≅ (Silicon-Aluminum Oxinitride) or ASimons≅ (Silicon-Magnesium Oxinitride), have also been experimented to control slag penetration.

3. Factors Governing Dissolution

Discussing the factors controling penetration and dissolution separately is at best artificial, since they interact together.

Dissolution is controlled by both the chemical reactivity and the specific area of the constituents. Smaller grains dissolve faster than the large ones. In refractories, the smallest grains are, by design, the bonding phases between the coarser grains (the aggregates). Since bonding occurs by liquid-solid sintering, the matrix materials have the lowest melting temperatures and the highest reactivity. In general, the dissolution of the bond phases (or the matrix) leads to debonding between the coarser grains, whose dissolution is inherently slower and is usually apparent only at the refractory hot face.

To minimize matrix dissolution, the first possible action is to alter its composition so as to raise its melting temperature, then to minimize the importance of segregated impurities in the system by upgrading its chemical purity. Examples of this are the use of tabular alumina instead of bauxite grains or the use of very high purity magnesia grains. A second way is by chemical addition rather than substraction. Finely powdered matrix additives may be included in the refractory mix to raise the matrix melting temperature. This must be used with caution since the matrix serves as a flux in sintering, hence such additions require precise temperatures to optimize the bonding and the mechanical properties of the sintering refractory product. The logic of reducing or eliminating low melting additives for improved corrosion resistance is compelling, however, the consequence is that the sintering temperature will have to be raised.

A thorough knowledge of the different binder systems (for non-fired refractory products) and the different bonding systems (after firing) goes a long way to explain the different dissolution characteristics of refractory products. Once more, the texture, and in particular the porosity, are key parameters to be considered. Coarse grain dissolution becomes a life-limiting factor only when penetration and matrix invasion are sufficiently minimized.

4. Factors Governing Oxidation

In carbon-bonded refractories, carbon and graphite oxidation is very detrimental, with material losses leading to porosity. Since carbon's specific gravity is lower than the oxide components in a refractory, 5% loss leads to a 7 to 10% porosity increase. To prevent carbon's oxidation, new additives have been successfully used, called anti-oxidants or oxygen inhibitors. Some general reviews on the role of anti-oxidants, are available, but fall outside the scope of this paper. Among the many additives (14-15) now available, the aluminum-based alloys are the most commonly used, combinations of the metallic additives are sometimes used in conjunction with various non-metals additives such as silicon carbide and some boron compounds. They are used to delay the carbon debonding effect and the graphite oxidation. They often act as pore-blockers and many of them have a net positive effect on the mechanical strength of the refractories.

CONCLUSION

Corrosion resistance should not be considered as an intrinsic property of refractories but a measure of a complex process involving many intrinsic and extrinsic parameters. Fundamentally, kinetics predominates, even if thermodynamics is to be considered. Pragmatically, the definition of all parameters and their evolution with time is as important as the definition of the refractory material itself. This may explain why so many different refractories can be used with some successes for a given application.

REFERENCES

[1]A. Muan and E.F. Osborn, "Phase Equilibria Among Oxides in Steelmaking", Addison-Wesley, Reading, Mass., 1965.

[2]H.M. Kraner, in "Phase Diagrams, Materials Science and Technology", Volume 6-II, A.M. Alper Editor, Academic Press Inc., New York, 1970, pp. 67-115.

[3]A.M. Alper et al., in "Phase Diagrams", Volume 6-II, Academic Press Inc., New York, 1970, pp. 117-146.

[4]J.R. Rait, "Basic Refractories, Their Chemistry and Performance", Iliffe, London, UK, 1950.

[5]J. White in "High Temperature Oxides", Part I, A.M. Alper, Editor, Academic Press Inc., New York, 1970, pp. 77.

[6]S.C. Carniglia and G.L. Barna, "Handbook of Industrial Refractories Technology", Noyes Publication, Park Ridge, New Jersey, 1992.

[7]A.D. Pelton et al., "FACT Thermochemical Databases for Calculations in Materials Chemistry at High Temperature", High Temp.Sci., 26, 1990, pp. 231-240.

[8]C.W. Bale et al., « Facility for the Analysis of Chemical Thermodynamics – User Manual 2.1", École Polytechnique de Montréal/McGill University, 1996.
[9]A.H. Heuer and V.L.K. Lou, "Volatility Diagrams for Silica, Silicon Nitride and Silicon Carbide and their Application to High-Temperature Decomposition and Oxidation", J.Am.Ceram.Soc. 73, (10), 1990, pp. 2789-2803.
[10]V.L. Lou et al., « Review-Graphical Displays of the Thermodynamics of High-Temperature Gas-Solid Reactions and their Applications to Oxidation of Metals and Evaporation of Oxides", J.Am.Ceram.Soc., 68, 2, 1985, pp. 49-58.
[11]E.S. Chen and O. Buyukozturk, "Modeling of Long Term Corrosion Behavior of Refractory Linings in Slagging Gasifiers", Am. Cer. Soc. Bull., Vol. 64 (7), 1985, pp. 995-1000.
[12]R. Crescent and M. Rigaud, in "Advances in Refractories for the Metallurgical Industries," Proc. 26th Annual Conference of Metallurgists, CIM, Montreal, 1988, pp. 235-250.
[13]American Society for Testing Materials, Vol. 15.01, Philadelphia, published annually.
[14]M. Rigaud in 8th CIMTEC Proceedings, ACeramics: Charting the Future,≅ P. Vincenzini (Ed.), Techna Srl., Faenza, Vol. 3A, 1995, pp. 399-414.
[15]A. Watanabe et al., "Behavior of Different Metals Added to MgO-C Bricks", Taikabutsu Overseas, Vol. 7 (2), 1987, pp. 17-23.

2000 Focused Sessions

St. Louis, MO, May 1, 2000

James P. Bennett, Program Chair
Albany Research Center
U.S. Dept of Energy
Albany, OR, USA

Dr. Jeffery D. Smith, Session Chair
University of Missouri – Rolla
Rolla, MO, USA

Charles Alt, Session Chair
LaFarge Calcium Aluminates
Chesapeake, VA, USA

APPLICATION OF THERMOCHEMISTRY TO REFRACTORIES

Akira Yamaguchi
Nagoya institute of Technology
Gokiso-cho, Showa-ku Nagoya, 466-8555 Japan

ABSTRACT

Carbon-containing refractories are composed of oxides, graphite and antioxidants such as metal, carbide, etc. The antioxidants often become a gas phase, diffusing the refractory and reacting with graphite, CO(g) and oxides at high temperature. During this process change the useful effect is brought about in the refractory. Knowledge of thermochemistry is very important in order to understand the effect and behavior of the gas phase. In this paper, thermochemistry is applied to the refractory in order to clarify the phenomenon of gaseous reactions that occur.

INTRODUCTION

The rate of corrosion reactions by molten slag, etc. is important problems in refractory composed of only oxides. Thermodynamics is not necessarily useful in elucidation of the reaction rate. In addition, since evaporation from the oxides is little and corrosion by gas phase hardly occurs, consideration of the gas phases is not necessity.

On the other hand, damage and microstructure changes in which a gas phase is concerned occur in carbon-containing refractories composed of oxide-graphite-antioxidant. Thermochemistry is a very effective tool to use in understanding the behavior of the gas phase.

In this paper, the method for clarifying the phenomenon in which a gas phase is concerned is described from the thermochemical viewpoint.

PROCESS CHANGE IN CARBON-CONTAINING REFRACTORIES AND THE BEHAVIOR OF THE ANTIOXIDANT

Fig. 1 is a schematic drawing which shows the behavior of an antioxidant and the binder in a refractory during heating and use. During the first heating process, binders like phenol resin decompose to form gases such as CO_2(g) CO(g), H_2O(g),

$CH_4(g)$, $H_2(g)$, etc. while the strength of the refractory decreases with this decomposition. The diffusion routes of gases in the refractory to the outside become open pores. The graphite is oxidized over about 700℃ and oxidation-damage of the refractory occurs. The bad phenomenon such as oxidation-damage, strength degradation and formation of pores as mentioned above are repaired by the action of added antioxidants. Besides, the addition of antioxidant gives other excellent properties to the refractory. The effects of the antioxidants are enumerated as follows.

(1) Decrease of porosity, namely, densification of the refractory.
(2) Reduction of CO(g) to C(s), suppressing the carbon diminution rate.
(3) Formation of a surface protective layer. Oxidation-damage and corrosion of the refractory are suppressed.
(4) Promotion of crystallization of amorphous carbon that derives from the binder, namely, an oxidation-resistant form of the carbon is improved.
(5) Increase of hot strength of the refractory.

Each antioxidant brings about varying degrees of the above effects. Next, the behavior and the effect of some antioxidants are discussed based on thermochemistry.

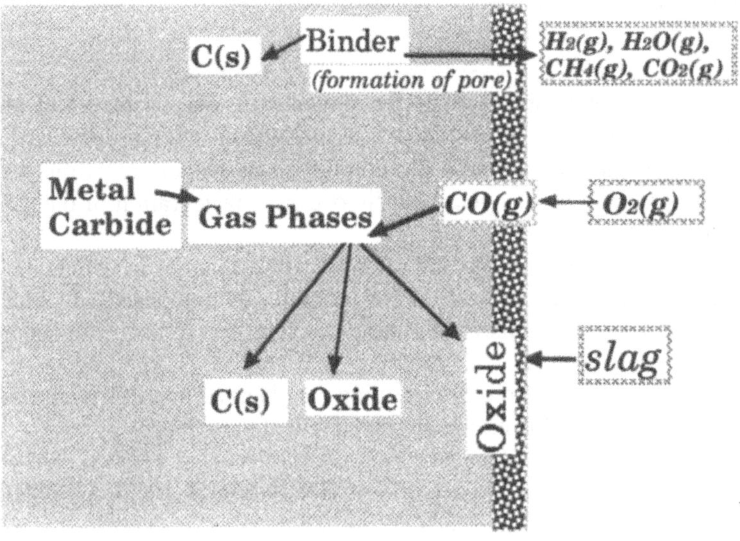

Fig. 1. Schematic drawing of behavior of an antioxidant in a carbon-containing refractory

OXIDATION OF CARBON

In the system C-O, there are many condensed phase and gas species; namely, C(s), CO_2(s,l) and O_2(l) as condensed phases, and C_2O(g), C_3O_2(g), CO(g), CO_2(g), O_2(g) and O(g) as gas phases. So in any given system, there are functional relations among each of the partial pressures of the gas species. Therefore, the gas species with high partial pressure under the given condition may be considered in order to evaluate the behavior of the gas phase.

When carbon is oxidized, various reactions as shown below can occur:

$$C(s) + O_2(g) = CO_2(g) \quad (1)$$

$$C(s) + CO_2(g) = 2CO(g) \quad (2)$$

$$2C(s) + O_2(g) = 2CO(g) \quad (3)$$

$$4C(s) + O_2(g) = 2C_2O(g) \quad (4)$$

------------------------ (X)

CO(g) and CO_2(g) are the main carbon containing gases created. Fig. 2 shows the partial pressure of CO(g), CO_2(g) and O_2(g) when their total pressure is 0.1MPa when in contact with C(s). Over about 1000°C, most of the gas is CO(g). The open pores in the refractory used at high temperature are considered to be filled by CO(g) at a pressure of about 0.1MPa, because the pressure in a furnace lined by a refractory is about 0.1MPa. Therefore, the antioxidant added to the refractory is considered to be exposed to an atmosphere of 0.1MPa CO(g).

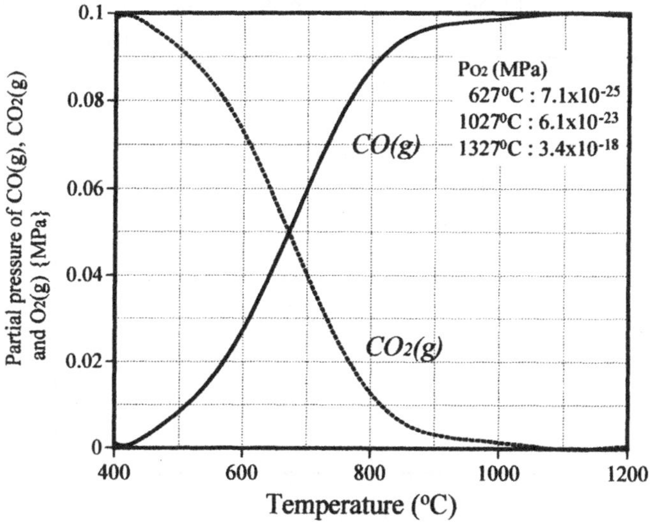

Fig. 2. Partial pressure of CO(g), CO_2(g) and O_2(g) in contact with of carbon. $\{P_{CO}+P_{CO_2}+P_{O_2} = 0.1MPa\}$

BEHAVIOR OF METAL

Metals such as Al, Mg, Si, etc. of which the oxide is stable at 0.1MPa of CO(g) in contact with C(s) are added to the refractory as antioxidants. The behavior of these metals is fundamentally same. Therefore, the behavior of Al is explained as an example. To begin with, the influence of carbon and oxygen on Al metal is examined based on thermochemistry. When a small quantity of carbon is added into a vacuum container in which only Al grains are present, it becomes C(g) and AlC(g). If additional amounts are added, Al_4C_3(s) condenses and C(s) deposits at ever higher additions. On the other hand, Al_2O_3 deposits if O_2(g) is added above a certain quantity. Fig. 3 shows the stable areas of the condensed species in the system Al-C-O for various partial pressures of C(g) and O_2(g), which are plotted in the vertical and horizontal axes, respectively, at 1600K. Al_4C_3(s) and C(s) coexist in equilibrium with Al_2O_3(s) at an oxygen partial pressure of –24.121 in log P_{O_2} and –3.814 in log P_{CO}. The partial pressure of CO(g) that is obtained from equation (5) is also shown in Fig. 3.

$$2C(g) + O_2(g) = 2CO(g) \quad (5)$$

Fig. 4 shows changes that occur among the condensed phases in the system Al-C-N for partial pressures of C(g) and N_2(g), which are obtained by similar method stated above.

The stability areas of Al_2O_3(s), Al_4C_3(s) and AlN(s) for log P_{co} and log P_{N_2} in contact with C(s) at 1600K are shown in Fig. 5, which is based on information in

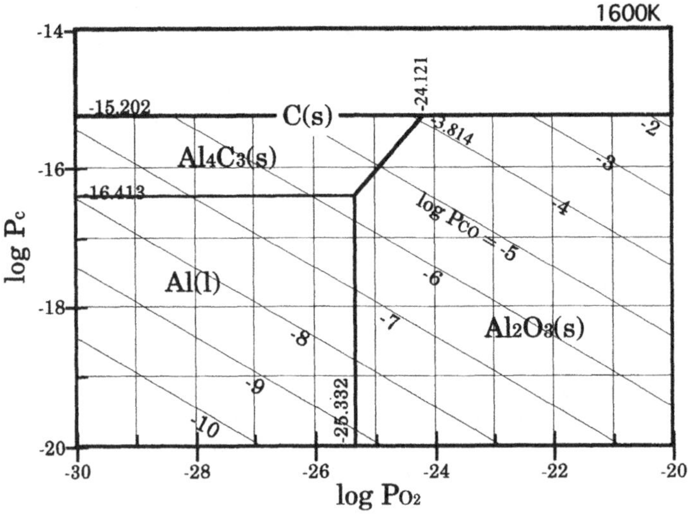

Fig. 3. Stable area of the condensed phases in the system Al-C-O for different partial pressures of C(g) and O_2(g) at 1600K.

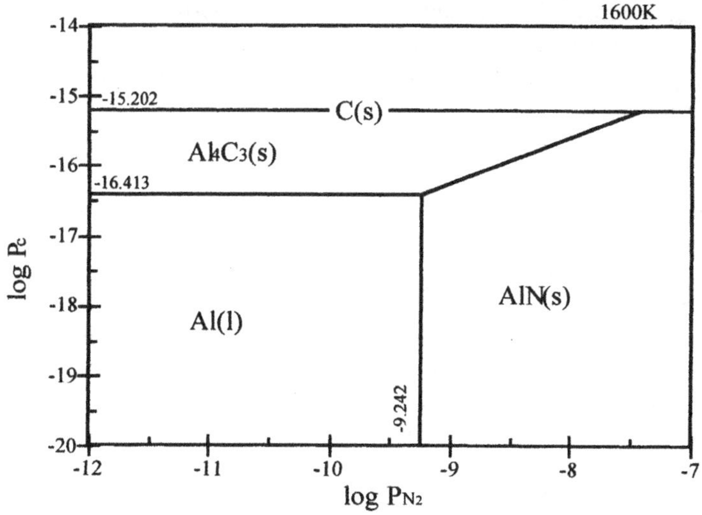

Fig. 4. Stable areas of the condensed phases in the system Al-C-N for partial pressures of C(g) and N_2(g) at 1600K.

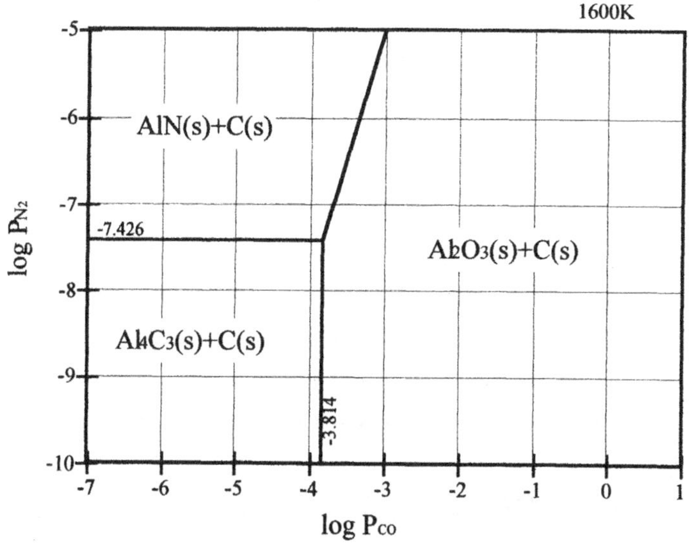

Fig. 5. Stable areas of the condensed phases in the system Al-C-N-O for partial pressures of CO(g) and N_2(g) at 1600K.

Fig. 3 and Fig. 4. Since Al_2O_3 is stable at 0.1MPa of CO(g) {log P_{CO} = 0} when in contact with C(s), Al added to the refractory changes to Al_2O_3 via Al_4C_3, AlN or directly. Fig. 6 shows the process changes of the condensed phases of Al metal added to the refractory.

However, the process changes of the Al is not clarified without considering the gas phase, because the gas phase is an important part in the process. Accordingly, the behavior of the gas phase is clarified by the application of thermochemistry.

As an example, the change of Al_2O_3 to Al_4C_3 via gas phase transition is discussed. The partial pressures of the gas species in the system Al-C-O in contact with carbon for partial pressure of CO(g) at 1600K is shown in Fig. 7. Al_2O_3 is stable at 0.1Mpa of CO(g), but Al_4C_3 is unstable. Therefore, Al_4C_3 reacts with CO(g), and its partial pressure drops until a value at which Al_4C_3(s) stabilizes. In this state, Al(g) and Al_2O(g) and having high partial pressures evaporate mainly from Al_4C_3(s) according to the equation of (6) and (7).

$$Al_4C_3(s) = 4Al(g) + 3C(s) \quad (6)$$

$$Al_4C_3(s) + 2CO(g) = 2Al_2O(g) + 5C(s) \quad (7)$$

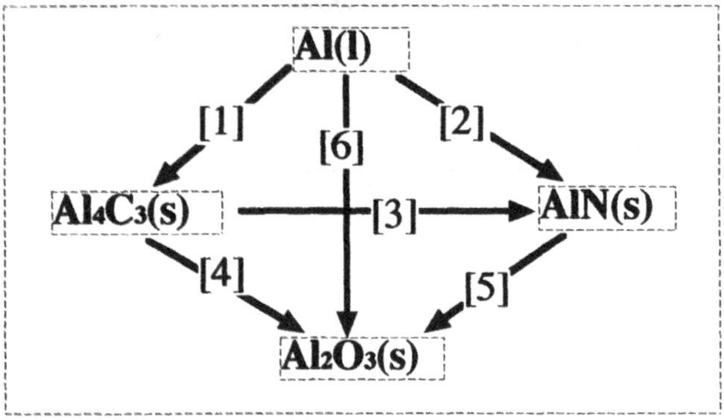

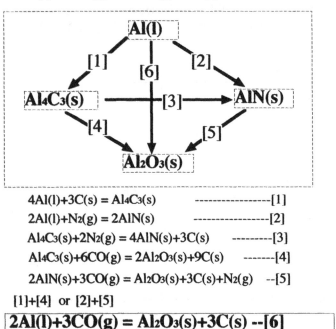

$4Al(l)+3C(s) = Al_4C_3(s)$ -----------------[1]

$2Al(l)+N_2(g) = 2AlN(s)$ -----------------[2]

$Al_4C_3(s)+2N_2(g) = 4AlN(s)+3C(s)$ ---------[3]

$Al_4C_3(s)+6CO(g) = 2Al_2O_3(s)+9C(s)$ -------[4]

$2AlN(s)+3CO(g) = Al_2O_3(s)+3C(s)+N_2(g)$ --[5]

[1]+[4] or [2]+[5]

$2Al(l)+3CO(g) = Al_2O_3(s)+3C(s)$ --[6]

Fig. 6. Changing process of the condensed phases of Al metal added to the carbon-containing refractory

Under these conditions, the partial pressures of Al(g) and Al_2O(g) near the surface of Al_4C_3(s) are values of X and X' as shown in Fig. 7.

When Al(g) and Al_2O(g) diffuse near the surface of the refractory, (the CO(g) partial pressure is about 0.1MPa at this point), these gases react with CO(g) and condense to form Al_2O_3(s) and C(s). The partial pressure of Al(g) and Al_2O(g) decrease to their equilibrium partial pressure according to the equation of (8) and (9).

$$2Al(g) + 3CO(g) = Al_2O_3(s) + 3C(s) \quad (8)$$

$$Al_2O(g) + 2CO(g) = Al_2O_3(s) + 2C(s) \quad (9)$$

Next, the equation (10) is obtained from {(6)+(8)×2} or {(7)+(9)×2}.

$$Al_4C_3(s) + 6CO(g) = 2Al_2O_3(s) + 9C(s) \quad (10)$$

It is understood from this equation that Al_4C_3(s) reduces CO(g) to C(s), and that Al_2O_3(s) condenses. The formation of Al_2O_3 brings about densification of the refractory. Si and Mg also reduce CO(g) to C(s), condensing to form SiO_2 and MgO.

BEHAVIOR OF CARBIDE

SiC, B_4C, ZrC, etc. are also added to the refractory as antioxidants. Since the behavior of these carbides is fundamentally almost the same in the refractory, only the behavior of SiC will be explained.

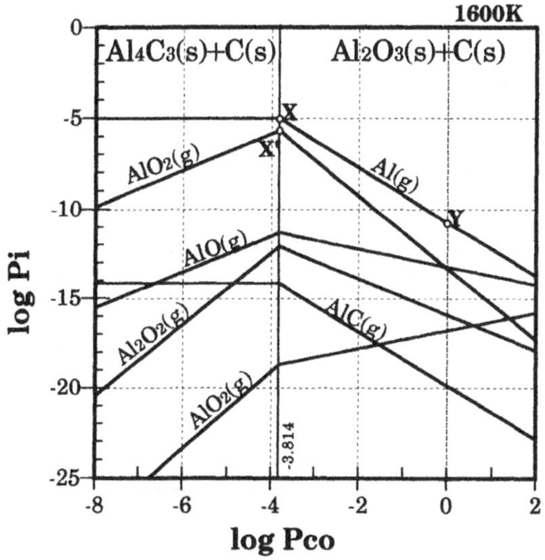

Fig. 7. Change of partial pressures of the gas species in the system Al-C-O in contact with carbon for partial pressure of CO(g) at 1600K.

Fig. 8 shows the stable areas of the condensed species in the system Si-C-N-O for partial pressures of N_2(g) and CO(g), which are plotted on the vertical and horizontal axes, respectively. This figure was made by the same method as Fig. 5.

Since SiC(s) is not stable at 0.1MPa of CO(g) {log P_{CO} = 0} at 1600K, it reacts with CO(g) to form SiO_2(s) and C(s). Consideration of the gas phase is important when SiC(s) changes to SiO_2(s). The mechanism of the change is described next.

There is Si(g), Si_2(g), Si_3(g), SiC(g), Si_2C(g), SiO(g) and SiO_2(g) as gas species in the system Si-C-O. Fig. 9 shows their equilibrium partial pressures for a CO(g) partial pressure at 1600K. SiO(g) with the highest equilibrium partial pressure is the mainly concern in reactions of these species. SiC(s) reacts with CO(g) to form SiO_2(s) according to the equation (11) because SiC(s) is unstable under 0.1MPa of CO(g),

$$SiC(s) + CO(g) = SiO(g) + 2C(s) \tag{11}$$

By this reaction, the partial pressure of CO(g) near the surface of SiC(s) crystals drops to a value shown by the point X in Fig. 9, at which SiC(s) is stable, and then SiO(g) partial pressure is a value of point X in Fig. 9. SiO(g) that diffuses near the surface of the refractory reacts with CO(g) to form SiO_2(s) according to equation (12). This occurs because the CO(g) partial pressure rises to 0.1MPa, as the equilibrium partial pressure of SiO(g) drops to Y in Fig. 9.

$$SiO(g) + CO(g) = SiO_2(s) + C(s) \tag{12}$$

The reaction is expressed by the equation (13) {=(11)+(12)}.

$$SiC(s) + 2CO(g) = SiO_2(s) + 3C(s) \tag{13}$$

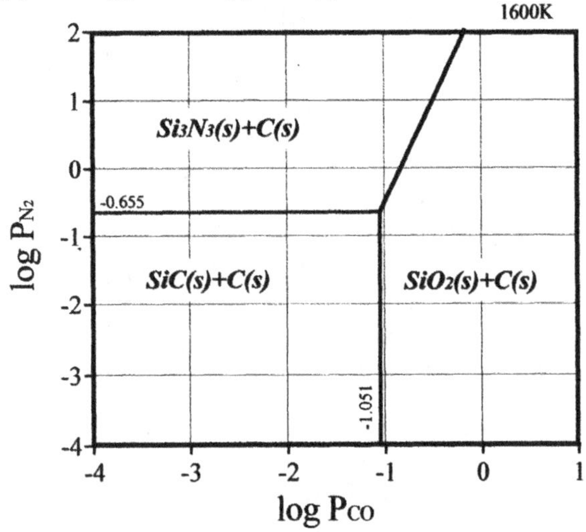

Fig. 8. Stable areas of the condensed phases in the system Si-C-N-O for partial pressure of CO(g) and N_2(g) at 1600K.

From this equation, it is proven that SiC(s) reduces CO(g) to C(s), suppressing the decrease of carbon. In addition, carbon, which is a constituent of the SiC, is deposited as C(s). The $SiO_2(s)$ that condenses near the surface of the refractory by equation (12) reacts with slag to form a surface protective layer. Because of this layer, the diffusion of oxygen into the refractory is also suppressed, and simultaneously, the corrosion by the slag is inhibited.

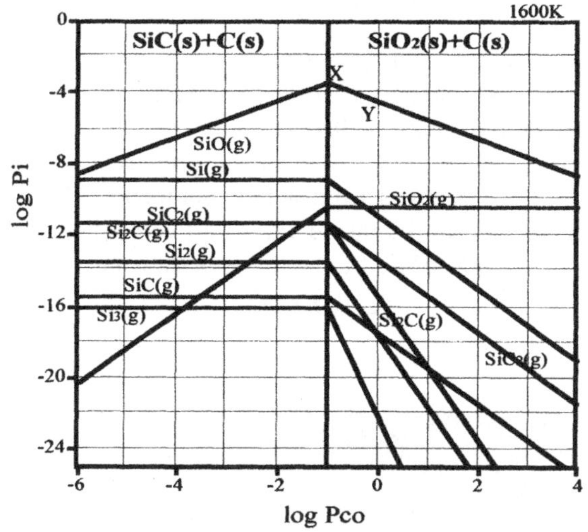

Fig. 9. Change of the partial pressures of gas species in the system Si-C-O in contact with carbon for partial pressure of CO(g) at 1600K.

BEHAVIOR OF ALLOY AND CONPLEX CARBIDE

Alloys and carbides with over two kinds of metal, namely, Al-Si alloy, Al-Mg alloy, Al-Ca-Mg alloy, Ca-Mg-Si alloy, Al_4SiC_4, or $Al_8B_4C_7$, etc. are can be added to the refractory. They improve refractory properties as follows.

(1) Al-Mg alloy

The formation of a dense MgO layer near the surface of MgO-C refractory will suppress oxidation damage and improve corrosion resistance. The addition of a Al-Mg alloy is one of the most effective methods for the formation of the layer. Fig. 10 shows the dense MgO layer formed in the used refractory with Al-Mg alloys. The effectiveness of the addition is described next from the viewpoint of thermochemistry.

The melting and boiling points of Mg metal are 640℃ and 1103℃, respectively. Mg evaporation becomes active over about 700℃, where Mg(g) is oxidized to form a dense MgO layer near the surface of the refractory during its first heating. It is necessary to keep growing the layer in order to make a useful protective layer. Otherwise it will be corroded by slag and disappear. Al metal contributes to the growth of this layer as described next.

MgO(s) reacts with C(s) to form Mg(g) and CO(g) in the refractory according the equation (14).

$$MgO(s) + C(s) = Mg(g) + CO(g) \qquad (14)$$

In a refractory without Al, the partial pressure of Mg(g) is a value of Y of Fig. 11, where a partial pressure of CO(g) is 0.1MPa.

The element Fe plays an important role for the growth of the dense layer as shown in Fig. 10. Iron ion changes from a Fe^{+3} compound {$MgFe_2O_4$} in the outside of layer to metal iron in the inside via a Fe^{+2} compound {(Mg, Fe)O} in middle of the layer. The change of Fe^{+2} to metal Fe is shown in the equation (15).

$$FeO(s) + Mg(g) = MgO(s) + Fe(s) \qquad (15)$$

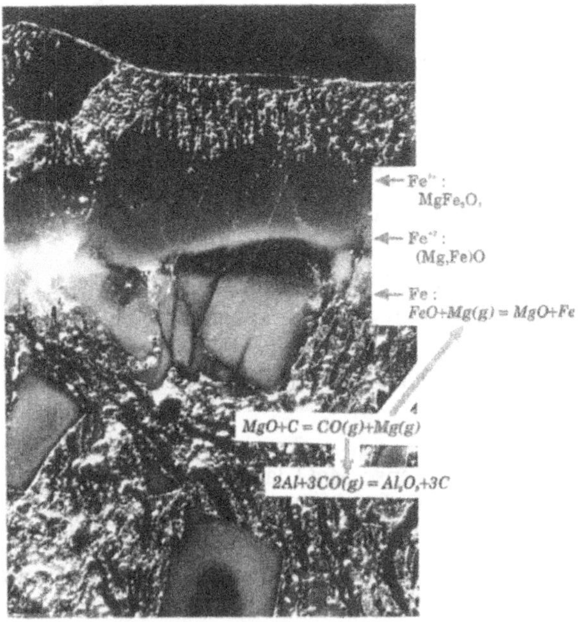

Fig. 10. Dense MgO layer formed in the used MgO-C refractory with Al-Mg alloy.

This reaction is promoted as the partial pressure of Mg(g) becomes higher, causing precipitation of MgO(s) to increase. When Al metal is added, the CO(g) partial pressure drops as described in section 4 and the Mg(g) partial pressure increases from Y to X as shown in Fig. 11. As a result, the growth of a dense MgO layer is promoted.

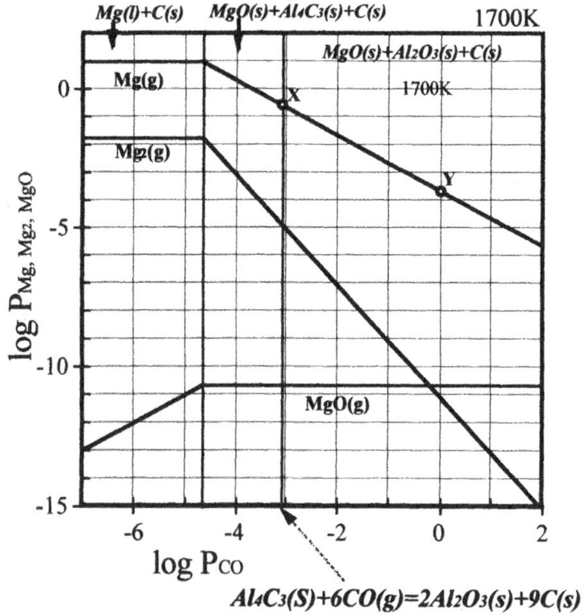

Fig. 11. Chang of the partial pressures of gas species in the system Mg-C-O in contact with carbon for partial pressure of CO(g) at 1700K.

(2) Al-Si Alloy and Al_4SiC_4

Fig. 12 shows the reaction products of Al, Al-Si alloy and Al_4SiC_4 where they are heated in CO(g) of 0.1MPa in the presence of carbon at various temperatures for 2 hrs and at 1500^{0}C for various times. Each of changes roughly as follow.

$Al \rightarrow Al_4C_3 \rightarrow Al_2O_3$

$Al\text{-}Si \rightarrow Al+SiC \rightarrow Al_4C_3+SiC \rightarrow Al_2O_3+SiC$

$Al_4SiC_4 \rightarrow Al_2O_3+SiC \rightarrow Al_6Si_2O_{13}$

In the case of an Al-Si alloy, SiC is formed first, then Al_4C_3 is formed. Al_2O_3 is formed from the mixture of SiC and Al_4C_3. On the other hand, in the case of Al_4SiC_4, Al_2O_3 and SiC are formed first, and Al_4C_3 is never formed. The reason why this order occurs is explained as follow:

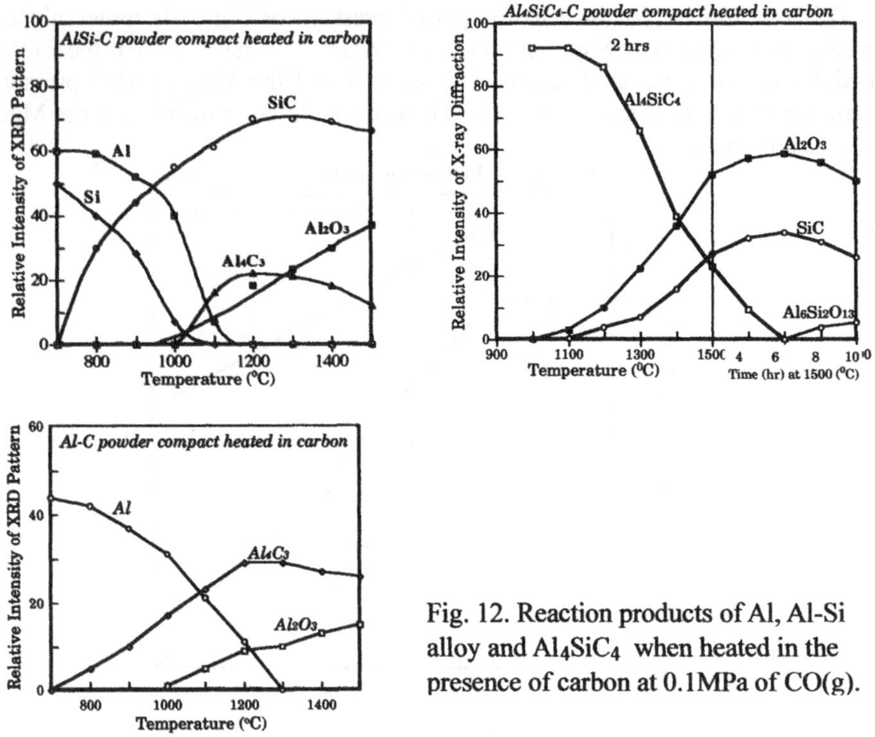

Fig. 12. Reaction products of Al, Al-Si alloy and Al_4SiC_4 when heated in the presence of carbon at 0.1MPa of CO(g).

Whether Al and Si first reacts with carbon to form a carbide is judged by comparing the free energy of formation of (Al_4C_3+Si) and that of (SiC+Al) according to the equation (16).

$$Al_4C_3(s)+3Si(s,l) = 3SiC(s)+4Al(s,l) \qquad (16)$$

Which carbide (Al_4C_3 or SiC) reacts with oxygen to form an oxide is determined by same method according to the equation (17).

$$Al_4C_3(s)+3SiO_2(s) = 2Al_2O_3(s)+3SiC(s) \qquad (17)$$

Fig. 13 shows the free energy change of the equations (16) and (17). Each of the energy changes based on the equation is negative. Accordingly, SiC is first formed from a mixture of Al, Si and carbon and Al_2O_3 is formed from a mixture of Al_4C_3, SiC and carbon.

Fig. 14 shows the stable condensed phases and the equilibrium partial pressures of the gas species in the system Al-Si-C-O for a CO(g) partial pressure in the presence of C(s) at 1700K. Al_4SiC_4 reacts with CO(g) as shown by the equation

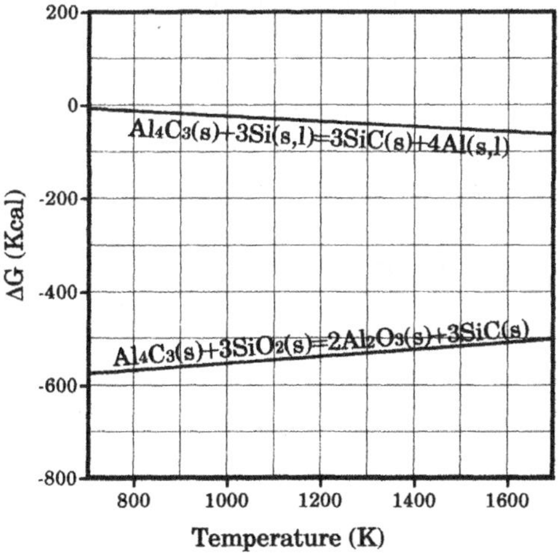

Fig. 13. Free energy change of the reactions.

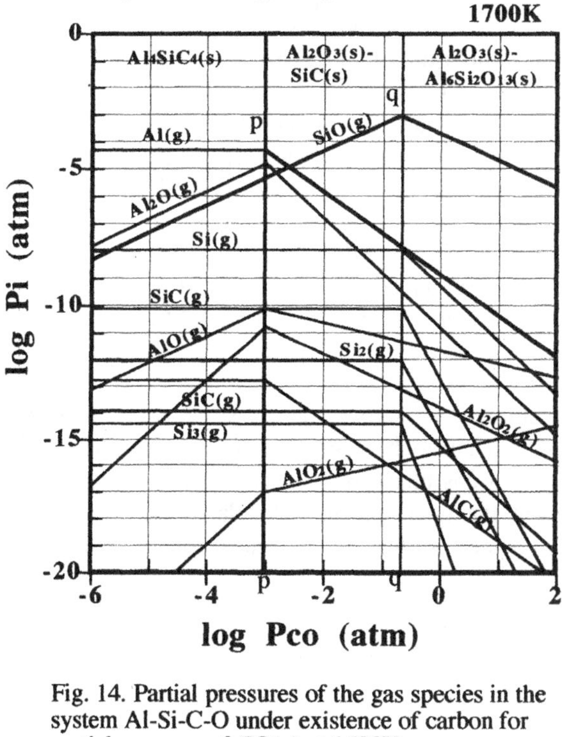

Fig. 14. Partial pressures of the gas species in the system Al-Si-C-O under existence of carbon for partial pressure of CO(g) at 1600K.

(18), because it is unstable at 0.1MPa of CO(g) at 1700K.

$$Al_4SiC_4(s) + 6CO(g) = 2Al_2O_3(s) + SiC(s) + 9C(s) \quad (18)$$

A gas phase is involved in this change. By the reaction of the equation (18), the partial pressure of CO(g) near the surface of Al_4SiC_4 grains drops to the value of P point where Al_4SiC_4 is stable. Since the equilibrium partial pressure of Al(g) and $Al_2O(g)$ are higher at the point P, both gas species are generated from Al_4SiC_4 as follows

$$Al_4SiC_4(s) = 4Al(g) + SiC(s) + 3C(s) \quad (19)$$

$$Al_4SiC_4(s) + 2CO(g) = 2Al_2O(g) + SiC(s) + 5C(s) \quad (20)$$

Al(g) and $Al_2O(g)$, which diffuse near the surface of the refractory where the CO(g) partial pressure raises, react with CO(g) to condense and form $Al_2O_3(s)$ and C(s) according to equations (21) and (22), and then the equilibrium partial pressure of Al(g) and $Al_2O(g)$ drops.

$$2Al(g) + 3CO(g) = Al_2O_3(s) + 3C(s) \quad (21)$$

$$Al_2O(g) + 2CO(g) = Al_2O_3(s) + 2C(s) \quad (22)$$

The equation (18) is obtained from {(19) + (21)x2} or {(20)+(22)x2}. When Al_4SiC_4 disappears, the CO(g) partial pressure rises to 0.1MPa. However, since SiC(s) is unstable at 0.1MPa of CO(g), SiC(s) reacts with CO(g) to give SiO(g) and C(s) and the partial pressure of CO(g) drops to the value of point q.

$$SiC(s) + CO(g) = SiO(g) + 2C(s) \quad (23)$$

The SiO(g) partial pressure is higher at the point q, but decreases with increasing CO(g) partial pressure. Therefore, SiO(g), which diffuses to the surface of the refractory, reacts with CO(g) to form $SiO_2(s)$ and C(s) according to the equation (24).

$$SiO(g) + CO(g) = SiO_2(s) + 2C(s) \quad (24)$$

The precipitated $SiO_2(s)$ is considered to form a protective layer with the precipitated $Al_2O_3(s)$ and /or slag near the surface of the refractory.

7. Summary

In the carbon-containing refractory, gas phase reactions play an important role in refractory behavior. Thermochemistry is important tool in learning the reaction mechanisms in which a gas phase is concerned. In this paper, the application of thermochemistry was discussed and used to clarifying the behavior and process changes involving metals, alloys and carbides added to the carbon-containing refractory.

CREEP MEASUREMENT AND ANALYSIS OF REFRACTORIES[1]

J. G. Hemrick
Mechanical Characterization and Analysis Group
High Temperature Materials Laboratory
Metals and Ceramics Division
Oak Ridge National Laboratory
Oak Ridge, TN 37831-6069

A. A. Wereszczak
U.S. Army Research Laboratory
Weapons and Materials Directorate
Aberdeen Proving Ground, MD 21005

ABSTRACT

Creep deformation of refractories can limit their service life if it is excessive or be problematic if it is misunderstood. Consequently, accurate and unambiguous engineering creep data are needed for furnace design in order to reliably predict dimensional stability of refractory superstructures. Existing creep testing and data interpretation practices that were developed for structural ceramics are surveyed and their applicability to refractory materials are discussed. Specimen geometry and environmental/experimental issues, compression vs. flexure vs. tension testing, and the utilization of existing creep models is visited. Lastly, other factors that can affect refractory superstructure dimensional stability (e.g., thermomechanical stress gradients, stress relaxation) are reviewed.

1 Research sponsored by the U. S. Department of Energy, Assistant Secretary for Energy Efficiency and Renewable Energy, Office of Industrial Technologies, Advanced Industrial Materials Program and the Glass Vision Team, under Contract DE-AC05-00OR22725 with UT – Battelle, LLC.

I. INTRODUCTION

Creep is defined as the deformation of a material under constant stress as a function of time and temperature [1]. Classically, a metallic material will go through three periods of creep prior to fracture, as represented by Fig. 1 [2]. Following the instantaneous elastic response to loading the material goes through a primary phase of creep where the strain rate transiently declines as a function of time. The secondary stage of creep, or steady-state creep, is characterized by a constant strain rate. This may be followed by a tertiary stage during which the strain rate increases and catastrophic fracture of the material ultimately occurs.

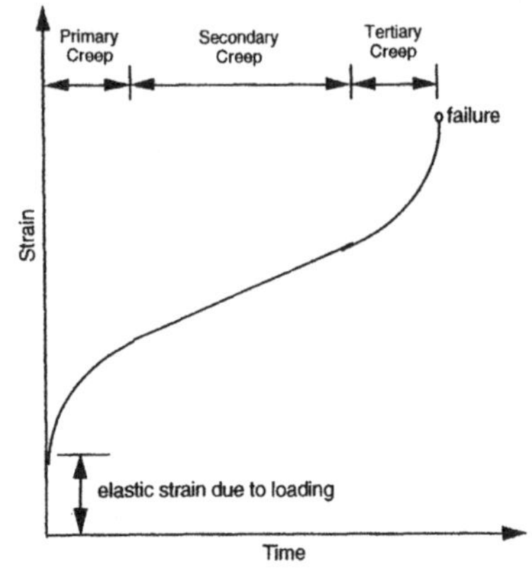

Fig 1. Classical creep curve for a generic metallic material showing the three creep regimes following loading.

The understanding of creep mechanisms is an entire field of science in of itself; however, all mechanisms essentially may be grouped into either lattice or boundary mechanisms [3]. Lattice mechanisms involve intra-granular diffusion and the movement of dislocations within the structure while boundary mechanisms occur through physical means such as diffusion along the grain boundaries or grain boundary sliding. Two primary mechanisms involve diffusion by the movement of vacancies in the structure. Nabarro-Herring creep occurs when vacancy flow is through the lattice (lattice mechanism) while Coble creep occurs when vacancy flow is along the grain boundaries (boundary mechanism). Typically, Coble creep is the rate-limiting mechanism of deformation in polycrystalline ceramics. Grain boundary sliding and cavitation are also typical boundary mechanisms which occur in polycrystalline ceramics, but are not as active in compression (unless very high magnitudes of stress are involved) as they are in tension.

The type of behavior described above and shown in Fig. 1 is known as classical creep and is typical for most metals; however, it is not necessarily representative of either tensile or compression creep in ceramic materials. Most polycrystalline ceramics will exhibit anisotropic creep behavior; namely, much more rapid creep deformation in tension than in compression for the same magnitude of stress [4-6]. Therefore, this must be respected when considering flexural testing results as both tensile and compressive creep deformations are occurring and there is a shift in the specimen's neutral axis with time.

One must be concerned with the creep of refractories due to their crucial role in structural applications such as in glass furnace superstructures and regenerators. For crown superstructures, a compressive loading inherently results from the arch-shaped construction, while in regenerator settings the loading is often simply due to the mass of vertically-stacked refractory components. These types of applications will subject the refractory materials to low stresses typically only on the order of 0.2 MPa (29 psi), but for long periods of time (up to a year or more) and at temperatures as high as 1650°C (3000°F) or above. This is contrary to the typically higher stresses and lower temperatures that components fabricated from normal structural materials are subjected to. Therefore, creep is an unavoidable event and must be accounted for in the design of the furnace crown in order to promote maximum service life of the furnace.

Standardized methods exist for both the measurement of creep in ceramic materials and for the measurement of creep specifically in refractories. ASTM C1291 [7] addresses the standard practice for measuring tensile creep strain, creep strain rate, and creep time-to-failure for advanced monolithic ceramics at elevated temperatures (typically between 1273 and 2073 K) using contact extensometry. It is assumed that the materials tested behave as macroscopically isotropic, homogeneous, continuous materials and that the materials possess a grain size of <50 μm. The results of testing can then be used to provide a measure of the load-carrying capability of a material as a function of temperature and time and therefore the service usefulness of the material. In this standard, an experimentally determined preheat and soak period is used to allow the entire test system to reach thermal equilibrium. This test is susceptible to external factors such as test environment (testing can be carried out in vacuum, inert atmosphere, humid or ambient air), chemical interactions between samples and fixturing, specimen bending, test temperature fluctuations, and machining flaws on samples which may influence the results of the test. Samples can be one of many shapes including square or rectangular cross-section dogbones and cylindrical button-head geometries. Example dimensions that have successfully been previously tested are given, but no set dimensional rules are specified.

ASTM C832 [8] is specifically tailored to the measurement of compressive creep of refractories. This standard addresses the measurement of uniaxial changes by using sensoring that is oriented parallel to the direction of loading. Testing is carried out by subjecting the refractory to a static compressive stress while being held at elevated temperatures. Similar to the standard for advanced ceramics, results can be used to characterize the load bearing capacity of a material that is uniformly heated such as a refractory in a blast furnace stove or glass furnace. Testing is specified to be carried out for 20 to 50 hours in ambient air on rectangular specimens with nominal dimensions 38 by 38 by 114 mm (1.5 by 1.5 by 4.5 in.) with the long dimension perpendicular to the pressing direction of the brick. A linear variable differential transformer (LVDT) is used to measure changes in the axial dimension of the sample and a preheat and soak is utilized to assure thermal equilibrium conditions for the test sample during testing.

The main differences between these two standards are the direction of loading (tensile verses compressive testing) and the method of sensoring. For the testing of refractories, it is stated elsewhere that (1) compressive testing should be performed since it is compressive stresses that are prevalent in refractory systems and (2) cylindrical specimens possess a cross section more suitable for uniform loading in compression [9]. Also note, as discussed, caution must be taken if tensile and compressive creep data are compared or if flexure data are analyzed. Polycrystalline ceramic materials creep more rapidly in tension than in compression and flexural creep data will be a convolution of both tensile and compressive deformation. The sensoring issue is a question of contact extensometry versus using LVDTs. As discussed in Section III, contact extensometry appears to be the more accurate method of measurement, but valuable information can be obtained by supplementing these measurements with those from a LVDT.

II. CREEP DEFORMATION MODELS

Time-dependent anelastic behavior such as creep and relaxation of materials are often modeled using linear viscoelastic theory. A motivation behind using such linear viscoelastic theory is that the solutions to the resulting differential equations are generally analytic. The forms of these mechanical analog models are a combination of linear springs and linear viscous dashpots with the following relationships [10].

$$\sigma = R\varepsilon \quad \text{(linear spring)} \qquad \text{(a)}$$

$$\sigma = \eta\frac{d\varepsilon}{dt} = \eta\dot{\varepsilon} \quad \text{(linear dashpot)} \qquad \text{(b)} \qquad (1)$$

where: σ = stress

R = linear spring constant or Young's Modulus

ε = strain

η = coefficient of viscosity

t = time

$\dot{\varepsilon}$ = strain rate

Several increasingly complex characteristic combinations of these basic elements have been defined. The first two are the familiar Maxwell and Kelvin models that consist of a single linear spring and a single linear viscous dashpot in series or parallel, respectively. While these models do represent the creep and relaxation of the material, they still have several limitations [10]. The Maxwell model shows no time-dependent recovery and does not represent decreasing strain rate under constant stress while the Kelvin model does not exhibit time-independent strain on loading or unloading and does not describe a permanent strain after unloading. A model of higher complexity and better representation is formed by joining a Maxwell and Kelvin model in series. This model is known as the Burgers Model [10]. The strain in this model is related to a spring in series with both a dashpot and a combination of a spring and dashpot in parallel. Through the use of Laplace transforms and their inverse, a constitutive equation between σ and ε can be determined and a resulting creep rate can be derived through differentiation. Yet, this model still is limited by its use of only a single time constant. Through the use of an equation incorporating a series of linear exponential terms and by the choosing of exponential terms such that they encompass several decades of time, one allows for multiple time constants while also achieving a superior fit to data modeled with only geometric time increases [11]. An example of this is shown in Fig. 2. Although the two fits are close, the exponential form fits the data quite well at all times, while this is not necessarily true for the Burgers model. This is also evident when looking at the corresponding R values for the two fits.

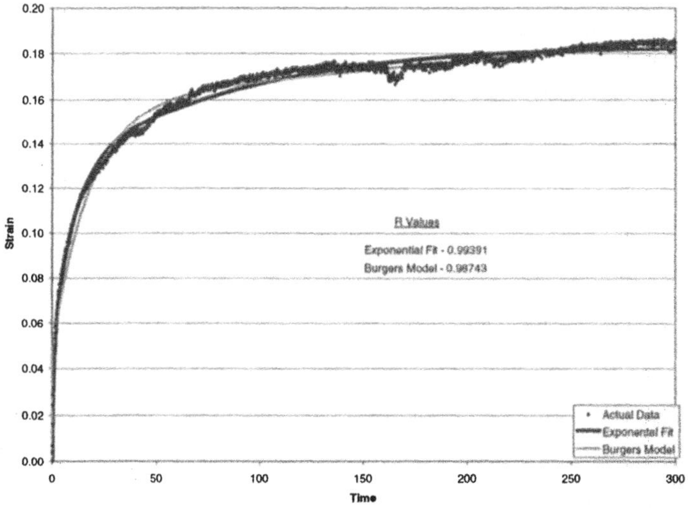

Fig 2. Comparison of multiple exponential fit and Burgers model fit for typical engineering creep strain data.

Traditionally, both tensile and compressive creep in ceramic materials have been characterized by an empirical creep equation that takes on the form of the Norton-Bailey-Arrhenius (NBA) equation as shown below [4].

$$\dot{\varepsilon} = A\sigma^{n} \exp\left(-\frac{Q}{RT}\right) \tag{2}$$

where: $\dot{\varepsilon}$ = strain rate (h^{-1})
A = dimensionless constant
σ = stress (Pa)
n = stress exponent
Q = activation energy (J/mole)
R = gas constant (8.314 J/(mole * K))
T = temperature (K)

Assuming that the same dominate (or rate-controlling) creep mechanism is active at all temperatures and stresses, the activation energy and value of the stress exponent can be found through regression of data at various temperatures and stresses. Examples of data fit using the NBA equation are shown in Fig. 3. The value of the stress exponent will then provide an indication of the rate-limiting creep mechanism [3]. A value of n = 1 indicates diffusion controlled creep, while a value of n = 2 is characteristic of grain boundary creep. Values of n = 3 and n = 5 are also possible indicating creep attributed to dislocation glide or climb of dislocations and creep due to intragranular dislocation controlled by climb of dislocations at dislocation pile-ups respectively. For polycrystalline ceramics the n-value is generally approximately equal to one for compressive creep (since dislocation activity is typically insignificant) and may be higher than one for tensile creep due to cavitation [3,4].

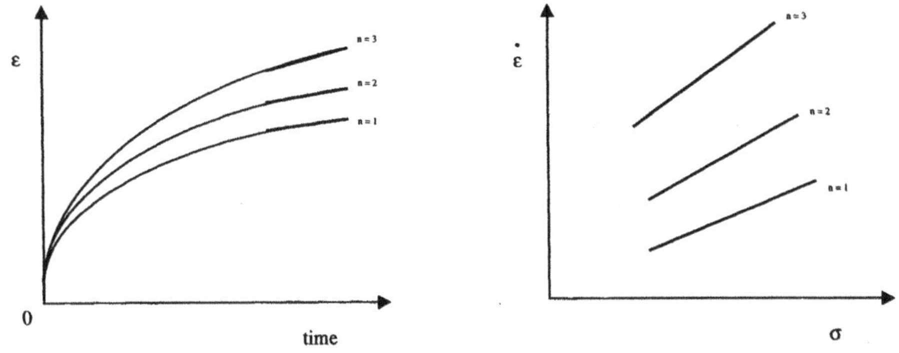

Fig 3. Generic data fit using the NBA equation. Note: data are for a single stress and temperature, but for different materials with varying stress exponents.

III. UNIQUE ISSUES CONCERNING THE MEASUREMENT OF CREEP OF REFRACTORIES

Since published engineering creep and high temperature modulus of elasticity data are essentially nonexistent for commercial refractories that are used for applications in the glass industry, there is a great need for testing to produce such data. Yet, obtaining such data can be experimentally difficult due to several unique issues concerning the measurement of creep in refractories. Therefore, care must be taken when performing creep measurements on this class of material.

Refractory materials are expected to perform at extremely high temperatures (in excess of 1600°C (2912°F) for many cases), therefore testing must be performed at equivalent temperatures to be meaningful. Due to these high temperatures, there must be a chemical compatibility between the test specimen and the material of which the test fixturing is made. For example, if a refractory containing high amounts of alumina is tested in contact with a SiC push-rod at these temperatures, a glassy alumino-silicate will form at the sample/push-rod interface degrading both the test specimen and the push-rod while also compromising the validity and accuracy of the resultant data. Therefore, it is wise to consider the phase diagrams concerning any materials one wishes to put in contact with one another in order to adapt one's fixturing to avoid such problems.

Another problem with measuring the creep of refractories is the difficulty in accurately characterizing the accumulated strain at loads corresponding to those that occur in service and for times similar to those of the life of a furnace campaign. A standard crown refractory will only see loads on the order of 0.2 MPa (29 psi), but may be subjected to a furnace campaign lasting many months or even years. At such small loads, very small amounts of accumulated strain are produced (on the order of 1% or less). Therefore, a method of strain measurement is needed that is both extremely precise and which possesses excellent long-term stability.

To achieve this, one must assure that thermal stability of the whole test system is in existence (including that of the furnace, load train, and extensometer) by using a means such as a dedicated water chiller to control overheating and temperature fluctuations of the furnace [12]. Also, any holes in the furnace insulation must be eliminated to assure "chimneying" does not occur as ambient air is pulled in one end of the furnace, passed through the heated chamber and forced out the other end (typically the top). Further, the method of strain measurement must be highly accurate and show high levels of stability at test temperatures while also being minimally insensitive to specimen and load-train-seating effects. For these reasons, contact extensometry is a favored technique. The extensometer can be enveloped in an air-cooled isothermal box that will provide a constant temperature for measurement during testing. Extensometery also allows for the large gauge lengths needed to accurately measure the small dimensional changes seen in large grained refractory materials. Finally, contact extensometry is not affected by contraction and seating of the load-train, which is sampled by other methods of strain measurement such as the use of a linear differential variable transformer (LVDT). LVDTs are useful as a secondary means of measurement, but it has been noted from previous experience and discussions in the literature that some problems may exist in monitoring creep when only a LVDT is used. These problems may include the above-mentioned sampling of the load train contraction in conjunction with the sample; along with rigid body motion effects, sampling of specimen interaction with air, and sampling of interaction between the push-rods and the sample [13]. Therefore, it is suggested that a contacting extensometer be employed in conjunction with a LVDT to insure isolation of the measured creep behavior of the sample apart from that of the load system.

IV. EXAMPLES OF CREEP DEFORMATION IN REFRACTORIES

The following sections will present a review of creep behavior in several refractory systems. For the testing of each material with the exception of the fused silica and the fusion-cast alumina, cylindrical test specimens were used with a 38.1 mm diameter, 76.2 mm length (1.5 in. dia., 3.0 in. length) and a primary axis parallel to the pressing direction of the original brick. Cylindrical specimens with a 9.0 mm diameter and 25.0 mm length (0.4 in. dia., 1.0 in. length) were used in the fused silica testing and for the fusion-cast alumina, the same sized cylinders were used as in the majority of the other testing, but they were taken from the homogeneous bulk region of the originally cast block. The given creep deformation is divided into two types based on the agreement of the behavior with classical creep theory as discussed above. Behavior that agrees with this theory is termed "classical", while behavior which deviates from the reviewed creep theory is termed "non-classical". Within each section, the relevance of the refractory type is briefly discussed along with the experimental conditions of the testing and a summary of the generated results.

IV.A. "Classical" Creep Deformation

Creep of CaO/SiO_2-containing MgO Refractories [14]

These refractories find use in soda-lime glass furnace regenerators due to their superior corrosion resistance to alkaline environments. In this study, five commercially available brands (designated "A", "B", "C", "D", and "E" to maintain their anonymity), commonly used in regenerator applications, were subjected to compressive creep testing at temperatures and stresses representative of service conditions. The creep rate for each brand was determined and explored as a function of stress and temperature.

Each brand was fired at temperatures in excess of 1535°C (2800°F) according to the vendors' technical literature and possessed a grain size between 20 and 100 μm with several grains making up each crystal. Testing was performed in ambient air using an electromechanical test machine in load control with α-SiC push-rods in conjunction with high purity (99.5%) alumina disks being used in contact with the test specimen to prevent reaction between the sample and the fixturing. A contact extensometer employing sapphire knives and a 40.00 mm (1.16 in.) gauge length was used for sensoring. All specimens were preloaded at 0.03 MPa (≈4 psi) during furnace heat up to keep all the load train components and the specimen in continuous contact and a soak of 15-20 hours at temperature prior to the application of the first stress was maintained. Tests were carried out at 1400°C (2550°F), 1475°C (2685°F), and 1550°C (2820°F) with specimens being loaded sequentially at three stresses; 0.1, 0.2, and 0.3 MPa (14.5, 29.0, and 43.5 psi) for approximately 75 hours at each stress.

The above testing generated curves similar to the one shown in Fig. 4 with an interesting phenomenon being observed in each test during the initial 15-20 hour soak as noted by the contraction which occurs in Fig. 4 during the first 19.8 hours despite the low applied stress.

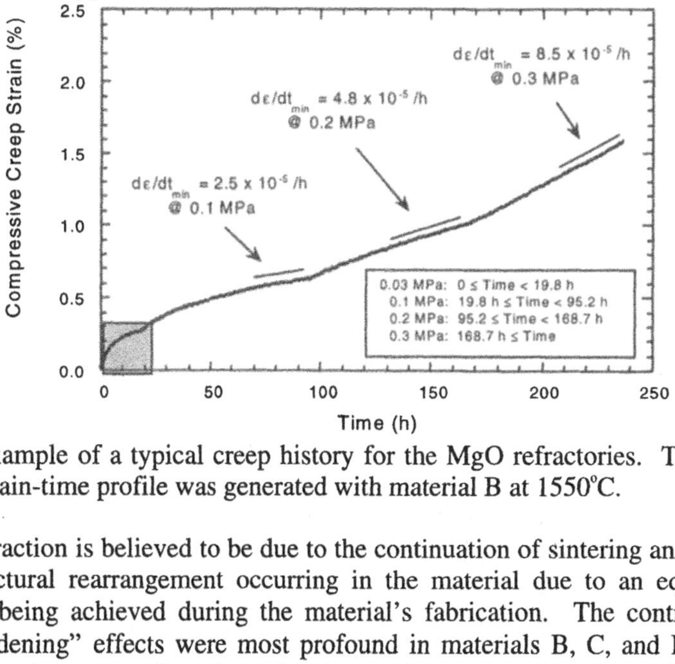

Fig 4. Example of a typical creep history for the MgO refractories. The shown strain-time profile was generated with material B at 1550°C.

This contraction is believed to be due to the continuation of sintering and/or some microstructural rearrangement occurring in the material due to an equilibrium state not being achieved during the material's fabrication. The contraction or "time-hardening" effects were most profound in materials B, C, and D and are believed to have contributed to the faster net creep rates observed in these materials. This would indicate that a superstructure that is several feet in size would in fact significantly contract solely due to the initial exposure of materials B, C, and D to temperatures in excess of 1475°C.

The general appearances of the creep curves were similar for each of the five brands and can be characterized by Fig. 4. Typically, the materials exhibited a period of primary creep lasting 20-40 hours followed by a period of longer duration characterized by a slowly decreasing creep rate. Tertiary creep was never observed. In many instances, a steady-state or constant creep rate was never achieved; consequently, the continuing transient nature of many of the creep rates resulted in a minimum (not steady-state) creep rate being defined.

The steady-state or minimum compressive creep rate ($d\varepsilon/dt_{min}$) was related to the applied compressive stress and temperature using Eq. 2. Multilinear regression was performed to determine the constants A, n, and Q for each brand of MgO refractory. The results for each of the five brands are illustrated in Fig. 5 with the values of parameters A, n, and Q tabulated in Table I.

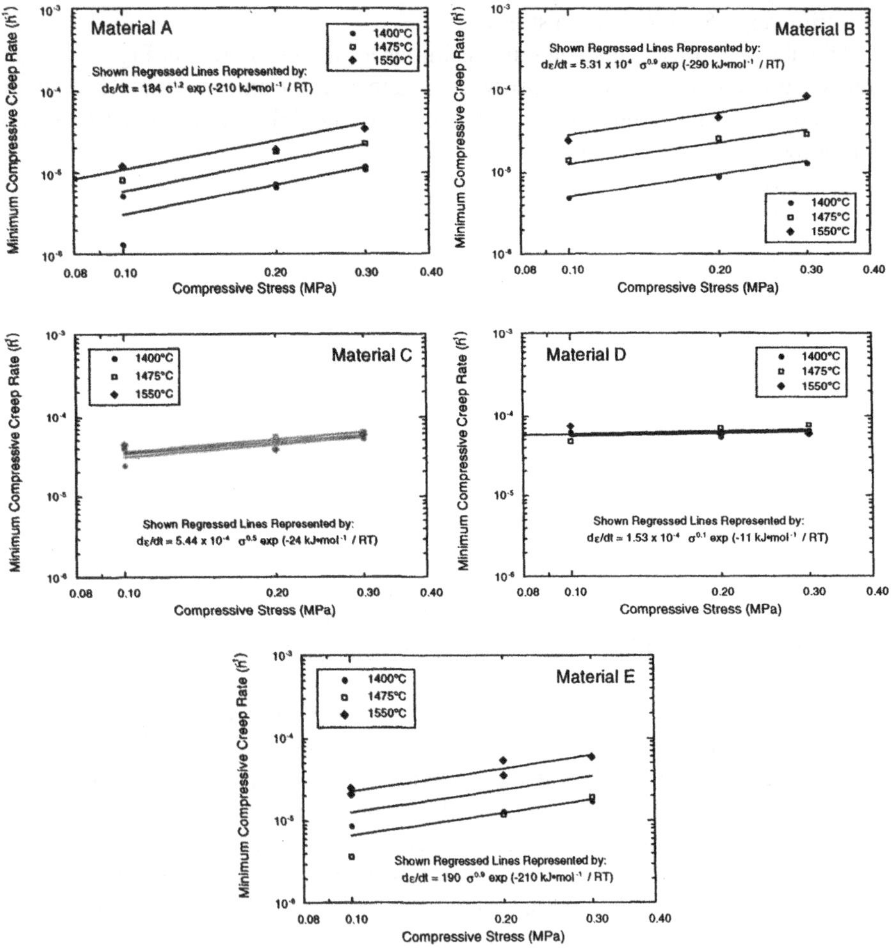

Fig 5. Creep rates as a function of stress for the five MgO brands.

Table I. Summary of values from power law multilinear regression fits of the five MgO brands.

Brand	A	n	Q
A	184	1.2	210 kJ/mol (50 kcal/mol)
B	5.31×10^{-4}	0.9	290 kJ/mol (69 kcal/mol)
C	5.44×10^{-4}	0.5	24 kJ/mol (5.7 kcal/mol)
D	1.53×10^{-4}	0.1	11 kJ/mol (2.6 kcal/mol)
E	190	0.9	210 kJ/mol (50 kcal/mol)

The parameters in Table I can then be used to directly compare the relative creep resistances of the five refractory brands at any hypothetical stress and temperature as shown in Table II and to interpolate between the shown temperatures and stresses.

Table II. Minimum creep rates of the five MgO brands at the hypothetical stress of 0.2 MPa and theoretical temperature of 1500°C (2730°F) as calculated using the parameters in Table I.

Brand	Minimum Creep Rate (h^{-1})
A	1.7×10^{-5}
B	3.6×10^{-5}
C	4.8×10^{-5}
D	6.2×10^{-5}
E	2.9×10^{-5}

Creep of Mullite Refractories [15]

Mullite refractories are also a candidate material for glass furnace crowns and find additional use in other non-glass-contact applications. This testing was also performed in ambient air using an electromechanical test machine in load control. A total of ten different mullite materials were tested comprising a combination of both fused-grain and non fused-grain materials. A load train similar to that used in the MgO testing above was employed for testing. All specimens were preloaded at 0.08 MPa (≈12 psi) during furnace heat up to keep all the load train components and the specimen in continuous contact and a soak of 15-20 hours at temperature prior to the application of the first stress was maintained. Tests were carried out at 1375°C (2500°F) and 1450°C (2640°F) with specimens being loaded sequentially at the stresses of 0.4 and 0.6 MPa (58.0 and 87.0 psi) for approximately 75 hours at each stress.

The above testing generated curves similar in appearance to the one shown in Fig. 6.

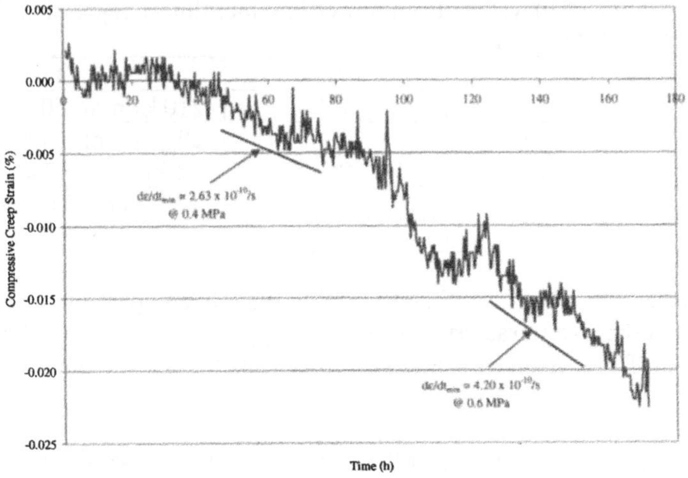

Fig 6. Example of typical creep history for the fused mullite refractories. The shown strain-time profile was generated at 1375°C.

The curves for the two different temperatures were similar with the 1450°C curve showing faster accumulation of strain that resulted in about twice as much total accumulated strain. As seen with the MgO refractories, the materials exhibited a period of primary creep lasting ≈20 hours followed by a longer duration of slowly decreasing creep rate. Tertiary creep was never observed. Since again a steady-state or constant creep rate was never achieved, the continuing transient nature of many of the creep rates resulted in a minimum (not steady-state) creep rate being defined. Creep rate versus stress at the two test temperatures is shown in Fig. 7.

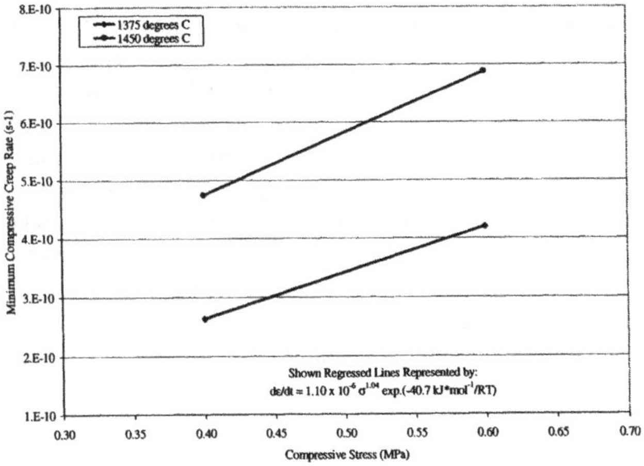

Fig 7. Creep rates as a function of stress for fused mullite refractory.

The steady-state or minimum compressive creep rate ($d\varepsilon/dt_{min}$) was related to the applied compressive stress and temperature using Eq. 2. Multilinear regression was performed to determine the constants A, n, and Q for the mullite refractory leading to Eq. 3.

$$\frac{d\varepsilon}{dt} = 1.10x10^{-6}\sigma^{1.04}\exp\left(\frac{-40.7kJ * mol^{-1}}{RT}\right) \quad (3)$$

<u>Creep of Fused Silica Refractories</u> [16]

While not used in glass furnace applications like the other refractories covered in this paper, fused silica refractories merit discussion as another type of refractory material that exhibits what has been termed "classical" creep deformation. These refractories are composed of amorphous or vitreous silica grains and find use in the metals casting industry as sacrificial "cores" during the casting process. In this testing, samples from six different core materials were tested. Each specimen was tested at temperatures of 1475 and 1525°C (2687 and 2777°F) and under stresses of 2.1, 4.1, and 6.2 MPa (300, 600, and 900 psi).

Behavior exhibited by the various specimens was similar with the behavior of the material at 1475 and 1525°C being characterized by the plots shown in Fig. 8 and Fig. 9 respectively.

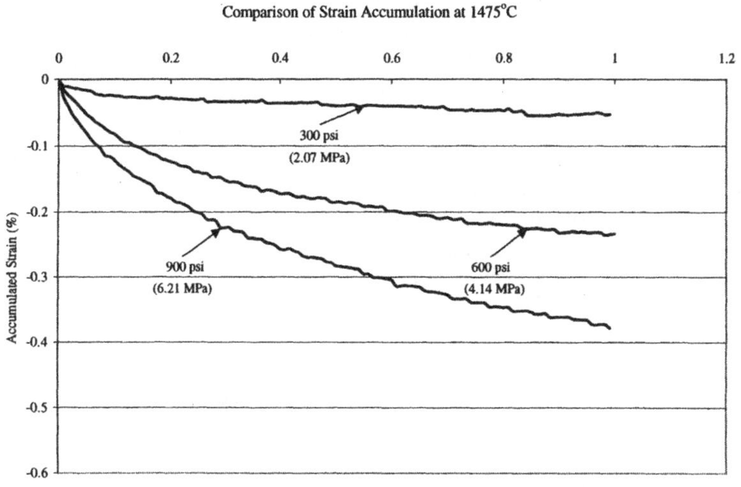

Fig 8. Strain accumulation for a fused silica refractory at 1475°C.

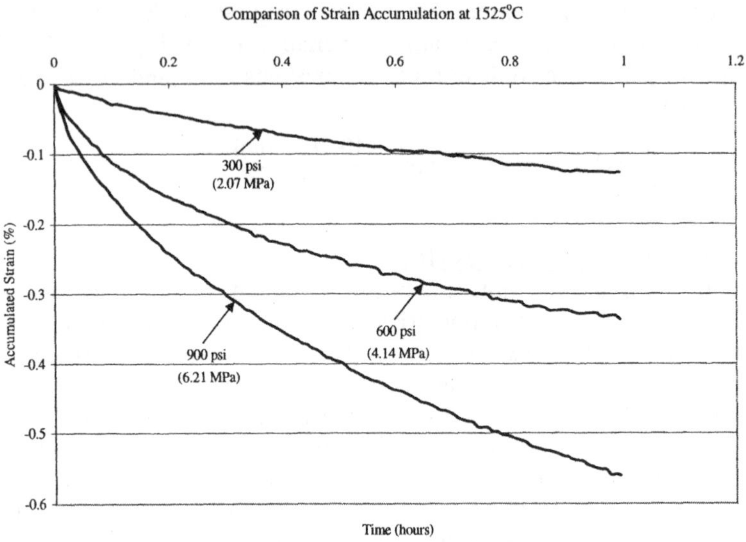

Fig 9. Strain accumulation for a fused silica refractory at 1525°C.

In these plots, a period of primary creep lasting ≈0.2 hours is followed by a relatively longer period of secondary creep. Again, no tertiary creep regime is seen. Behavior at both temperatures is similar with strain accumulating faster at higher temperatures and more total accumulated strain being present. Increasing stress has a similar effect. Since again a steady-state or constant creep rate was never achieved, the continuing transient nature of many of the creep rates resulted in a minimum (not steady-state) creep rate being defined. Creep rate verses stress at the two temperatures of interest is shown in Fig. 10.

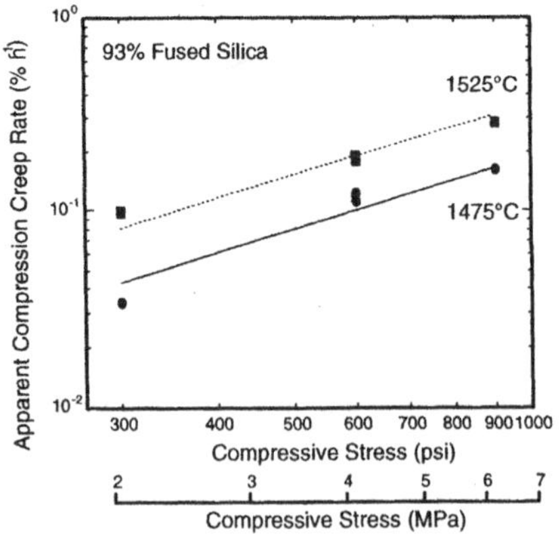

Fig 10. Creep rates as a function of stress for fused silica refractory

The steady-state or minimum compressive creep rate ($d\varepsilon/dt_{min}$) was related to the applied compressive stress and temperature using Eq. 2. Multilinear regression was performed to determine the constants A, n, and Q for the fused silica refractory leading to Eq. 4.

$$\frac{d\varepsilon}{dt} = 0.235\sigma^{1.20} \exp\left(\frac{-330\text{kJ} * \text{mol}^{-1}}{RT}\right) \tag{4}$$

IV.B. "Non-classical" Creep Deformation

Creep of Conventional Silica Refractories [17]

Conventional silica refractories are polycrystalline and have long been used in the superstructures of traditional air-fuel fired glass furnaces and are still considered by many for application in newer oxy-fuel furnaces. Six conventional silica refractories shown were subjected to temperatures and stresses characteristic of current superstructure applications. All examined brands contained a minimum of 95.5% silica by weight as reported by the manufacturers and were "Type A" [18] conventional silicas with the exception of one "Type B" [18] silica which was also tested. The creep rate for each refractory was determined and explored as a function of stress and temperature.

Testing was performed in ambient air using either a pneumatic power digitally controlled system or a hydraulic power analog controlled system. To prevent reaction between the sample and the push-rods, α-SiC push-rods in conjunction with α-SiC disks were used in contact with the test specimen. A contact extensometer employing α-SiC knives and a 40.00 mm (1.16 in.) gauge length was used for sensoring in the tests involving the pneumatic system while a scissors-type mechanical extensometer with SiC arms concurrent with a laser scanner was used in the tests involving the hydraulic system. Consistency between the two sensoring systems was verified prior to the commencement of the test program.

All specimens were preloaded at 0.04 MPa (≈6 psi) during the 4 hour furnace heat up to keep all the load train components and the specimen in continuous contact and a soak of 25 hours at temperature prior to the application of the first stress was maintained. Tests were carried out on each refractory at 1550°C (2820°F), 1600°C (2910°F), and 1650°C (3000°F) and at sequential stresses of 0.2, 0.4, and 0.6 MPa (29.0, 58.0, and 87.0 psi) for 75 hours at each stress using a temperature/stress history similar to that shown in Fig. 11. Tests were only performed on the "Type B" material at 1550°C (2820°F) and 1600°C (2910°F) due to its excessive creep deformation seen at 1600°C.

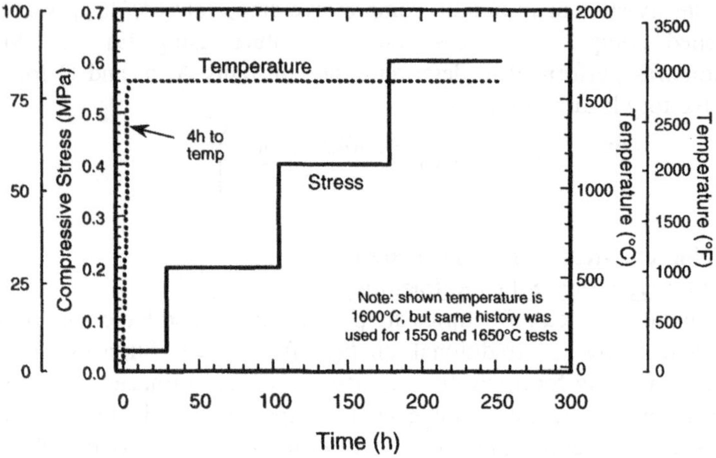

Fig 11. Temperature and stress history used in the creep testing of conventional silica refractories.

The above testing generated curves similar to the ones shown in Fig. 12 with the general appearances of the creep curves being similar for the five "Type A" silicas.

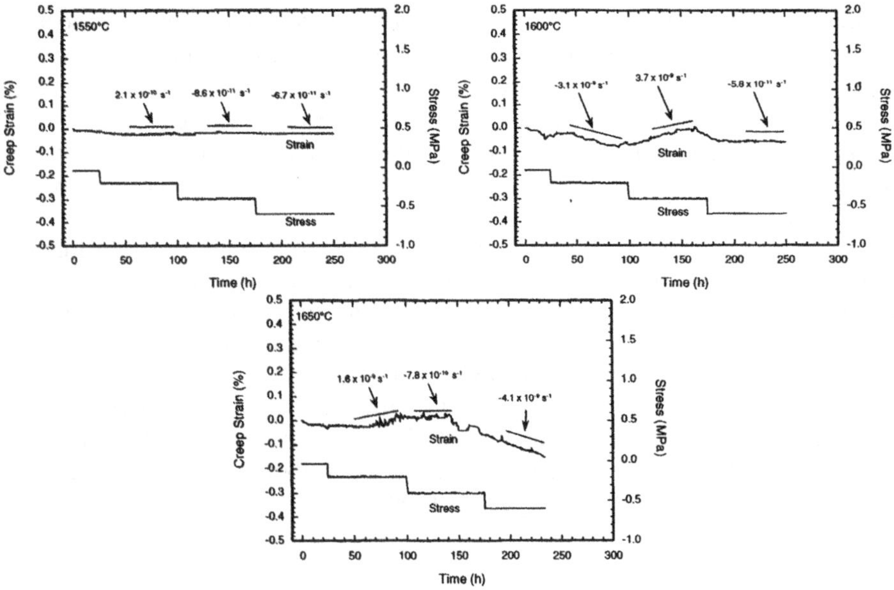

Fig 12. Typical creep behavior exhibited by "Type A" conventional silica refractories.

Typically, these materials exhibited less strain at 1550°C and higher amounts at 1600°C and 1650°C but the overall amount of compressive creep was negligible at these temperatures and stresses. None of these five materials exhibited statistically significant superior creep resistance to the others as shown in Table III.

Table III. Creep rates (h^{-1}) of conventional silica refractories at various temperatures and stresses.

	1550°C			1600°C			1650°C		
Material	0.2 MPa	0.4 MPa	0.6 MPa	0.2 MPa	0.4 MPa	0.6 MPa	0.2 MPa	0.4 MPa	0.6 MPa
A	7.6E-7	-3.1E-7	-2.4E-7	-1.1E-5	1.3E-5	-2.1E-7	5.8E-6	-2.8E-6	-1.5E-5
B	7.9E-7	1.6E-6	-9.0E-7	-1.2E-6	-1.2E-6	-2.1E-6	-6.1E-6	-1.5E-6	-2.1E-7
C	3.6E-7	-2.2E-7	1.3E-7	1.6E-5	1.3E-5	-2.3E-6	6.8E-6	2.2E-5	1.4E-5
D		6.8E-7	3.5E-7	1.5E-6	1.9E-6	1.2E-6	1.4E-6	5.8E-7	9.4E-7
E	3.4E-8	-5.8E-7	4.0E-7	2.1E-6	9.0E-7	1.4E-6	5.4E-6	9.0E-6	1.6E-5
F	1.0E-6	6.1E-7	4.0E-7						

Note: Materials A-E were "Type A" conventional silica refractories, while material F was a "Type B" conventional silica material as defined by reference [18].

The "Type B" material showed negligible compressive creep at 1550°C; however, its creep resistance was found to be much less than that of the other five "Type A" silica brands at 1600°C with contraction of more than 20% as shown in Fig. 13.

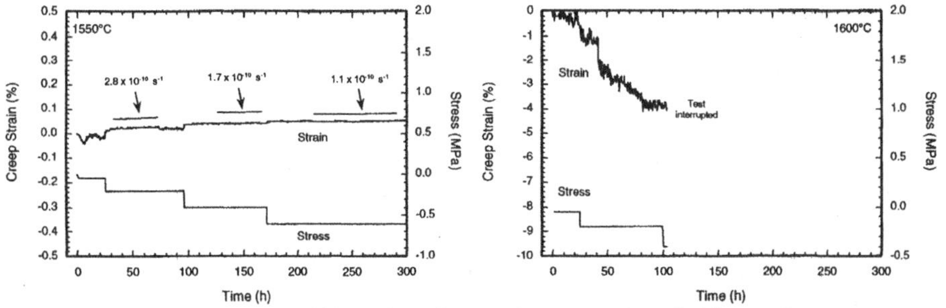

Fig 13. Creep behavior exhibited by "Type B" conventional silica refractory.

Due to this limited creep resistance, the "Type B" refractory was not tested at 1650°C although its creep resistance at this temperature was still expected to be inferior to the other five silica refractories.

Attempts were made to relate the steady-state or minimum compressive creep rate ($d\varepsilon/dt_{min}$) to the applied compressive stress and temperature using Eq. 2. This analysis was impossible due to concurrently active mechanism(s), other than creep, which resulted in larger (or oppositely anticipated) dimensional changes than were produced by creep. Thus, this effect limited the ability to identify or interpret the lesser-active creep mechanism in these silica refractories. Multilinear regression which was performed using the determined creep rates as a function of stress and temperature yielded undefined values for the constants in the equation. Therefore, models other than the NBA equation (or equivalent "creep" models that represent creep rate as a function of temperature and stress) are needed to represent the long-term dimensional stability of conventional silica superstructures that are subjected to compressive stresses less than 0.6 MPa and temperatures between 1550 and 1650°C.

Creep of Fusion-Cast Alumina Refractories at Low Stresses [19]

With the transition of glass furnaces from traditional air-fuel firing to oxy-fuel, alternative refractories are being sought for application in glass furnace superstructures. One such refractory under considerable consideration is fusion-cast alumina due to its lack of bond interfaces and low porosity, high refractoriness, good spall resistance, and bulk structural uniformity.

Fusion-cast alumina refractories from two manufactures were tested at 1450 and 1550°C (2640 and 2820°F) and stresses of 0.6, 0.8, and 1.0 MPa (87.0, 116.0, and 145.0 psi). Testing was carried out in ambient air using an electromechanical test machine in load control. Fusion-cast alumina push-rods in conjunction with high purity (99.5%) alumina disks were used in contact with the test specimen to prevent reaction between the sample and the fixturing. A contact extensometer employing sapphire knives and a 100.00 mm (3.94 in.) gauge length was used for sensoring. All specimens were preloaded at full load prior to the start of each test to keep all the load train components and the specimen in continuous contact during furnace heat up and to eliminate errors in the data due to seating of the load column. Tests were carried out for approximately 200 to 300 hours at each stress and temperature. Data from each test was smoothed prior to analysis by fitting to the linear exponential function discussed earlier. All fits were found to have a correlation coefficient of 0.97 or greater.

Selected behavior exhibited by this material at 1450 and 1550°C is shown in Fig. 14 and Fig. 15 respectively.

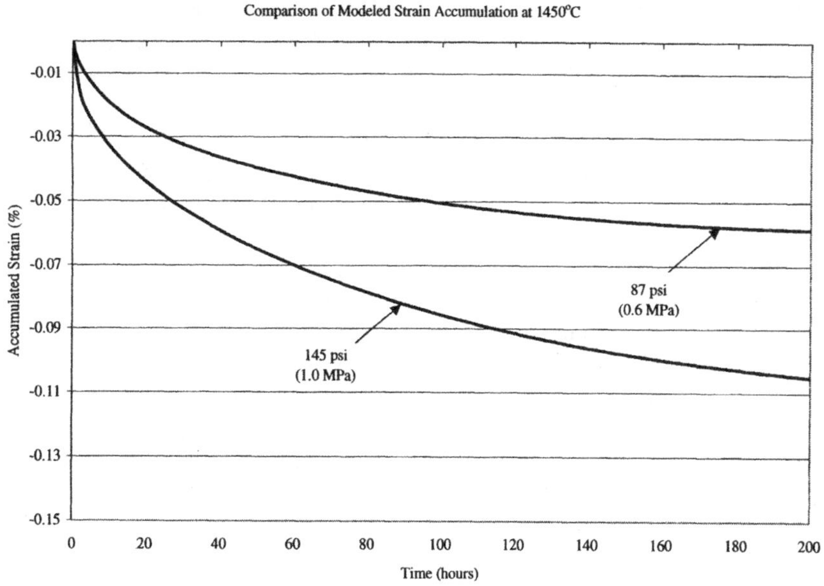

Fig 14. Modeled strain accumulation for fusion-cast alumina refractories at 1450°C.

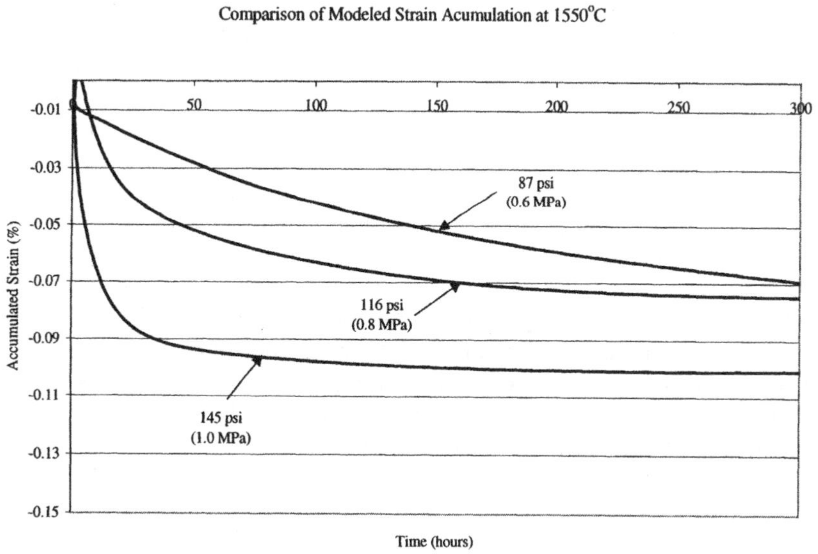

Fig 15. Modeled strain accumulation for fusion-cast alumina refractories at 1550°C.

The behavior of the samples from the two different suppliers was deemed statistically equivalent, therefore a single curve for each test condition is shown representing both materials. The behavior seen at these low stresses takes on a "non-classical" form where there is initially a large amount of creep strain accumulated over the first ≈50 hours, then the accumulated strain levels off asymptotically. The initial rate of strain accumulation is faster at 1550°C than at 1450°C, but the overall accumulated strain at each stress is similar for both temperatures.

Similar to the case of the conventional silica behavior above, trouble was encountered when trying to relate this data to Eq. 2. Therefore, steady state creep rates could not be obtained due the dimensional changes that leveled off asymptotically with time. Thus, this effect limited the ability to identify or interpret the active creep mechanism in these refractories. Again, models other than the NBA equation (or equivalent "creep" models that represent creep rate as a function of temperature and stress) are needed to represent the long-term dimensional stability of this material and its performance in superstructures that are subjected to compressive stresses less than 1.0 MPa and temperatures greater than 1450°C.

V. SUMMARY

The design and analysis of glass furnace refractory superstructures has created a need for accurate and unambiguous engineering creep data. This requires that new methods be adopted to address the unique issues and conditions relevant to superstructure refractories by modifying existing creep testing and data interpretation practices that were developed for structural ceramics. Through this type of analysis, the dimensional stability of refractory superstructures can be reliably predicted and furnace design capabilities can be substantially enhanced. It must be remembered though, that while classical models such as linear viscoelastic theory and the well-known Norton-Bailey-Arrhenius equation are applicable to many refractory systems, others exhibit "non-classical" behavior that requires new models to be explored which do not represent creep rate as a function of temperature and stress. Therefore, work must be ongoing to further our understanding of both "classical" and "non-classical" refractory systems and to further support glass furnace designer's efforts.

ACKNOWLEDGMENTS

In addition to work performed by the authors, results discussed in this paper are the culmination of work done by T.P. Kirkland and K. Liu at Oak Ridge National Laboratory, W.F. Curtis of PPG Industries, and E. Krug of Certech.

The authors wish to also thank K. Liu, M. Ferber, P. Angelini, J. Swab, and P. Huang for reviewing the manuscript and for their helpful comments.

REFERENCES

[1] D. W. Richerson, Modern Ceramic Engineering: Properties, Processing and Use in Design, Marcel Dekker, Inc., New York, NY, 1992.

[2] D.J. Bray, "Creep of Refractories: Mathematical Modeling," *Advances in Ceramics: New Developments in Monolithic Refractories*, Vol. 13, 69-80 (1985).

[3] A.H. Chokshi and T.G. Langdon, "Characteristics of Creep Deformation in Ceramics", *Materials Science and Technology*, Vol. 7, 577-584 (1991).

[4] A.A. Wereszczak, M.K. Ferber, T.P. Kirkland, A.S. Barnes, E.L. Frome, and M.N. Menon, "Asymmetric Tensile and Compressive Creep Deformation of Hot-isostatically-pressed Y_2O_3-Doped-Si_3N_4", *Journal of the European Ceramic Society*, Vol. 19, 227-237 (1998).

[5] W.E. Luecke, S.M. Wiederhorn, B.J. Hockey, R.F. Krause Jr., and G.G. Long, "Cavitation contributes substantially to tensile creep in silicon nitride", *Journal of the American Ceramic Society*, Vol. 78, No. 8, 2085-2096, (1995).

[6] M.K. Ferber, M.G. Jenkins, and V.J. Tennery, "Comprison of tension, compression, and felxure creep for alumina and silicon nitride ceramics", *Ceramic Engineering Science Proceedings*, Vol. 11, No. 7-8, 1028-1045 (1990).

[7] "Standard Test Method for Elevated Temperature Tensile Creep Strain, Creep Strain Rate, and Creep Time-to-Failure for Advanced Monolithic Ceramics", ASTM C1291-95e1, Vol. 15.01, American Society for Testing and Materials, West Conshohocken, PA (2000).

[8] "Standard Test Method of Measuring the Thermal Expansion and Creep of Refractories Under Load", ASTM C832-99, Vol. 15.03, American Society for Testing and Materials, West Conshohocken, PA (2000).

[9] C.A. Schacht, "Recommended Additional Material Data for Evaluating the Mechanical Strength of Refractories", Ceramic Engineering Science Proceedings, Vol. 9, No. 1-2, 32-38 (1988).

[10] W.N. Findley, J.S. Lai, and K. Onaran, Creep and Relaxation of Nonlinear Viscoelastic Materials. Dover Publications, Inc. New York, 1989.

[11] J.D. Lubahn, "The Role of Anelasticity in Creep, Tension, and Relaxation Behavior", *Transactions of the American Society of Metals*, Vol. 45, 787-838 (1953).

[12] H.T. Lin, P.F. Becher, and M.K. Ferber, "Improvement of Tensile Creep Displacement Measurements", *Journal of the American Ceramic Society*, Vol. 77, No. 10, 2767-2770 (1994).

[13] A.R. Prunier Jr., "Sources of Systematic and Random Errors in Differential Push-Rod Dilatometry", Proceedings of the 11th International Thermal Expansion Symposium, Pittsburgh, PA, (1995).

[14] A.A. Wereszczak, T.P. Kirkland, and W.F. Curtis, "Creep of CaO/SiO_2-containing MgO refractories", *Journal of Materials Science*, Vol. 34, 215-227 (1999).

[15] A.A. Wereszczak, K.C. Liu, M. Karakus, B.A. Pint, T.P. Kirkland, and R.E. Moore, "Compressive Creep Performance and High Temperature Stability of Mullite Refractories", in preparation, ORNL Technical Report.

[16] A.A. Wereszczak, K. Breder, M.K. Ferber, T.P. Kirkland, C.J. Rawn, E. Krug, C.L. LaRocco, R.A. Pietras, and M. Karakus, "Dimensional Stability and Creep of Silica Core Ceramics used in Investment Casting of Superalloys", in preparation.

[17] A.A. Wereszczak, M. Karakus, K.C. Liu, B.A. Pint, R.E. Moore, and T.P. Kirkland, Compressive Creep Performance and High Temperature Dimensional Stability of Conventional Silica Refractories. Oak Ridge National Laboratory Internal Report (ORNL/TM-13757), 1999.

[18] "Standard Classification of Silica Refractory Brick", ASTM C416, Vol. 15.01, American Society for the Testing of Materials, West Conshohocken, PA (1998).

[19] J.G. Hemrick, "Creep Behavior and Physical Characterization of Fusion-Cast Alumina Refractories", in preparation, University of Missouri – Rolla Ph.D. Dissertation.

CORROSION OF REFRACTORIES IN GLASS-MELTING APPLICATION

SM Winder (UKSS Inc., Grand Island, NY, 14072, USA)
KR Selkregg (Monofrax Inc., Falconer, NY, 14733, USA)

ABSTRACT

Refractories exposed to glass-melting environments may experience corrosion by interaction with molten glass, and also vapor phases, molten batch components and condensed species above the melt. Since refractory selection influences the melter service life and the propensity to create defects, melter productivity can be highly dependent upon the refractories employed. This paper describes the chemistry and microstructure of selected superstructure refractories, both before and after Oxyfuel glass-melting service, and relates them to performance.

1. INTRODUCTION

Refractories used in glass-melting furnaces are exposed to a wide range of corrosion mechanisms. The optimum refractory selection for a given furnace will rely upon the glass composition to be melted, the service conditions, and the required quality of the glass product. Since service conditions vary with position in any given furnace, a range of refractory types may be utilized. Thus, refractories applied in contact with molten glass may require different properties than those applied in superstructure above the molten glass bath. Cooler glass melt, in refining zones and working ends, close to the furnace exit, has little chance to 'digest' any defects. Here, refractories are not necessarily subjected to the most hostile environment, but must corrode in a benign manner - without contaminating the glass melt with heterogeneities.

Generally, high quality glass production requires application of premium refractories in order to minimize defect generation. Additionally, the current move to potentially cleaner and efficient Oxyfuel combustion also increases the potential for corrosion of certain superstructure refractories. This paper will focus on the important aspects of corrosion for selected fusion-cast refractories that are currently applied in production of high quality glasses under Oxyfuel combustion. This requires an understanding of the basic forces driving corrosion.

1.1. Physical Processes Of Refractory Corrosion

The amount of corrosion that a particular refractory may undergo will be determined by both its' properties and the environment to which it is exposed. The chemical endpoint of long-term refractory exposure may be predicted by thermodynamic calculation, and consideration of available phase equilibrium diagrams. The rate at which the refractory will approach that endpoint must then depend upon the kinetics of the corrosion processes involved in the degradation.

In the absence of forced convection, refractories in contact with molten glass corrode due to dissolution, which occurs by diffusion. As exposure time proceeds, the concentration gradients formed from the refractory / melt interface decrease, and a boundary layer forms which decelerates the diffusive process. Where forced convection is present, due to the 'pull' of glass in the normal production process, due to stirring, or due to the action of bubbles, the corrosion rate increases. Thermal and density convection effects also intensify any degradation. Surface and interfacial tensions drive flow which results in particularly severe reaction at the 3-phase (refractory / glass-melt / furnace atmosphere) contact area known as the 'melt line', 'metal line' or 'flux line'. Additional perturbations along the melt line, due to bubble activity, further accelerate the refractory decomposition. The reader is directed to the literature[2] for a more detailed treatment of melt-contact refractory corrosion phenomena.

Superstructure refractories do not contact the molten glass bath but are exposed to the hostile environment of the glass-melting atmosphere. Reactions may occur with the products of Oxyfuel combustion (ideally ~67%H_2O + 33%CO_2), vapor phases formed above the liquid surface ($NaOH$, KOH, PbO, $NaBO_2$, etc) and also batch dusts (containing SiO_2, alkali and alkaline earth compounds, B_2O_3, etc.). Products of condensation may lead to vigorous 'rat holing' corrosion at Silica refractory brick joints. The potential reactant vapor phases are further discussed in Section 1.2. below, for various glass types, and greater detail is available in the literature[3],[4],[5].

Impingement of atmospherically borne dusts upon superstructure refractories is unavoidable, particularly in the area of the batch chargers. The range of melter designs, and Oxyfuel burner designs available, means that some furnaces are more problematic than others. This mode of material transport enables mobility of inherently less volatile glass batch components, which could not otherwise interact to such a major extent with superstructure refractories. Thus, relatively large concentrations of Silica, Zircon, alkaline earths, and other compounds, are able to contact the superstructure refractories and react with them.

1.2. Thermodynamic Aspects Of Refractory Corrosion

Vapor phases are derived due to reactions between the furnace atmosphere (mostly H_2O and CO_2 for Oxyfuel firing) and the surface of the glass bath. The major corrosive vapor species present, and their relative amounts, have been previously ascertained by thermodynamic minimum free energy calculations (using HSC Chemistry For Windows, Outokumpu)[3],[4]. *Soda-Lime glasses* (commonly used for production of flat glass and containers) form NaOH and KOH as the major corrosive vapor phases. Modern *TV Panel glasses* generate high concentrations of KOH and NaOH, and both *TV Funnel glasses* and *Lead Crystalware glasses* create PbO, KOH and NaOH. *Borosilicate glasses* produce HBO_2, $NaBO_2$ and KBO_2 as major vapor phase species, in different amounts depending upon the particular glass composition being melted. Silica, and any alkaline earth batch components present, generate compounds with relatively low vapor pressures under the oxidizing conditions prevalent in most melters.

The H_2O present in the furnace atmosphere may also react with the refractories, commonly to generate vapor pressures of NaOH. Reference to available phase equilibrium diagrams[6] can aid in understanding of reactions between corrosive vapor phases and the range of refractories available for glass-melting superstructure application. Consider Figure 1., a simplified form of Figure 501[6].

Figure 1. Schematic Of Liquid Formation In The SiO_2-Al_2O_3-Na_2O System

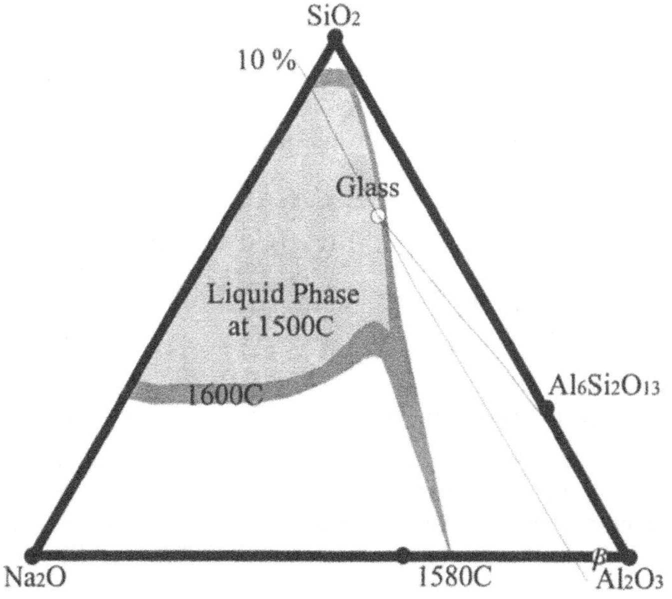

At equilibrium, addition of ~10wt% Na_2O to pure Silica or Mullite ($Al_6Si_2O_{13}$) results in considerable liquid formation at 1500°C. In the case of Silica, this Na_2O addition causes complete liquefaction with no residual solid. In the case of Mullite, this Na_2O addition effects complete decomposition of Mullite to Corundum (α-Al_2O_3) and a NaSiAlO liquid. The liquid freezes upon rapid cooling to form a stable Glass phase at the marked composition in the diagram. Substituting NaOH for Na_2O, the presence of OH^- decreases silicate liquid viscosity, thereby increasing corrosion rates. Reaction of Na_2O with pure Alumina, or β-Alumina ($Na_2O.9Al_2O_3$), does not form a liquid phase at 1500°C.

If the temperature is increased to 1600°C, only ~5% Na_2O addition is required for complete liquefaction of Silica, Figure 192[6]. With increasing temperature, Mullite forms more liquid phase and less solid Corundum than at 1500°C. Alumina may form a liquid component at 1580°C, or higher, due to addition of Na_2O greater than the proportion in $Na_2O.9Al_2O_3$, Figure 4282[6].

1.3. Kinetic Aspects Of Refractory Corrosion

In the absence of barrier layers or convective forces, which may alter reaction kinetics, refractory corrosion rates may generally be treated as determined by simple diffusive behavior. Consider a semi-infinite homogeneous refractory slab containing a pre-defined concentration (C_o) of diffusant. Expose it to service allowing a specific surface concentration (C_s) of diffusant to contact one surface at time t_o, and maintain that concentration at the surface to allow diffusion into or out of the slab for a set time period.

If the diffusant moves through the slab, its concentration ($C_{x,t}$) at any position from the surface (x) or time (t) may be defined from a knowledge of the effective diffusion coefficient (D) using the following equations[1]:

1. For diffusion into the slab,

$$(C_{(x,t)} - C_o)/(C_s - C_o) = erfc(x/2\sqrt{Dt}) \qquad [1]$$

2. For diffusion out of the slab,

$$(C_{(x,t)} - C_o)/(C_s - C_o) = erf(x/2\sqrt{Dt}) \qquad [2]$$

Where *erf* is the Gaussian error integral and the complementary error function *erfc* = *1-erf*. Both equations include the term $\sqrt{Dt}$, which is the mean diffusion distance attained during the time interval t. The net movement of diffusant into or out of the refractory slab will depend primarily upon the chemical potential gradient, which defines the chemical driving force for diffusion.

Equations [1] and [2] only apply when C_s is constant and when D is constant throughout the slab, so they are only valid for isothermal situations and homogeneous, isotropic, cases in which D is independent of $C_{(x,t)}$. If the diffusant reacts with the refractory to alter the effective diffusion coefficient (D), the mean diffusion distance attained will often increase, but may theoretically decrease for a case in which D is decreased. The temperature dependence of D is usually described by an Arrhenius relationship $D = D_o \exp(-Q/RT)$, in which D_o is a temperature independent constant known as the frequency factor, Q is the activation energy, R is the molar gas constant, and T is the absolute temperature.

Figure 2. Time Dependence Of In-Diffusion With A Temperature Gradient

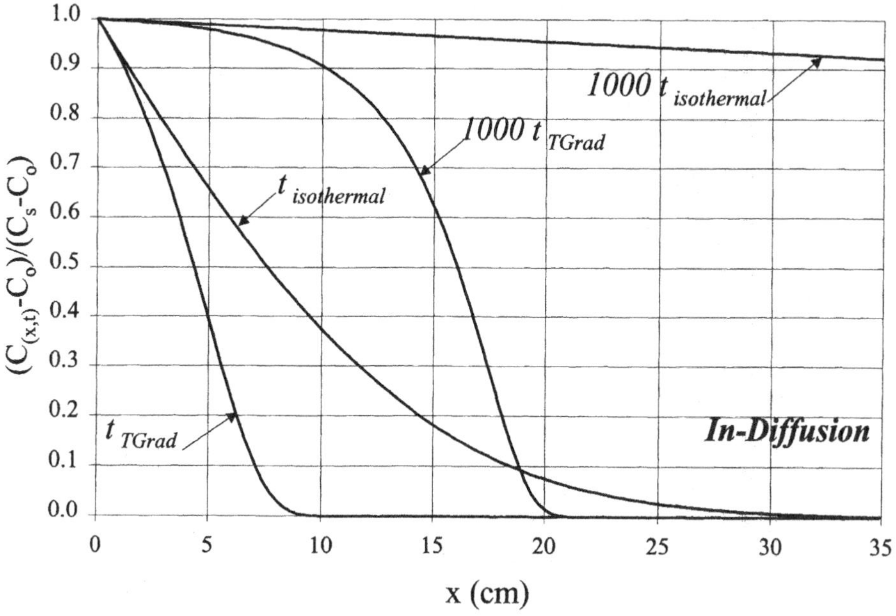

A refractory block in service with a temperature gradient imposed through it experiences an effective diffusion coefficient that changes rapidly as a function of depth. This effect may be appreciated by comparison of the plots labeled $t_{isothermal}$ and t_{TGrad}, both obtained over time t, for an isothermal case and with imposition of a temperature gradient respectively, calculated for a hypothetical case and presented in Figure 2. The time dependence of diffusion in both circumstances may be appreciated by consideration of the plots labeled *1000* $t_{isothermal}$ and *1000* t_{TGrad}, obtained at a much longer time = 1000 t.

For isothermal service, or situations with small temperature gradients, homogenization tends to completion over long time periods and the mean concentration of diffusant (C_m) may be approximated using trigonometric functions[1] of the form:

3. For out-diffusion,

$$(C_m - C_s)/(C_o - C_s) = 8\pi^2 \exp(-\pi^2 Dt/4x^2) \quad [3]$$

In Summary:

- A combination of thermodynamic calculations and reference to phase equilibrium diagrams can be utilized to predict which pure compounds will degrade, due to liquid phase formation, in a given environment.
- In reality, multiphasic refractories contain sintering aids, or other minor impurities that promote liquid formation at lower temperatures. These may severely degrade performance, altering corrosion mechanisms and rates.
- Carefully designed laboratory studies may be useful for comparing performance of similar refractories containing different amounts of impurity phases, and defining reaction kinetics under isothermal conditions.
- Recognition of the temperature dependence of diffusion, the formation of boundary layers, and generation of convective effects are also necessary for an understanding of corrosion kinetics occurring in service. Indeed, operation of these phenomena in glass-melting practice may alter overall corrosion rates by orders of magnitude as time proceeds. This can exercise more influence on the amount of refractory degradation experienced, at any given position in a furnace, than the particular prevalent corrosion (or diffusion) mechanism.

2. EXPERIMENTAL PROCEDURES

Laboratory testing allows for idealized comparison of superstructure refractory behavior due to exposure under Oxyfuel or Airfuel conditions. Here, it is possible to control all parameters so that the effect of changes to atmospheric chemistry and flow rate can be carefully isolated. This type of testing has been used to explore superstructure refractory corrosion mechanisms and kinetics under isothermal conditions. Testing over longer times, in the presence of thermal gradients and contaminating batch dusts, is more easily accomplished by 'in-situ' testing - utilizing available space in operating industrial melters.

2.1. Short-Term Laboratory Superstructure Refractory Corrosion Testing

Laboratory testing has been used to evaluate the short-term (from 50 to 240hrs) performance of a range of 'bonded' and 'fusion-cast' refractory types, exposed to vapor species resulting from both Airfuel and Oxyfuel combustion, with glass melts of different compositions. The laboratory test apparatus presented in Figure 3. was used to gain an understanding of relative refractory performance due to exposure to combustion products and vapor phases only.

Figure 3. 'Non-Accelerated' Laboratory Test For Superstructure Refractories Developed At Monofrax Inc. (Schematic)

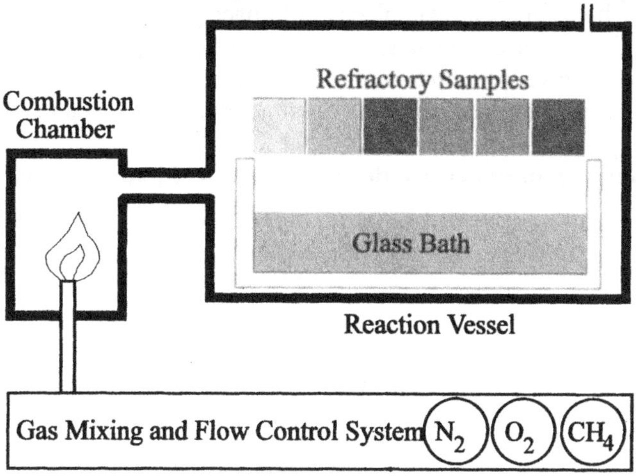

Oxygen and Methane gases, regulated by mass-flow control, were burnt at a nozzle in an insulated combustion chamber in order to create Oxyfuel combustion products. Nitrogen gas was added to the combustion chamber, through a separate dispersion system, when simulating Airfuel conditions. The combustion products were then passed through an electrically heated muffle, which formed the reaction vessel and held two large refractory crucibles filled with cullet of the glass composition of interest. Refractory brick samples of size ~50x50x115mm (~2x2x4.5in) were positioned above the glass baths for isothermal reaction with the glass-melting atmosphere. Temperature was monitored as a function of length and width within the reaction vessel.

2.2. Long-Term 'In-Situ' Superstructure Refractory Corrosion Testing

Following laboratory testing, selected refractory samples were installed in operating industrial glass-melting furnaces. Samples have now been recovered following long-term exposure in several units melting different glass compositions. In each case, refractories were exposed in different furnace zones, where they were subjected to dissimilar temperatures and to varying amounts of batch dust contamination.

Initially, rod-shaped samples were simply suspended into the furnace atmosphere through holes in the crown refractories. Following these early tests, a procedure was developed for exposing samples to the local furnace atmosphere while avoiding contamination from any liquid 'run-down' resulting from superstructure refractory corrosion. Samples were mounted in custom sized protective sample holders (designed with drip ledges and provision for thermocouples) and then hung through holes in the crown, placed in furnace ports, or inserted into existing peepholes. Figure 4. is a schematic example of a typical peephole installation. So far, refractory core samples up to ~70mm in diameter and 350mm long have been tested in operating industrial Oxyfuel Glass-melters for times up to ~17 months.

Figure 4. Generic Specimen Holder Developed At Monofrax Inc. (Schematic)

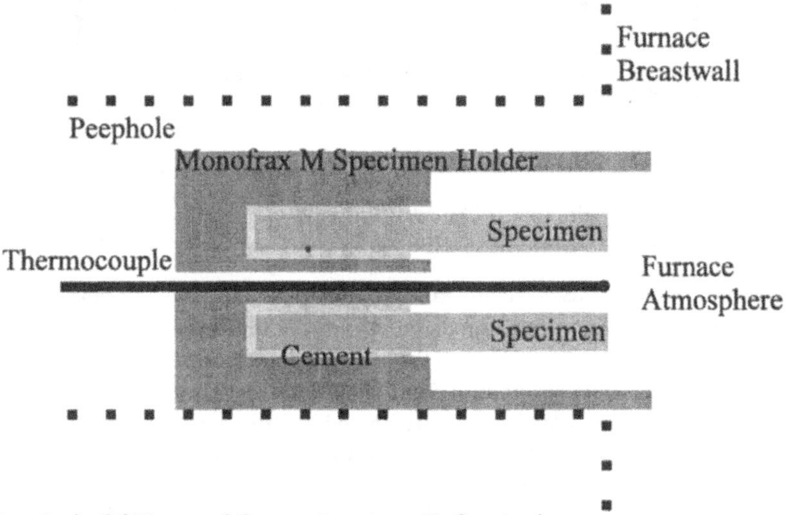

2.3. Analysis Of Exposed Superstructure Refractories
Refractories were investigated using Optical Microscopy, Scanning Electron Microscopy with Energy Dispersive X-Ray Spectroscopy (SEM/EDS), Dynamic Secondary Ion Mass Spectroscopy (DSIMS), X-Ray Fluorescence Spectroscopy (XRF) and X-Ray Diffraction (XRD), as necessary. Altered refractory surfaces were analyzed following exposure to corrosion. Microstructures were also characterized from cross-sectional samples as a function of position below the exposed surfaces. Cross-sections of unexposed and exposed refractories were evaluated, and this allowed for normalization of post-exposure chemistries by pre-exposure chemistries - in order to compare degree of chemical alteration.

3. RESULTS

Results from analysis of untreated refractories and post-service studies of superstructure refractories exposed to various glass-melting environments are presented. 'Bonded' refractories are treated in a limited fashion only, since the intention is to differentiate the relative performance of the various classes of premium 'fusion-cast' refractories for Oxyfuel application.

3.1. Characterization Of Unexposed Superstructure Refractories

'Bonded refractories', fabricated by pressing and firing technology, generally tend to contain a matrix of 'bond phase' which surrounds coarser grains or agglomerates - holding them in place. The 'granular' component is usually relatively dense, often containing dispersed micro-scale closed porosity. The 'bond' component is usually characterized by a large volume of coarse open porosity and other less refractory phases derived from any sintering aids and impurities. These refractories are relatively homogeneous as a function of position and frequently contain up to 20% apparent porosity, see Figure 5.

Figure 5. The Structure Of Unexposed 'Bonded' Silica Refractory

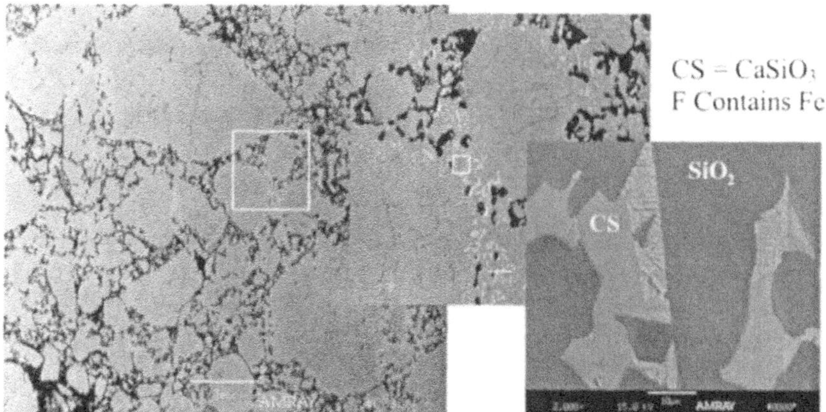

This Silica Contains $CaO.SiO_2$ And An Fe-Rich Phase At Grain Boundaries.

'Fusion-Cast refractories', fabricated by arc-melting and casting into molds, generally command a cost premium. They usually exhibit chemical and structural inhomogeneity as a function of position in the casting. This segregation is intrinsic to the solidification process, but can be minimized to some extent by use of differing casting techniques. 'Regular cast' materials contain large shrinkage voids whereas 'void-free' materials have had the part of the block containing the shrinkage void removed. Apparent porosity is typically quoted at less than 1%.

Figure 6. Unexposed 'Fusion-Cast' Refractory Structures

Monofrax CS3 AZS

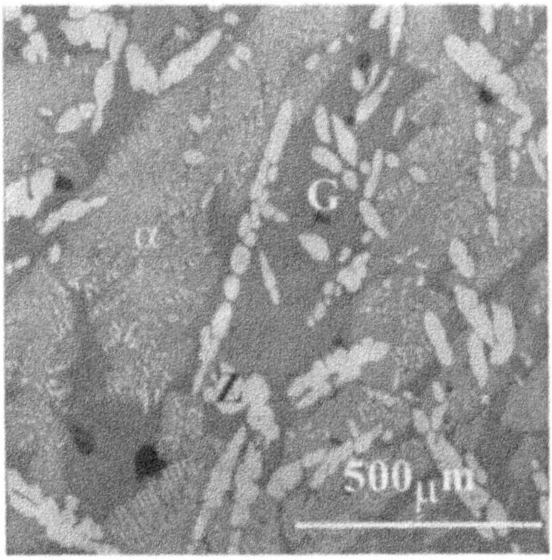

α=Al_2O_3, Z=ZrO_2, G=Glass

Monofrax M αβ-Alumina

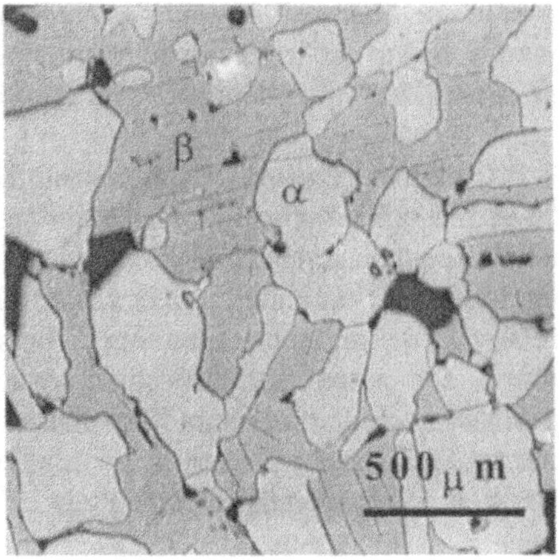

α=Al_2O_3, β~($Na_2O.8.2Al_2O_3$)

Monofrax H β-Alumina

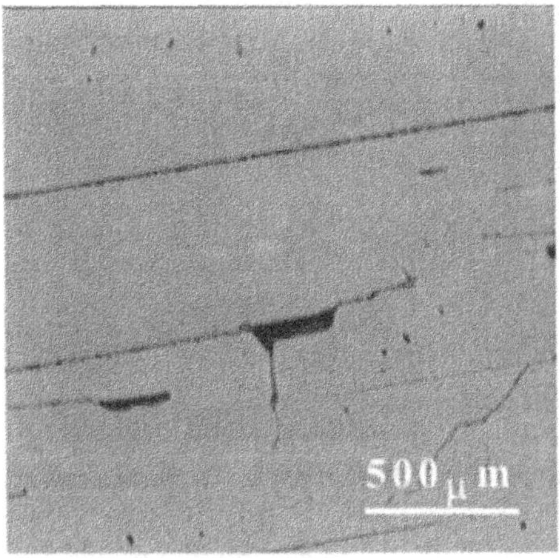

β-Alumina ~($Na_2O.9.2Al_2O_3$) in 'H'

Monofrax L MgO.Al_2O_3 Spinel

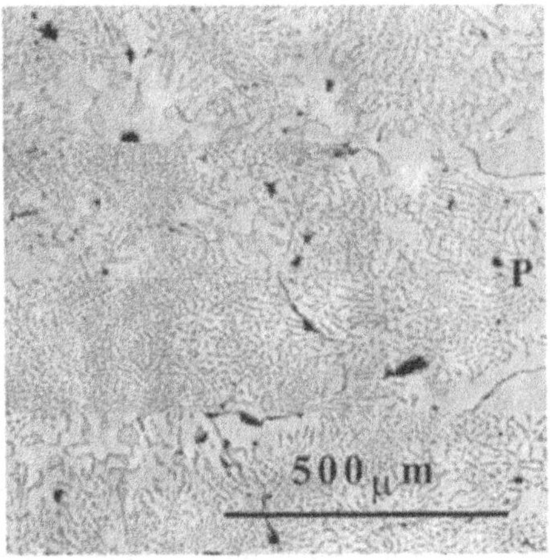

Periclase (P)=MgO

Alumina-Zirconia Silicate (AZS) refractory is available in a number of grades, principally distinguished by Zirconia content (~33 – 40wt%). All fusion-cast AZS contains both primary Zirconia dendrites and Alumina (Corundum) plates within a large volume of NaSiAlO glassy matrix phase. See the top left image in Figure 6. The Alumina plates contain coprecipitated Zirconia crystals, which undergo a volume expansion on cooling from casting temperatures - which often fractures the Alumina plates. The glassy matrix provides relatively high strength at room temperature, but softening of the matrix decreases strength (and creep resistance) at service temperatures, see Figure 7.

Figure 7. Temperature Dependence Of MOR Strength For Fusion-Cast Refractories, Related To Glassy Phase Content

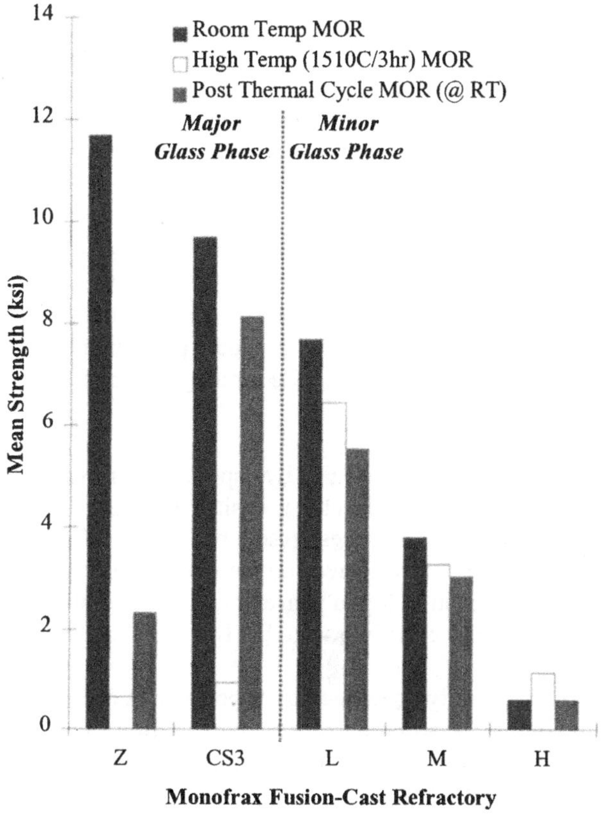

Monofrax Z **Fusion-Cast Refractory Has ~93% ZrO_2 With ~7% Glassy Matrix Phase.**

Monofrax CS3 AZS **Has ~33% ZrO_2, ~ 46% Crystalline Al_2O_3, And ~21% Glassy Matrix Phase.**

Monofrax L, M & H **Refractories Are Essentially Free Of Interconnected Glassy Phase Volume.**

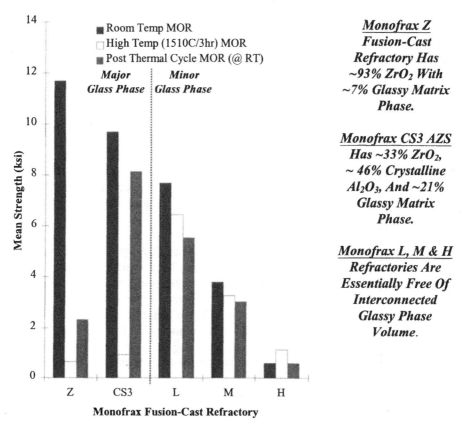

$\alpha\beta$-Alumina (Monofrax M) refractory is comprised essentially of ~40mol% α-Al_2O_3 and ~60mol% β-Na_2O.~8Al_2O_3 crystalline phases with a minor Nephelitic boundary phase content, as seen in the top right image of Figure 6. This material exhibits excellent high temperature strength retention and creep resistance due to its essentially biphasic crystalline nature. The combination of α-Al_2O_3 in a compliant β matrix phase bestows this extremely refractory material with relatively 'tough' composite load extension behavior in high temperature service.

β-Alumina (Monofrax H) refractory is essentially a monophasic material comprised of large β-Na_2O.~9Al_2O_3 grains. See the bottom left image of Figure 6. The large inherent flaw population in this material endows stable performance under conditions of thermal shock.

MgO-Alumina Spinel (Monofrax L) refractory is comprised essentially of ~60mol% MgO.Al_2O_3 Spinel grains with ~40mol% Periclase (P = MgO, Figure 6. bottom right image) and a minor silicate boundary phase. The Periclase is distributed bimodally as a fine coprecipitated structure locked within the Spinel grains, or as large primary MgO dendrites. The latter are particularly evident near external block surfaces, due to preferred orientation arising from rapid cooling. MgO shrinks more than the Spinel phase upon cooling from fabrication temperatures. This relatively fine-grained refractory exhibits high strength with excellent high temperature retention and extremely high creep resistance.

3.2. Refractories Exposed To Short-Term Vapor Corrosion Testing

Several refractory grades may be subjected to superstructure testing in the laboratory apparatus at one time. Test conditions, and the particular refractories tested, are usually dictated by the interests of the glass producer providing the cullet to be melted. Several examples from short-term laboratory superstructure testing are presented below:

'Bonded refractories' are generally easily penetrated by vapor species (NaOH, KOH etc.) due to their large volume of interconnected porosity. Invasive vapors may then undergo dissolution into any liquid phases present within the refractory structure. This usually causes degradation by decreasing the viscosity of any silicate liquids and expansion of the existing liquid phase volume by dissolution of the refractory grain structure. Examples are presented in Figures 8. and 9.

'Fusion-Cast refractories' are less easily penetrated by vapor phases due to their intrinsically lower open pore volume. However, the presence of liquid silicate phases (at service temperatures) makes some fused refractories more vulnerable to degradation by invasive vapor phases than others.

Fusion-Cast Alumina-Zirconia Silicate (AZS) refractory may undergo considerable alteration. The corrosion mechanism involves dissolution of alkali vapor species into the AZS matrix phase, causing dissolution of the crystalline Alumina structure and expanding the liquid phase volume on the refractory surface. Dissolved OH^- further decreases liquid phase viscosity, and the resultant Alumino-Silicate liquid corrosion product contains both dissolved and crystalline Zirconia, creating 'run-down' with high defect potential. Increased NaOH concentration under Oxyfuel results in significantly more corrosion and defect potential than experienced under comparative Airfuel combustion, see Figure 10.

Figure 8. The Structure Of Bonded Silica Refractory Following Laboratory Oxyfuel Superstructure Testing With Soda-Lime Glass, 1600°C/75hrs

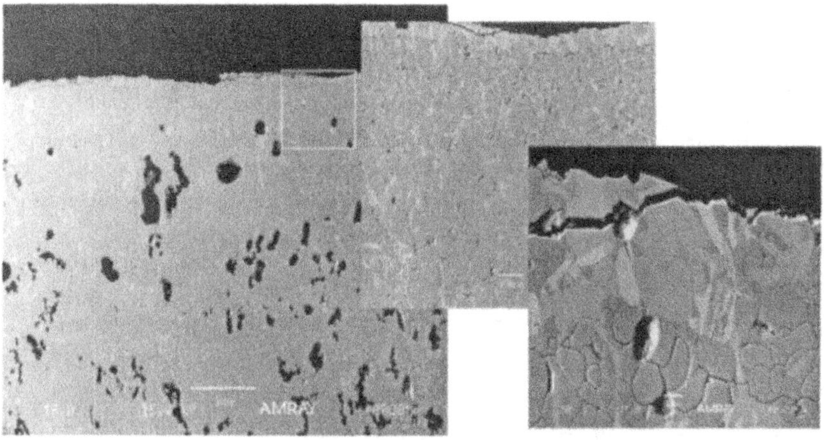

Reaction Of NaOH With Intrinsic Grain Boundary Liquid Phase Caused Progressive Dissolution Of The SiO_2 Grain Structure.

Figure 9. Bonded Mullite Refractory Following Laboratory Superstructure Testing With Soda-Lime Glass, 1510°C/50hrs

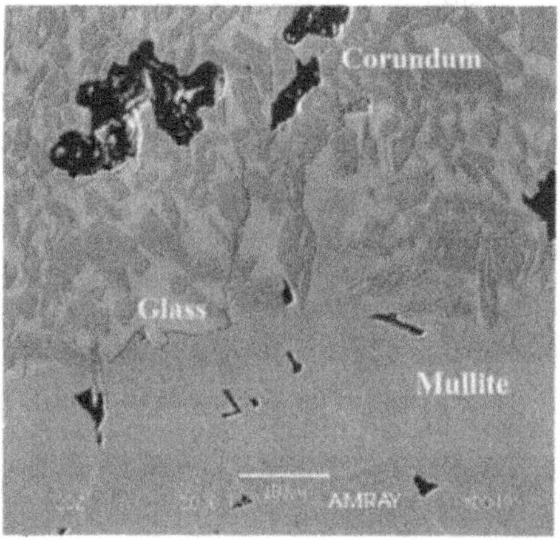

As Expected From Theory, Alkali Penetrated The Mullite Refractory Reacting With The Intrinsic Grain Boundary Liquid Phase And Dissolving The Crystalline Grain Structure. This Generated A Large Liquid Volume. The Mullite Refractory Decomposed To Form Corundum (α-Al_2O_3) And A NaSiAlO Liquid Phase, Which Cooled To A Stable Glass Following The Test.

Figure 10. Corrosion Of AZS Superstructure Refractory

Alkali In-Diffusion Due To Testing With Soda-Lime Glass, 1510°C/100hrs:

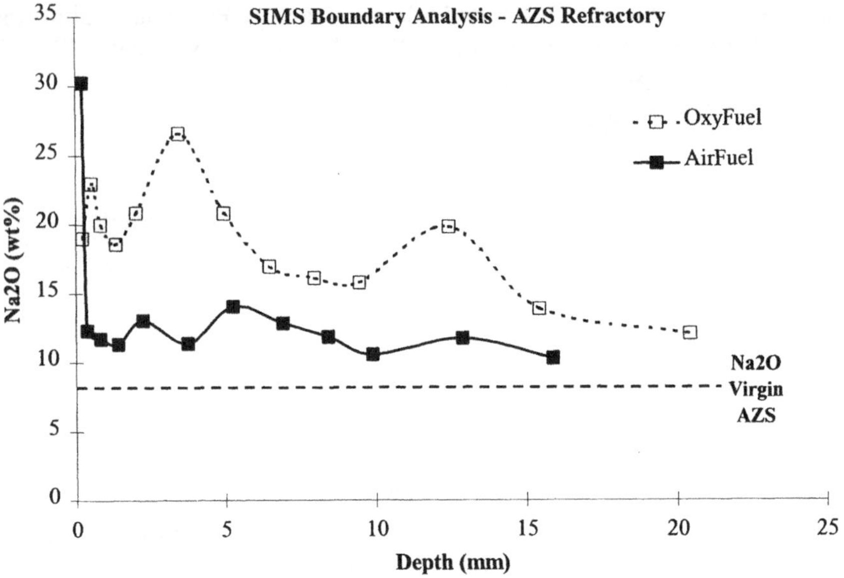

Effect Of Alkali In Decomposing Crystalline Alumina Structure (Schematic):

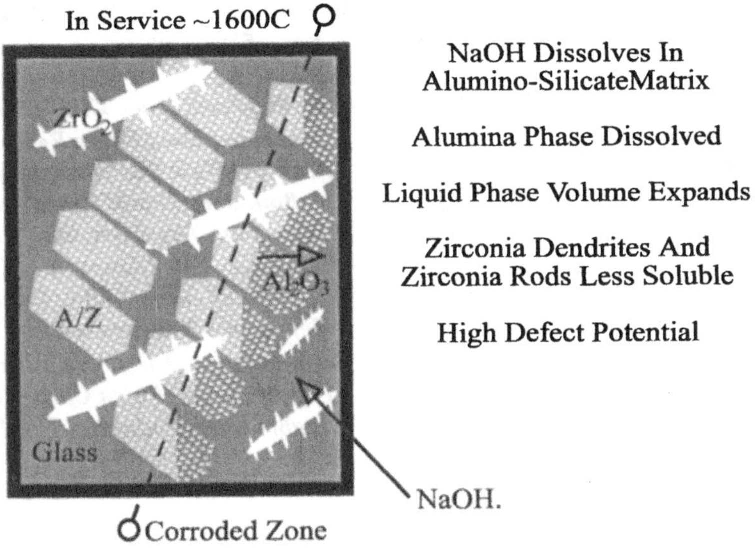

Fusion-Cast Alumina refractories are intrinsically more stable (than Silicate-rich materials) in the presence of vapor phases present in glass production. In fact, both theory and empirical observation prove that most glass compositions do not create enough alkali vapor pressure to stabilize the β-Alumina phase, see Figure 11. This diagram is constructed for Soda-Lime glasses, with addition of excess Na_2O to a baseline Flint Container glass in order to produce a range of glass compositions with variable Na_2O content (along the horizontal axis). The vertical axis allows determination of the equilibrium NaOH vapor concentration (from theoretical calculation) derived above the glasses, as indicated by the dotted isotherms. The solid isotherm at 1470°C is labeled with refractory stability results from empirical observation made in the laboratory corrosion test. The thick, almost vertical, line at ~26% Na_2O represents the equilibrium vapor pressure above β-Alumina phase. Thus, glass compositions to the left do not provide enough NaOH to stabilize β-Alumina, and α-Al_2O_3 formation must result.

Figure 11. Thermodynamic Calculations And Empirical Observations Of Phase Stability Fields For Fusion-Cast Alumina Refractories

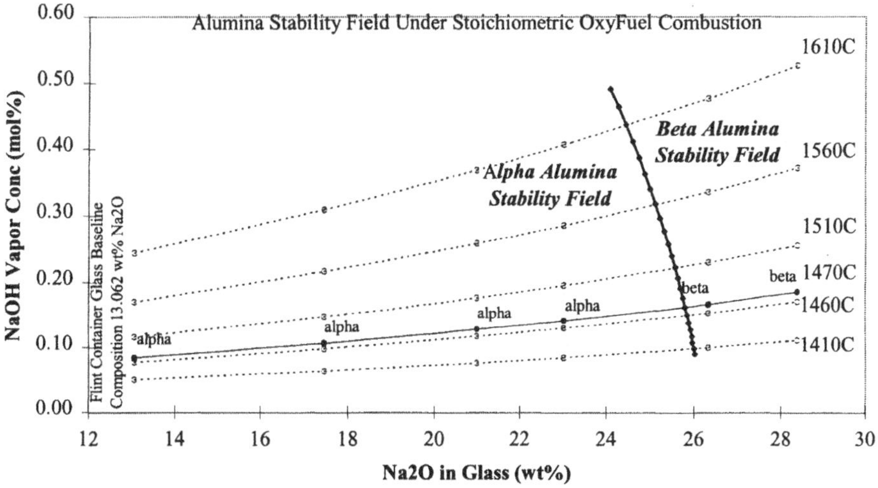

Vapor Phase Exposure Over Most Glass Compositions Results In Formation Of An α-Alumina Layer On Fusion-Cast Alumina Refractory Surfaces.

An equivalent diagram could be constructed for addition of K_2O, assuming direct substitution of K for Na in the β-Alumina. However, since KOH is more prevalent than NaOH for equal amounts of alkali, the total glass-melt alkali content necessary to stabilize β-Alumina must then decrease to some extent.

Figure 12. Short-Term 'In-Situ' Testing Of αβ-Alumina Superstructure Refractory Exposed 14 Days To PbO-Alkali-Silicate Glass-Melting

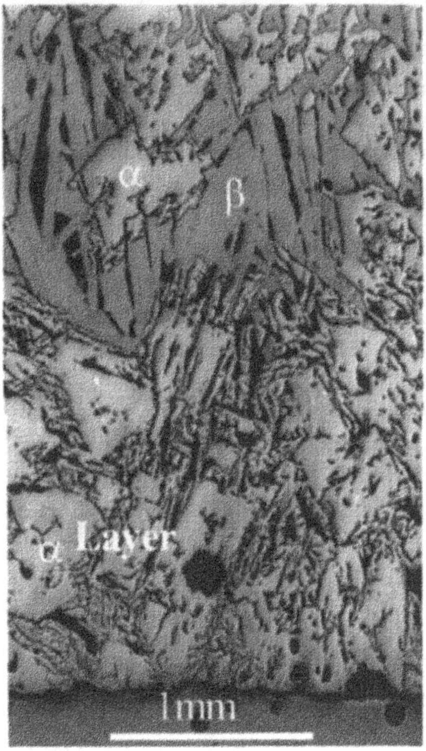

PbO Is The Major Vapor Phase Present In This Environment, But Reaction With The Refractory Was Not Detected.

Fusion-Cast αβ-Alumina And β-Alumina Refractories Contain Na_2O In Solution In The β-Alumina Phase. This Is Available To React With H_2O In The Glass-Melting Furnace Atmosphere. The Consequent Formation Of NaOH At The Refractory Surface Leads To Loss Of Na From The Refractory In Service, See Figure 11.

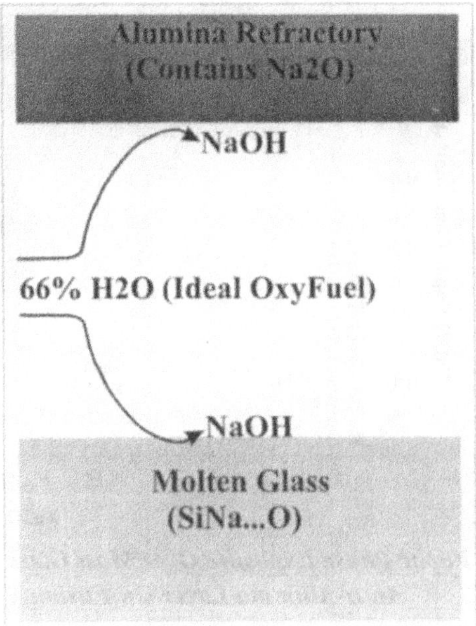

The Observed Formation Of An α-Alumina Layer In "In-Situ' Industrial Testing Confirms Both Theoretical Calculations And The Validity Of ***'Non-Accelerated'*** *Laboratory Vapor Phase Corrosion Testing.*

Some Test Houses Perform Laboratory Testing Using Excess Alkali In Order To 'Accelerate' Corrosion Kinetics. This Practice May Move The Experiment Into The Beta-Alumina Stability Field, Thereby Creating An Erroneous Result (Beta Phase Formation) And An Invalid Test.

αβ-Alumina (Monofrax M) refractory and **β-Alumina (Monofrax H) refractory** both lose Sodium (as NaOH) in vapor corrosion testing against most glass compositions, forming a continuous, dry, benign α-Al_2O_3 surface layer.

When service conditions impose KOH in-diffusion, concurrent loss of NaOH from the refractory results in net loss of alkali - leading to formation of an α-Al_2O_3 surface layer that is devoid of any β-Alumina phase. Underneath the Corundum surface layer, α/β structure of normal appearance may be found, but further analysis usually indicates some substitution of K for Na in the β-Alumina phase. The depths of both the α-Al_2O_3 surface layer and any KOH penetration increase with increasing test time. It should be noted that invalid test techniques (using excess alkali to 'accelerate' corrosion kinetics) are often performed in the β-Alumina stability field - leading to erroneous results. Industrial 'in-situ' testing confirms the validity of 'non-accelerated' laboratory testing, as seen in Figure 12.

3.3. Refractories Exposed To Long-Term 'In-Situ' Corrosion Testing

Selected fusion-cast refractories were exposed to very heavy batch dust contamination in an Oxyfuel TV Panel glass melter exhaust port for~70 days. The results are presented below in Figures 13, 14, 15 and 16.

Figure 13. Refractories Exposed To Batch Dust In A TV Panel Glass Furnace Exhaust Port ~70 Days

Fusion-Cast Refractories:
Monofrax CS3 AZS
Monofrax Mx98 αβ-Alumina
Monofrax M αβ-Alumina
Monofrax H β-Alumina
Monofrax L MgO.Al_2O_3 Spinel

Post-Test Condition:
Very Heavy Batch Dust Deposition
Liquid Phase Surface Deposit

Fusion-cast Monofrax CS3 AZS and Monofrax M αβ-Alumina refractories were also exposed to minor batch dust contamination in peepholes at the charging end of an Oxyfuel PbO-Silicate 'Crystalware' melter, for both 111 days and 411 days. Results are presented in Figure 17. Additionally, a sample of αβ-Alumina refractory was subjected to minor batch dust contamination over an even longer-term exposure, of 498 days, in the crown of an operating Oxyfuel TV Funnel glass melter, see Figure 18. All results are summarized for each refractory below:

Fusion-Cast Alumina-Zirconia Silicate (AZS) refractory exposed to heavy TV Panel batch dust for ~70 days was covered by a deposition layer of thickness ~1mm. Upon cooling, this layer was comprised of Zirconia clusters (from decomposed Zircon batch) in a (K,Na,Sr,Ba,Zn,Zr,Al,Si,O) glassy matrix. Inter-diffusion (in service) between the liquid layer and the refractory caused significant alteration of the AZS microstructure to a depth of ~11mm below the interface with the contamination. The presence of Zirconia dendrites in a Nephelitic matrix phase marked the original AZS surface. Here, complete dissolution of Alumina plates occurred, although Corundum volume then increased with depth below the interface. The Nephelitic-type matrix chemistry ended at ~11 mm depth, with a residual glassy matrix (altered by the presence of K_2O and excess dissolved Alumina) present at greater depths.

Figure 14. Fusion-Cast AZS / Heavy TV Panel Glass Batch Dust ~70 Days

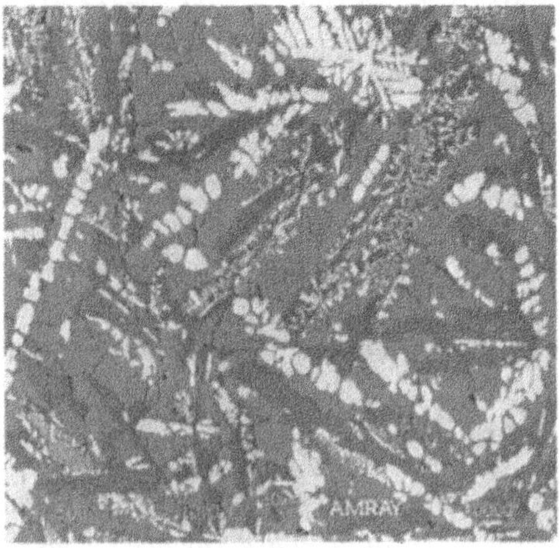

AZS Significantly Altered
Alumina Dissolution
Liquid Phase + ZrO_2 Surface

M_2O/MO Penetration ~11mm
Below Deposited Material

Least Resistant Fusion-Cast
Refractory In This Trial

High Defect Potential

Other exposure of AZS for ~111 days, and in a separate trial for ~411 days, in a PbO-Crystalware melter led to a reaction with vapor phases and a relatively minor amount of batch dust deposition. A loosely adherent scale, comprised of glassy phase with Alumina and Zirconia crystals, was identified on the outer surface. The considerable liquid phase expansion on the AZS surface was associated with the presence of K_2O (KOH), and to a minor extent, PbO. This again caused dissolution of the crystalline Alumina phase within the AZS, which further increased the total liquid volume.

EDS 'Linescan' data from the 411 day exposed sample, is represented by solid circle markers (with solid line) in the chart of Figure 17. This analysis was obtained by scanning a line through a length of microstructure at a given depth below the surface. The open circle markers similarly represent data from the 111 day exposed sample. Each such data point therefore represents a mean analysis of all crystalline and glassy phases in the microstructure at a particular depth. In each case, the mol% content of (M_2O+MO) obtained in this manner was summed and then normalized by the amount of Na_2O present in unexposed refractory at the same depth. In this normalized chart, a Y axis value of 1 indicates essentially unaltered refractory structure whereas higher numbers represent factors of increase in the corresponding oxide content. Thus, the 411 day exposed sample is seen to contain ~3x more (M_2O+MO), throughout the 25mm total analyzed depth, than unexposed material.

As expected, K and Pb were identified in the AZS glassy matrix phase following both exposures. K_2O concentrations in the expanded glass phase volume were high throughout the entire analyzed thickness of ~50mm (after 411 days ~11.0mol% at 0.05mm and ~10.7mol% at 50.8mm depth), accompanied by lower but consistent PbO concentrations (after 411 days ~0.4mol%).

αβ-Alumina (Monofrax M) refractory exposed to heavy TV Panel batch dust for ~70 days was also covered by a thick deposition comprised of Zirconia clusters in a matrix composed of a Nephelitic type phase and a Barium-rich glass. Inter-diffusion (in service) between the liquid layer and the refractory caused some Al_2O_3 dissolution into the liquid. The refractory structure to a depth of ~100μm below the interface with the surface contamination was affected by in-diffusion of alkali and alkaline earth components of the TV Panel glass. This resulted in conversion of the original α-Alumina and (Na,O)β-Alumina to a modified (Na,K,Sr,Ba,O)β-Alumina phase. The Ba and Sr content of the new β-Alumina phase decreased with distance from the exposed surface. However, the K_2O content of all β-Alumina crystals remained significant, even in the bulk of the sample away from exposed surfaces.

Exposure of αβ-Alumina refractory for ~111 days, and in a separate trial for ~411 days, in the PbO-Crystalware melter allowed reaction with vapor species and a relatively minor amount of batch dust deposition. After 411 days exposure the refractory was sheathed with a thin (~300μm thick) outer surface zone comprised of modified (K,Na,Zn,Si,Fe,O)β-Alumina plates and a grain boundary glass phase (liquid during exposure). Despite the presence of this liquid phase at grain boundaries, the outer surface appeared essentially dry.

Figure 15. Fusion-Cast Aluminas / Heavy TV Glass Batch Dust ~70 Days

αβ-Alumina **β-Alumina**

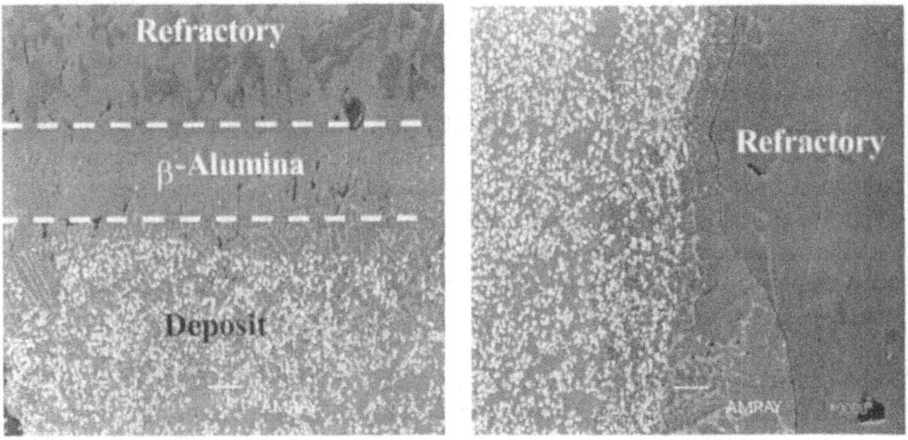

Liquid Deposit On Surface Caused Some Dissolution Of Refractory
M_2O+MO Liquid Penetration Below Liquid Deposition:
'M'~100μm 'H'~300μm
Original Surface Structure Converted To (Na,K,Sr,Ba,O)β-Alumina

Figure 16. Fusion-Cast Spinel / Heavy TV Panel Glass Batch Dust ~70 Days

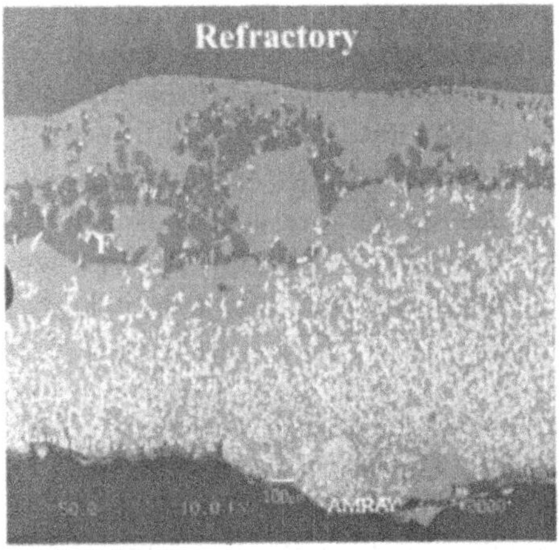

Liquid Phase Surface
Minor Dissolution Of MgO
Forsterite (F) Formed.

No Penetration Detected Below Liquid Deposition.

Best Resistance To Heavy Batch Dust Corrosion

Figure 17. - Surface Structure And Chemical Changes Within Refractories After Long-Term Exposure To PbO-Silicate Glass-Melting Atmosphere

Monofrax CS3 AZS (111 Days)	Monofrax M αβ-Alumina (111 Days)

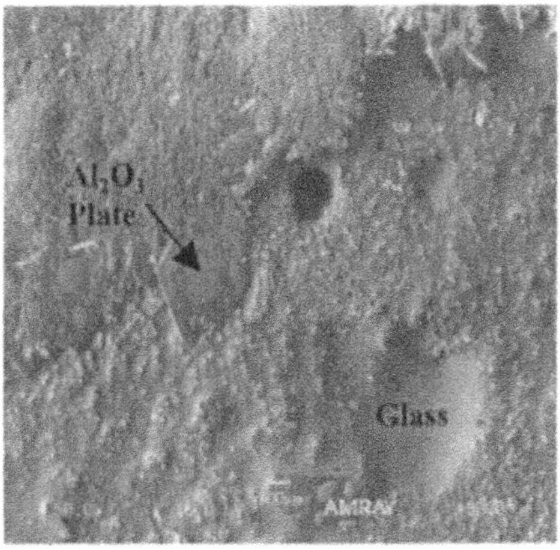

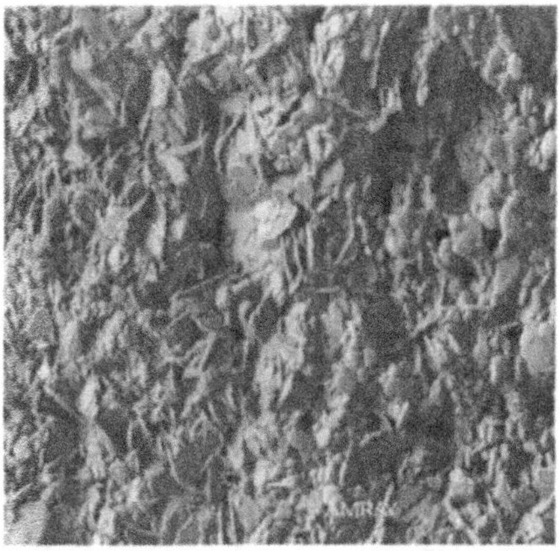

AZS Liquid Phase Surface	*αβ-Alumina Essentially Dry Surface*

Normalized Linescan EDS Analysis:

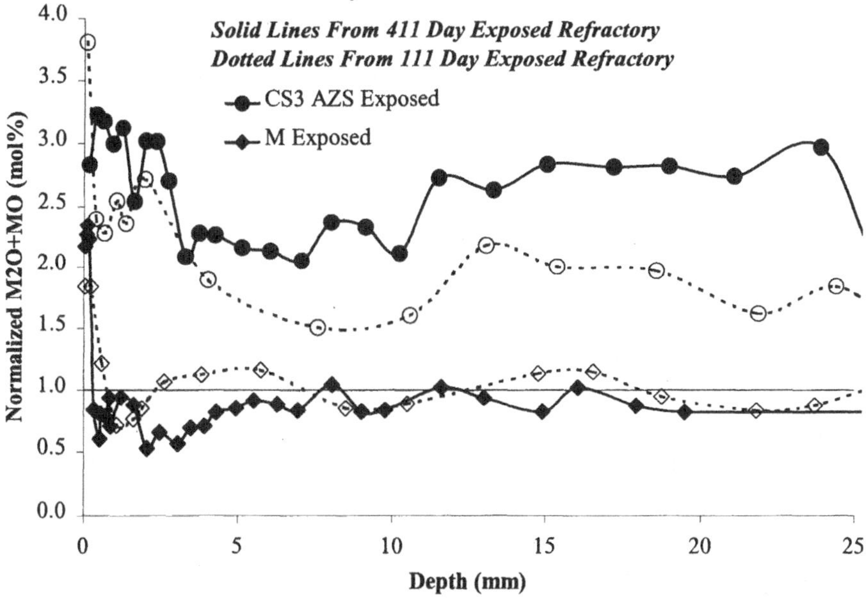

Figure 18. αβ-Alumina Refractory Exposed To Oxyfuel TV Funnel Glass Melting~498 Days In Crown Above Batch Chargers

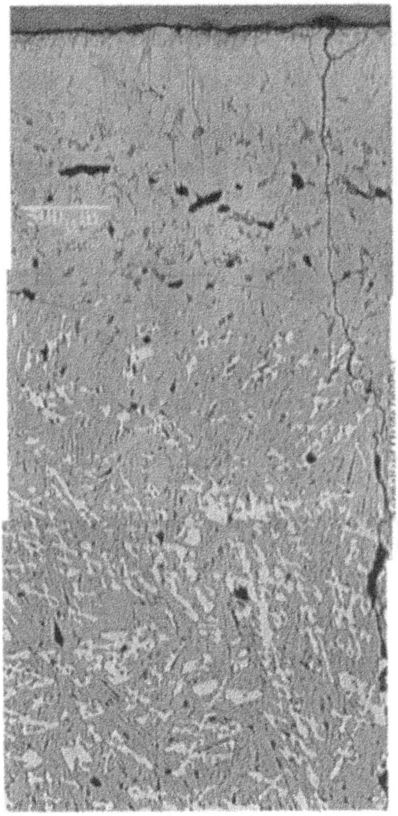

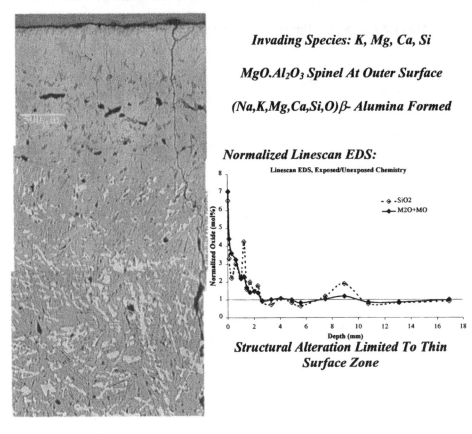

Invading Species: K, Mg, Ca, Si

$MgO.Al_2O_3$ Spinel At Outer Surface

(Na,K,Mg,Ca,Si,O)β- Alumina Formed

Normalized Linescan EDS:

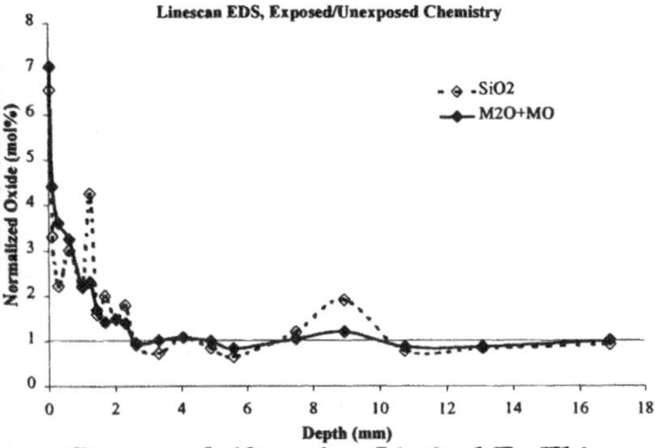

Structural Alteration Limited To Thin Surface Zone

The next (~0.3-1.3mm deep) zone was highly enriched in α-Al_2O_3 phase, with only minor modified β-Alumina content, and a continuing excess of grain boundary glass (liquid).

The underlying (~1.3-7mm deep) zone contained more porosity with increasing β-Alumina content and a diminished but continuing presence of Pb-bearing glassy (liquid) phase. The increased porosity may be a consequence of NaOH loss and conversion from β-Alumina to the denser α-Al_2O_3 phase. The grain boundary phase decreased significantly in this zone and rapidly diminished in volume below ~7mm total depth. Zn and K were present in both the glassy and modified β-Alumina phases, but Zn could not be detected below ~7mm depth. K was essentially constant in concentration to a depth of at least 50mm. Thus, below ~7mm depth the structure is essentially regarded to be that of normal (unaltered) αβ-Alumina refractory with some substitution of K_2O for Na_2O in the β phase.

Normalized EDS linescan data from exposed αβ-Alumina refractory is represented in the chart of Figure 17 by solid diamond markers (411 day exposure) and open diamond markers (111 day exposure). Below the benignly altered surface zone, the data demonstrates relatively little change in the overall nature of this Alumina refractory over the 411 day experimental time period.

Another sample of αβ-Alumina refractory was exposed to batch dust contamination in a TV Funnel glass melter for ~498 days. The outer surface was covered with an ~100μm thick layer containing $MgO.Al_2O_3$ Spinel crystals in a minor volume of glassy matrix of variable composition close to Gehlenite ($2CaO.Al_2O_3.SiO_2$). Below this outer surface the refractory structure was composed of modified (Mg,Na,Ca,K,Si,O)β-Alumina plates with an apparently glassy Ca,Na,K,O-Alumino-Silicate grain boundary phase. The MgO content of the boundary phase was undetectable, and the MgO content of the β-Alumina phase decreased rapidly as a function of depth, from ~13.5mol% at 0.11mm depth to undetectable at ~1.44mm depth (where α-Al_2O_3 crystals became evident).

Normalized linescan EDS data is presented in the chart of Figure 18, a rapid decrease in CaO, SiO_2 and MgO was noted as a function of depth. The solid diamond markers represent total (M_2O+MO), and the open diamond markers represent SiO_2, as a function of depth. Thus, the surface of the exposed sample is seen to contain considerably more (M_2O+MO) and SiO_2 than unexposed material, and the underlying refractory structure is essentially indistinguishable from unexposed material below a depth of ~2.6mm. However, it should again be noted that K_2O penetrated the sample to a considerable depth, but this gain in one alkali species was counteracted by a loss of Na_2O from the original structure - so that the net sum of alkali and alkaline earth components remained essentially unchanged below ~2.6mm depth. PbO was not detected anywhere within the exposed surface samples, although PbO condensation was detected further down the thermal gradient on the joint surfaces.

β-Alumina (Monofrax H) refractory exposed to heavy TV Panel batch dust for ~70 days behaved in a similar manner to αβ-Alumina (Monofrax M) refractory. The glaze layer consisted of Zirconia crystals embedded in a volume of silicate liquid phase containing dissolved Alumina - which was contributed by corrosion of the refractory. Again, the surface liquid phase developed a Nephelitic type crystal and a Barium-rich glass upon cooling, and a new β-Alumina phase formed at the interface due to reaction with alkali and alkaline earth cations from the batch dust. Compositions were similar to those observed in αβ-Alumina, with significant K_2O concentrations throughout the structure and alkaline earth concentrations decreasing as distance from the interface increased, but this time penetrating to a depth of ~300μm compared to ~100μm for Monofrax M.

$MgO.Al_2O_3$ Spinel (Monofrax L) refractory exposed to heavy TV Panel batch dust for ~70 days also became coated with a thick liquid-rich layer. However, this time the surface deposit exhibited a different structure than those observed upon the Alumina or AZS refractories (see Figure 13.), and apparently caused only minor reaction with the underlying refractory (see Figure 16.). The small amount of corrosion of the Spinel refractory involved free-MgO from the refractory going into solution in the surface liquid, with consequent formation of Forsterite ($2MgO.SiO_2$) crystals. The refractory structure below the interface with the surface contamination appeared to be completely unaffected, with no signs of penetration or gross in-diffusion of TV Panel melt cations.

4. DISCUSSION

Vapor Phase Corrosion of silicate superstructure refractories occurs in idealistic laboratory tests in the absence of atmospherically borne batch dust. This degradation mechanism may also be the prevalent corrosive component in certain zones of operating industrial furnaces, or for short 'in-situ' test periods where batch dust impingement has not accrued significantly. However, in long-term 'in-situ' testing, most superstructure refractory samples are subjected to some degree of batch dust contamination.

Glass-melting in the presence of vapor phases and the absence of batch dusts generally exposes superstructure refractories to alkali vapor species, although other species including Pb, Zn, Sb, or B may also be present. Vapor pressures of alkaline earths, Silica, Alumina etc. are normally relatively low. Under such conditions, refractories with a significant silicate component are subject to degradation. The dissolution of alkali into the silicate phases leads to liquid phase expansion by dissolution of the refractory structure. Any significant volume of open porosity accelerates the kinetics of attack, so that low density bonded refractories may exhibit more rapid degradation than less porous material. Dense fusion-cast Alumina refractories are not degraded under such conditions, except for loss of alkali from a relatively shallow zone immediately adjacent to the exposed surface, resulting in formation of Corundum.

'Accelerated' Testing achieved by addition of excess alkali does not allow realistic simulation of conditions in real operating furnaces. Previous work[7] has proven that the typical accelerated testing, currently employed widely within the glass industry, subjects the test refractories to different vapor species in very different concentrations than found in industrial glass-melting environments. In some cases this testing may lead to incorrect refractory selection due to reaction in the wrong phase stability field.

Atmospherically Borne Batch Dusts may expose superstructure refractory surfaces to considerable concentrations of Silica and other compounds that would not otherwise be problematic. This usually leads to liquid phase formation and inter-diffusion between the refractory and deposited surface layer. Again, any open porosity allows accelerated refractory degradation. Whenever batch dust contamination occurs, there must also be a component of vapor phase transport contributing to the total refractory degradation.

The formation of silicate liquid phases on fusion-cast Alumina refractory surfaces allows for some local dissolution of the refractory structure and preferential formation of modified β-Alumina phases rather than Corundum. It is possible that the original refractory structure alters to form this modified β-Alumina by solution and reprecipitation through the liquid phase at grain boundaries. However, this structural modification is an effective sink for containing invasive M_2O and MO species, thereby preventing the formation of even more excessive liquid phase volume. In comparison, AZS and other Silica-rich refractories absorb incoming species by production of large liquid volumes, which further expand by dissolution of the original crystalline refractory structure. This causes visible 'run-down' with potential for defect formation in the molten glass bath.

Compare the behavior of fusion-cast AZS and αβ-Alumina refractories as presented in Figure 17. Exposure to relatively light batch dust contamination in a PbO-Crystalware melter allowed comparative discrimination of the relative performance of these refractories. The AZS underwent considerably more alteration than the Alumina refractory during 111 days of 'in-situ' testing (open circles / dotted line). Again, this degradation included dissolution of the Alumina structure within the AZS and liquid phase expansion. After 411 days of testing the AZS (solid circles / solid line) degraded significantly further with greater in-diffusion of glass making components to ~3 times the original (M_2O+MO) concentration at the limit of analysis ~25mm below the exposed surface.

In comparison, the αβ-Alumina refractory maintained an essentially dry surface, although liquid phase was identified at underlying grain boundaries, decreasing in volume to a depth of ~7mm. Below this depth the refractory structure was effectively unaltered except for substitution of K for some Na in the β-Alumina phase. Further testing of fusion cast αβ-Alumina in a TV Funnel furnace confirmed this superior mode of behavior, as may be seen from Figure 18.

Testing in the presence of heavy batch dust deposition (Figures 13, 14, 15 and 16) again proved that AZS underwent more severe alteration, with more defect potential, than αβ-Alumina refractory. β-Alumina (Monofrax H) refractory behaved in a similar manner to the αβ-Alumina (Monofrax M) material except that penetration depths were greater due to the intrinsically larger grain and flaw size of the β-Alumina product. αβ-Alumina refractory exhibits higher strength, is intrinsically more refractory with higher creep resistance, and is even successfully utilized in glass contact at temperatures below ~1300°C. In comparison β-Alumina is restricted to relatively dry superstructure application due to its higher solubility in most silicate melts.

Considering the similar cost of AZS, 'M' and 'H' type materials, αβ-Alumina is currently considered the optimum superstructure refractory for most glass-melting applications where fusion-cast is considered. This material has therefore been widely and successfully applied in crown and superstructure construction for Oxyfuel melting of a variety of glass compositions including Soda-Lime, TV Panel, TV Funnel and Borosilicate.

Looking forward, Monofrax L fusion-cast Spinel refractory exhibits even higher strength, refractoriness, creep resistance and corrosion resistance than Monofrax M. This material may represent the ultimate superstructure option for high temperature glass-melting furnace zones subjected to heavy batch dust deposition, but to date there are no proven service applications of this material in glass-melting superstructures.

5. CONCLUSIONS

5.1. Thermodynamics

Glass melt reacts with atmospheric H_2O to derive vapor species (KOH etc.).
Oxyfuel combustion (~67%H_2O) therefore generates considerably more vapor species than Airfuel combustion (~19% H_2O).
Atmospherically borne batch dusts allow superstructure refractories to react with compounds that exhibit low vapor pressures.
All refractories react with glass-melting atmospheres, but to different degrees:

- Silicate refractories react with alkali vapor producing liquid phases.

- Fusion-Cast Alumina refractory generates a dry (α-Al_2O_3) surface in the presence of vapor phases, but batch dusts generate liquids resulting in growth of a thin β-Alumina surface.
- Fusion-Cast $MgO.Al_2O_3$ Spinel refractory offers the best performance against heavy batch dust exposure.
- Reactions causing refractory degradation can be theoretically predicted.

5.2. Kinetics

Industrial glass-melting operations create transient, non-ideal, environments due to the presence of convection currents, boundary layers, batch dust deposition, thermal and chemical gradients, etc.

Laboratory (or computer) simulation of the glass-melting environment is not simple, therefore reliable derivation of corrosion kinetics data is difficult.

Refractory Corrosion Mechanisms Are Well Characterized And Understood - Allowing Sensible Refractory Selection. Lifetime Prediction Is More Difficult.

ACKNOWLEDGEMENTS

Thanks to the various glass producers who participated in this work by welcoming 'in-situ' furnace testing in their industrial melters. Thanks also to S Gould of Monofrax Inc. for contributions in design and operation of the laboratory corrosion test apparatus and 'in-situ' test sample holders.

REFERENCES

[1]. WD Kingery, HK Bowen and DR Uhlmann, *Introduction To Ceramics – Second Edition*, Wiley, NY, (1976), 219-227.

[2]. W Trier, *Glass Furnaces Design Construction And Operation*, (Translated by KL Lowenstein), Society of Glass Technology (UK), (1987), 169-173.

[3]. SM Winder, A Brach and A Gupta, *'Refractory Selection For OxyFuel Glass-Melting Furnace Superstructure'* Presented At The HVG Colloquium On 13th March 1997, Mainz, Germany.

[4] J Kynik, SM Winder And KR Selkregg, *'Superstructure Refractory Selection For Oxyfuel Melting Of Lead-Alkali-Silicate Glass'*, Presented At The Conference On Advances In Fusion And Processing Of Glass, Toronto, 28th - 31st July 1997.

[5]. SM Winder, KR Selkregg and A Gupta, *'An Update On Selection Of Refractories For Oxyfuel Glass-Melting Service'* Proceedings Of the 59th Conference on Glass Problems, Amer. Ceram. Soc. (1999), 81-106.

[6]. Various Authors, *'Phase Diagrams For Ceramists, Volumes I-V'*, Amer. Ceram. Soc. (to 1985).

[7]. SM Winder and JR Mackintosh, "Testing Oxyfuel Furnace Crown Materials", Glasteknisk Tidskrift, 1, [52], 20. (1997).

OXYFUEL FIRING EFFECTS ON REFRACTORIES

R. E. Moore, M. Velez, M. Karakus, W. D. Headrick
University of Missouri-Rolla
Ceramic Engineering Dept.,
Rolla, MO 65409-0330

ABSTRACT

Oxygen is used as a boost to fuel/air combustion and in oxyfuel burner technology in various melting and heating processes conducted in refractory lined vessels. Advantages to oxyfuel technology include reduction of NOx and emissions, reduced fuel consumption, increased product throughput and, in some cases, higher product quality. This paper reviews the major applications of oxyfuel technology, focusing on the effects on the refractory linings, especially in steel, glass and aluminum melting equipment.

I. CORRODANTS IN THE MELTING OF SODA CONTAINING GLASSES

With the clean fuels used in contemporary airfuel furnaces, corrodants derived from them can be largely ignored. This is not true generally as oil continues to be used in some quarters of the industry, in which case the corrodants can include sulfur, alkalis and even oxides of Fe, V, and Ni, which are extremely reactive with SiO_2. Whether the oxygen for combustion is derived from air or from O_2 itself, there will be plenty of water produced in the combustion of the hydrocarbons. The water is available to combine with volatiles emanating from the glass surface or coming from the fuel in the case of dirty fuels. When Na_2O is the only alkali in a conventional glass being melted in a typical continuous melter, there are a number of factors which combine to determine the actual concentration of NaOH in the interface between the combustion space and the crown refractories:

1. Na_2O content in the glass
2. Na_2O from batch carry over
3. Local gas velocities/flow patterns
4. Oxidizing/reducing condition of the flame

The NaOH concentration varies over the total surface area of the crown. Ideally that concentration should be known, but the accurate and rapid measurement of NaOH concentrations in the combustion space, especially adjacent to the crown refractory surface, is not straightforward. The EPA method is being used at the University of Missouri-Rolla (UMR) following the practice by lab and industry investigators [1]. Measured levels of NaOH in commercial flat glass tanks vary from 50-400 ppm, but the values are not well quantified due to problems with condensation in the sampling conduits. At UMR we have minimized this problem and are controlling and measuring NaOH ppm in oxyfuel simulations in the range of 150-250 ppm, but this concentration level requires the addition of Na_2O to the bottle glass batch. The enhancement correlates to the achievement of substantial attack of silica crown brick. Melting of the unenhanced glass only produced a NaOH concentration of about 50 ppm and this points to the difficulty of simulation of the atmospheres of a commercial tank melter.

When the corrodant is NaOH in either an airfuel or an oxyfuel melter employing a silica refractory, the attack of the silica is straightforward but not so simple as in the case of dissolution or compounded formation. There are two general modes of damage:

1. Formation of a high alkali glassy phase in the surface regions of the silica brick
2. Concentration of NaOH at depth in the brick points followed by local attack termed ratholing.

The glassy phase formed actually is a modification of the glassy phase preexistent in the brick. Classical studies have focused on the complete characterization of the altered refractory, i.e. the volume and distribution of the reaction glass formed and its chemistry changing with depth. Suffice it to say that when the NaOH concentration builds in an oxyfuel melter, the resulting enhanced concentrations of NaOH can result in the development of large amounts of high Na reaction glasses which aggressively dissolve the silica, a process aggravated by temperature, especially by flame contact.

Conversion of an air/fuel combustion system to an oxyfuel system results in an altered corrosion scenario for all the brick exposed to the combustion space glasses. The first conversions led to the following consequences:

1. Enhanced concentrations of alkalis in the form of NaOH and KOH
2. Elevated concentrations of H_2O vapor
3. Higher superstructure temperatures
4. Gas flow patterns and flame conditions apparently more prejudicial to the refractories.

Table 1 presents some measured flue gas compositions for fuel/air and fuel/oxygen fired melters. It should be noted that the results such as these could vary widely between two designs of similar furnaces melting the same glass.
Figure 1 shows concentration profiles of Na_2O in silica brick with depth after exposure of only 20 hours to respective hot face temperatures of 1600, 1450 and 1350°C when the NaOH concentration was enhanced to 650-vol. ppm NaOH.
At 1600°C the higher rates of diffusion serve to triple the amount of Na_2O at depths greater than ≈ 1/8 inches (3.5 mm).

Table 1. Measured flue gas composition in combustion chambers for oxyfuel and air-gas fired soda lime glass furnaces [2].

	Oxy-fuel fired furnaces		Air-gas fired glass furnace
Na_2O content in glass	16 wt%	13 wt%	13 wt%
Gas Component (vol%)			
O_2	2-4	1-2	2
CO_2	31	33	12
H_2O	56	53	18-20
N_2	10-13	12-13	68
NaOH (ppm)	220-260	140-200	50-60

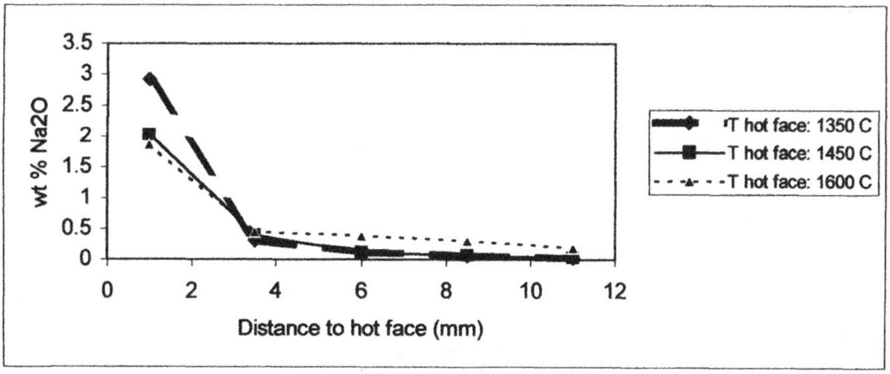

Figure 1. Concentration profiles of Na_2O in bonded silica samples, after 20 h exposure to 650 vol. ppm NaOH [2].

Figure 2 presents a theoretical equilibrium, concentration versus temperature curves for NaOH concentration reflecting air fuel (60 ppm) and oxyfuel (300-vol. ppm). The model predicts no uptake difference for CaO. The data suggest much high diffusion rates with temperature for the two oxide species. The mechanism of diffusion of species is an open consideration whether interface or transport control is dominant. Figure 3 represents theoretical curves for these controls plus a mixed control situation. It would be hard to verify control mechanisms in diffusion medium as complex as a highly heterogeneous silica brick. This figure is a plot of the relative corrosion rate of silica brick as a function of relative soda (NaOH) concentration in the region adjacent to the crown surface. This is schematized only but based on simulation trials. Little data have been published for silica alternative, but such data are being developed at the University of Missouri-Rolla.

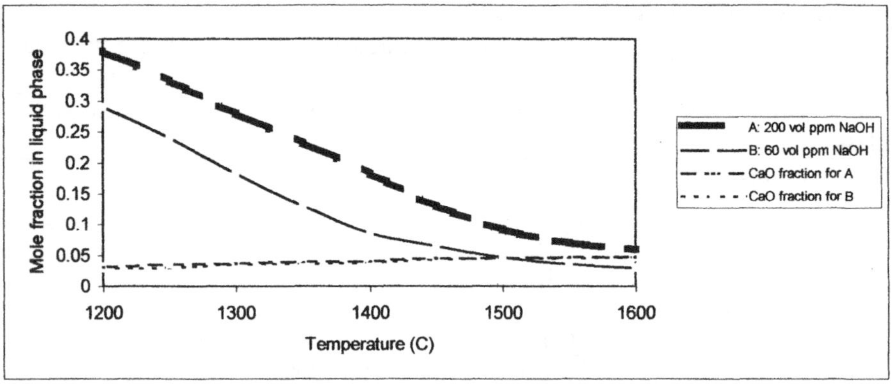

Figure 2. Calculated NaOH and CaO concentrations of glassy reaction products due to interaction between silica and NaOH-containing atmospheres, as a function of temperature, assuming thermodynamic equilibrium [2].

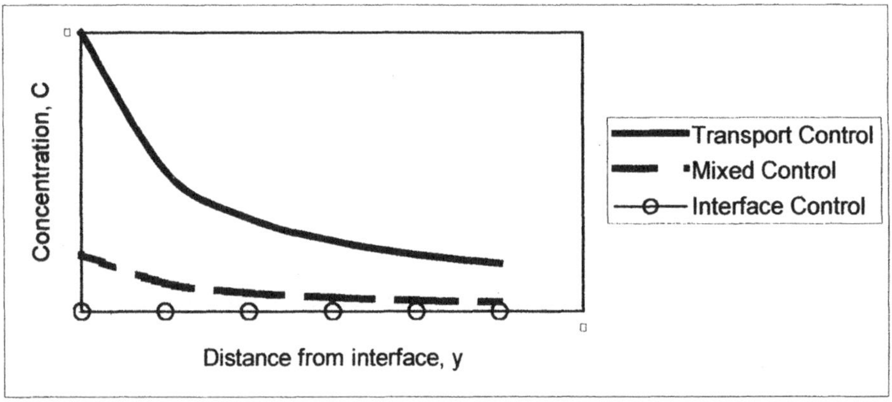

Figure 3. Schematic showing concentration distribution for various types of control of dissolution process.

There are three principal motivations for converting to oxyfuel-melting systems. These are as follows:

1. Vapor corrosion is aggressive due to the specific compositions being melted, i.e. high alkali glasses
2. Higher melt temperatures are desired
3. The need to reduce NOx emissions

The higher melt temperatures may result in higher defect potentials and will require higher performance silica brick (higher purity, lower porosity) or other brick compositions with higher corrosion and creep resistance.

II. OPERATING WITH SILICA BRICK IN OXYFUEL FURNACES

There are conversions wherein silica brick continue to serve. Measures to reduce NaOH vapor attack in these melters include:

1. Reducing NaOH levels in the furnace atmosphere
2. Reducing Na-uptake by the silica refractories
3. Optimizing temperature profiles of the silica crown

According to Leblanc [3] countermeasures to silica attack in oxyfuel furnaces should include:

1. Melting Practice: (a) Optimal oxyfuel flame, low velocity, good luminosity, blanket form close to the glass; (b) Avoid reducing conditions; and (c) High quality oxygen supply
2. Refractory practice/design: (a) Maintain a temperature above 1750° F (953° C) throughout the crown brick; (b) Avoidance of holes, cracks or open joints to preclude vapor passage; (c) Use of a monolithic seal over the entire crown; and (d) Carefully programmed heat-up to avoid openings of joints.

As mentioned earlier, selecting higher purity lower porosity brick is an obvious way to reduce Na uptake [2]. Figures 4 and 5 show schematics of seal and insulation systems geometries which will reduce NaOH movement and condensation. Sealers have been very effective in elimination of ratholing. Characteristics of sealer systems should include:

1. Chemical compatibility with crown and insulation
2. Bondable to the crown itself
3. Possess creep and chemical durability to serve as a crown in case of partial crown loss

4. Can be installed so as to effect a monolithic-sealing mass over the entire crown

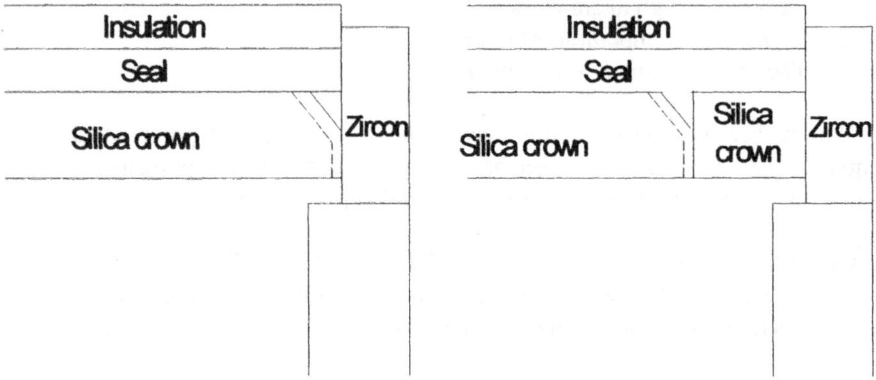

Figure 4. Schematics of seal and insulation systems geometries which will reduce NaOH movement and condensation.

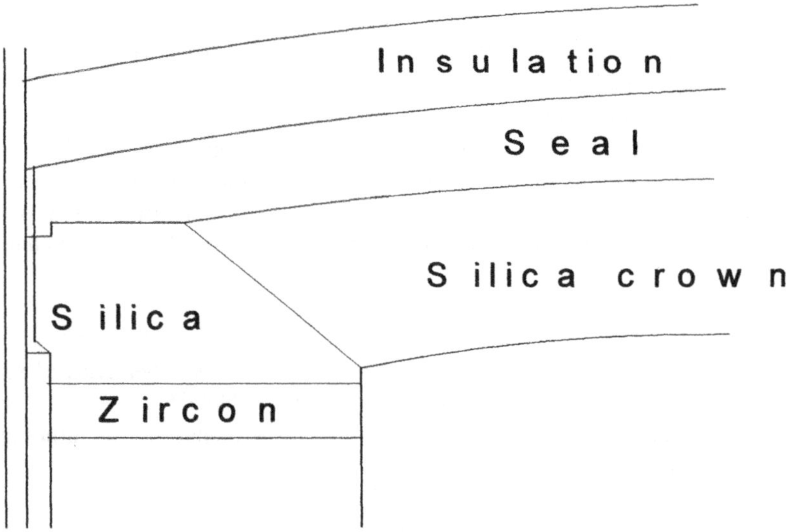

Figure 5. Schematics of seal and insulation system geometry which will reduce NaOH movement and condensation.

III. GENERIC CRITERIA FOR CROWN BRICK SELECTION IN OXYFUEL FURNACES

There are reports that fused AZS and M-type (α-β Al_2O_3) shapes are being used to substitute for silica brick in oxyfuel fired melters and that a number of brick compositions, bonded or fused, are being promoted for consideration [4,5]. The cost effectiveness of silica brick is very difficult to match. The cost of the brick and the costs of installing and supporting the much heavier crown refractories must be factored into the decision. There are instances in which a ten-fold increase in brick cost is prohibitive and conversions did not go ahead. However, the motivations to convert fuel technologies cited earlier are powerful, in particular, the move to higher (= faster) melting temperatures. Three major generic criteria for selection of refractories for the more demanding oxyfuel furnace atmospheres and/or for higher temperature melting include:

Rule 1. The refractory must be physically stable in a given furnace
Rule 2. The refractory must be physically stable when exposed to chemical species in any combination
Rule 3. Reaction kinetics must be minimized

In addition, there are the obvious criteria of cost, availability, and defect potential.

Figure 6 is a schematic of alkali resistance versus refractory cost per unit volume. The comparison between silica and AZS/alumina is priced with a ratio as high as 25. Most fused oxides and mixed oxides will probably be located in the 10-25 range. Bonded fused grains and high sintered aluminas, spinel types, etc. will possibly be located below 10. It is this group in the "opportunity" region of Figure 6 that need to be fully evaluated. This raises the question about methods of evaluation as the standard tests developed for silica and silicates are not applicable. New tests and simulations need to be designed and proven for oxyfuel tank refractories. Velez has reviewed the early work with oxyfuel simulators [6,7]. Long exposure times are needed to assure that alterations to the most resistant crown substitutes do not preclude long term service. No postmortem studies of the fused products have been published to date.

Table II provides a comparison between silica, a bonded spinel and two fused products. The property list serves as a fuller criteria compiling all the important properties and characteristics with some of them like defect potential being critical.

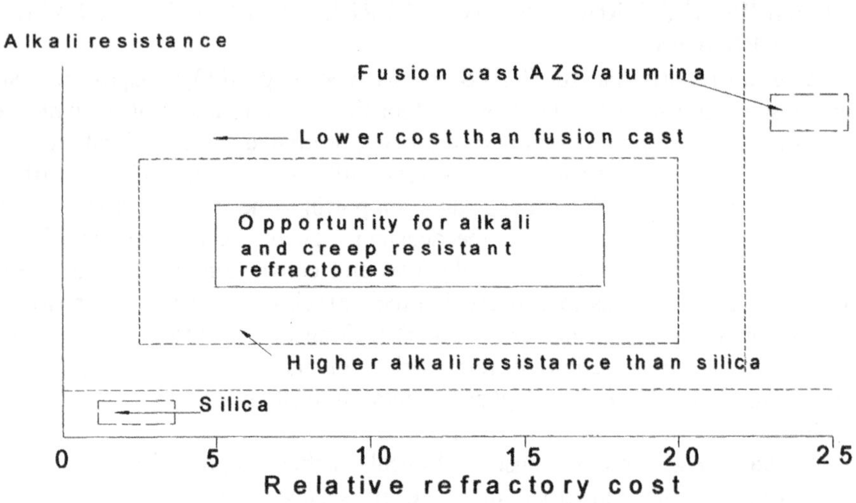

Figure 6. Schematic diagram showing the alkali resistance versus refractory cost per unit volume and the comparison between silica and AZS/alumina in price [8].

Table II. Comparison Superstructure Refractories.

Property	Product			
	Silica	Spinel	Fused Cast	Fused Cast
Cost per m^3 (compared to silica)	1	7	10	11
Density (g/cc)	1.79	2.99	3.72	3.4
Corrosion resistance	Poor	Excellent	Medium	Excellent
Construction type	Standard bonded end arch	Standard bonded end arch	Ground block, needs supporting steel	Ground block, needs supporting steel
Thermal conductivity (W/mK @ 1200° C)	2.0-2.3	3	5	5.19
Thermal shock	Poor < 500° C	Good (5 cycles 950° C to water)	Limited due to low porosity	Limited due to low porosity
Thermal expansion	Non-linear	Linear	Non-linear	Linear

Alternatives to silica brick discussed in the literature include fused AZS, fused alumina types, sintered spinel, rebonded spinel and fused spinel. Only the first two are reported to be in service in full crown constructions. Spinels are showing potential in various tests and trials. There are rebonded products on the market and they feature [9]:

1. Exceptional alkali resistance
2. Shapes available in standard arch shapes and sizes
3. Low linear thermal expansion
4. Resist ratholing
5. Low defect potential; are soluble in soda lime glasses.

These claims are made for melting conventional soda lime glasses at typically furnace temperatures.

Thermal conductivity trends are shown for several refractory crown materials in Fig. 7 and suggest that attention must be paid to the distinct behavior of the alternatives compared to silica in regard to thermal losses. Figure 8 depicts data for linear thermal expansion for the four products making clear that it will be much simpler to manage heat-ups with alternatives. Creep data comparisons are made in Fig. 9 and the data suggest no problems with creep up to 1600° C.

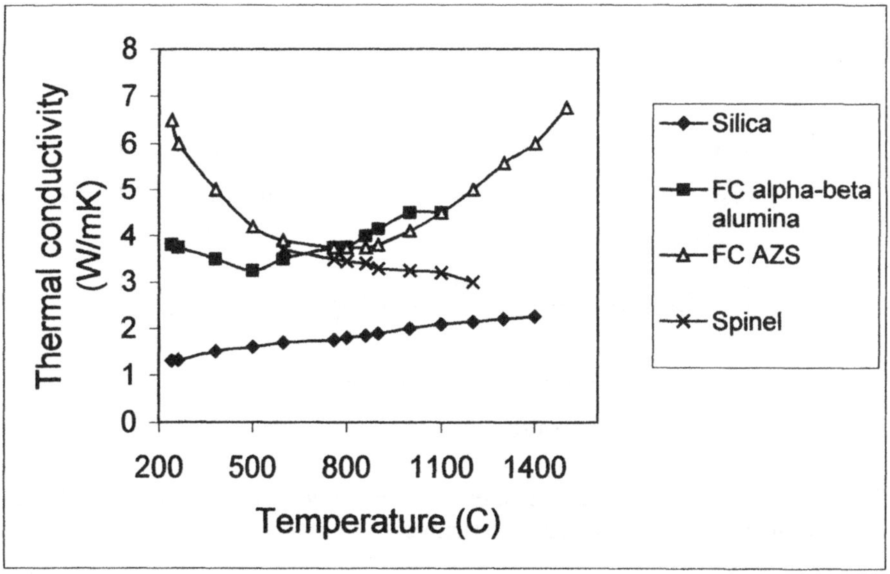

Figure 7. Thermal conductivity of spinel refractories in comparison to crown silica and fused cast alternatives.

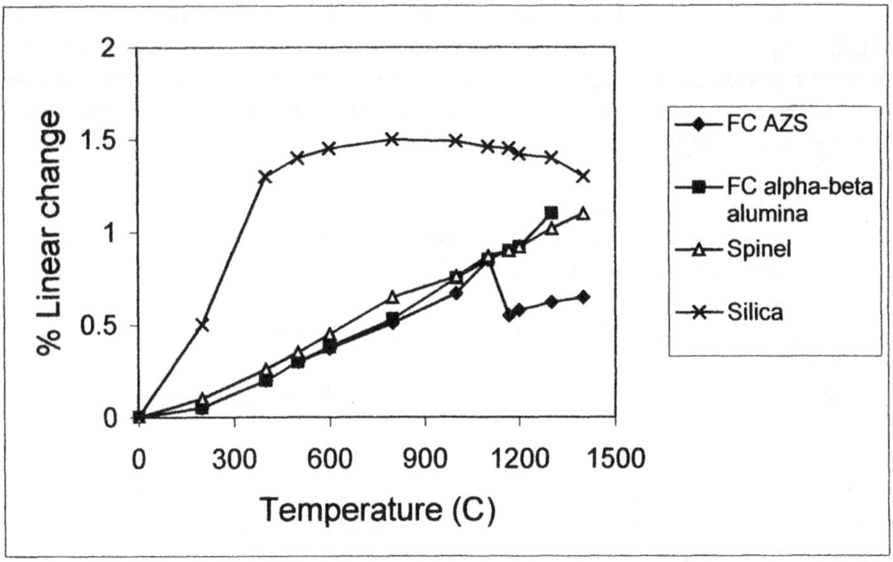

Figure 8. Linear thermal expansion of spinel refractories in comparison to crown silica and fused cast alternatives.

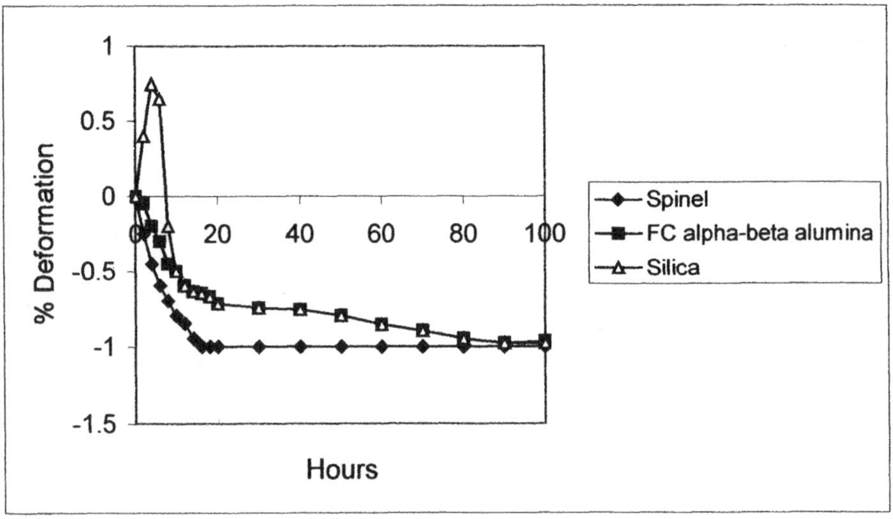

Figure 9. Creep of spinel refractories in compression at 1600° C (0.2 MN/m^2) versus crown silica and fused cast alternatives.

IV. SUMMARY AND CONCLUSIONS

Silica brick will continue to serve in crown construction in many melters due to the excellent creep resistance, low conductivity and low cost. The fused products used for glass contact appear to be functional alternatives for crowns but may not be cost effective in many instances. The next-in-line products appear to be bonded and fused spinels and other fused grains with new bond systems. Some sintered products based on alumina and alumina-zirconia mixtures may present some advantages in terms of shape availability density and cost. Melters for specialty glasses will be used in commercial scale trials of alternatives. Conventional glass melters will await these results and results obtained from the exposure of silica alternatives in simulators. Some monolithic products are being adapted and evaluated in oxyfuel and air/fuel simulations.

V. REFERENCES

1. S. S. Tong, J. T. Brown, L. H. Kotacska, "Determination of Trace Impurities in a Furnace Atmosphere at Operating Temperature," *Ceram. Eng. Sci. Proc.,* 18[1] 208-215 (1997).
2. A. J. Faber, O. S. Verheijen, "Refractory Corrosion Under Oxy-fuel Firing Conditions," *Ceram. Eng. Sci. Proc.,* 18[1] 109-119 (1997).
3. J. Leblanc, "Impact of Silica Attack on Soda Lime Oxyfuel Furnaces," *Glass Technol.,* 37[5] 153-155 (1996).
4. L. H. Kotacska, T. J. Cooper, "Testing of Superstructure Refractories in a Gas-Oxy Atmosphere Against High-Alkali Glasses," *Ceram. Eng. Sci. Proc.,* 18[1] 136-145 (1997).
5. A. Gupta, S. M. Winder, "Ongoing Investigation of Oxy-fuel Firing Impact on Corrosion of Nonglass Contact Refractories," *Ceram. Eng. Sci. Proc.,* 17[2] 112-120 (1996).
6. M. Velez, J. Smith, R. E. Moore, "Refractory Degradation in Glass Tank Melters, A Survey of Testing methods," *Cerâmica (Brazil),* 43[283-284] 180-184 (1997).
7. C. Carmody, M. Velez, L. Carroll, W. D. Headrick, R. E. Moore, "Refractory Corrosion in Oxy-fuel Systems," *Ceram. Industry,* 149[13] 61-63 (1999).
8. S. Baxendale, "Refractories Technology – New Capabilities for the Glass Industry," *The Refractories Engineer,* pp. 16-18, March 2000.
9. C. Windle, "Rebonded Magnesia-Alumina Spinel, The Oxyfuel Solution," *The Refractories Engineer,* pp. 12-14, March 2000.

COLD-SETTING CORDIERITE CASTABLES

Esteban F. Aglietti*, Nora E. Hipedinger** and Alberto N. Scian*
CETMIC (Centro de Tecnología de Recursos Minerales y Cerámica)
Camino Centenario y 506, C.C. 49 - (1897) Manuel B. Gonnet. Argentina

ABSTRACT

The heat-setting and cold-setting properties of some compositions are used to design chemically bonded refractories. Cordierite and cordierite-mullite are extensively used as refractory materials. This investigation consisted in the development and characterization of cold-setting cordierite-mullite castables. The castables were formulated with cordierite-mullite aggregates (scrap refractory material) with adequate grain size distribution and a matrix composed by silica, alumina and magnesium oxide. The setting bond was obtained using phosphoric acid (PH) or monoaluminum phosphate (MAP). Aggregates plus matrix were cast in steel molds after adding water and PH or MAP, obtaining PC and MC castables, respectively. Tixotropic properties were not observed. The setting time was between twenty minutes and five hours and it was controlled mainly through the type of MgO and the MgO/phosphate ratio. The reaction phase $MgHPO_4 \cdot 3H_2O$, newberyite, was observed but amorphous phases were predominant. Dilatometry of crude PC and MC probes showed a shrinkage of 1.1 % and 1.6 %, respectively, at 1350°C from demolded dimensions. The main final crystalline phases at 1100-1350°C were cordierite and mullite. A XRD peak of d= 0.408 nm appeared at 1100°C, indicating that c-$AlPO_4$ was formed by reaction of alumina with PH or MAP. At 1350°C the aluminum phosphates tended to disappear. The matrix calcined without phosphates did not convert properly, then cordierite formation during heating was favored by phosphate liquid phases. Physicochemical properties of the PC and MC castables as density, porosity, thermal expansion and flexural strength were determined.

INTRODUCTION

Cordierite ceramics are extensively used as refractory materials due to their low thermal expansion and outstanding thermal shock resistance. Cordierite based

* CONICET - Universidad Nacional de La Plata, Argentina.
** CIC - Universidad Nacional de La Plata, Argentina.

materials are employed mainly as kiln furniture. The low intrinsic strength of cordierite can be compensated by the presence of mullite even though some increase in the thermal expansion coefficient results. Cordierite-mullite materials exhibit a high thermal and strength resistance.

Mixes of talc, plastic clay and alumina produce bodies with a high content of cordierite, accompanied in some cases with spinel and mullite [1, 2, 3]. Refractory cordierite materials are based on a cordierite matrix and aggregates of different compositions and sizes, such as calcined clay, mullite content aggregates, alumina and others. These mixes are then pressed, rammed, cast, etc., producing cordierite based bodies after heating. For refractory concrete applications, adequate aggregate size distributions are required to obtain materials with the best service properties (density, strength, expansion, etc.), depending also on the application or installation method employed.

Mendoza et al. [4] studied the mechanical and physical properties of cordieritic refractory concretes starting from clay, talc, alumina and feldspar, using semi-dry pressing and vibration casting as fabrication techniques. Another author [5] developed a low cement cordierite castable using alumina, fine silica and cordierite grains.

The refractory industry has used the bonding properties of phosphate materials for about eighty years because of the high fusion temperature of many phosphates. The heat-setting and cold-setting properties of some compositions are used to design chemically bonded refractories [6,7]. Gonzalez and Halloran [8] studied the reactions between phosphoric acid and aluminum oxides.

The reaction responsible for the cold setting characteristic of the magnesium-phosphate cement is an acid-base reaction or, more accurately, the reaction between magnesium compounds and phosphates in aqueous solution [9]. The two components that have drawn most of the attention regarding quick-setting are magnesia and acid ammonium phosphate [10-14].

In this investigation the development and characterization of cold setting cordierite-mullite castables were studied. The setting bond was reached using the reaction of phosphoric acid and monoaluminum phosphate with magnesia.

The principal parameters of setting and hardening of the castables are analyzed in relation with the new phases formed in the system. The evolution with the temperature of the flexural strength, expansion coefficient, porosity and other physicochemical properties was observed.

EXPERIMENTAL

The cold setting castables were formulated with cordierite-mullite aggregates (scrap refractory material) with adequate grain size distribution from mesh 8 to 50 (ASTM). These aggregates constituted 60 wt.% of the total composition. The fine fraction (matrix) was formulated with silica, alumina and magnesium oxide. The matrix was 40 wt.% of the total and was designed to form cordierite during heating. The SiO_2 employed was silica fume EMS 965 from Elkem Materials Inc. The alumina utilized was a calcined type (Alcan-type S3G) with a mean particle size of 5μm. The magnesium oxide was refractory grade with a particle size under 75 μm. The amount of MgO was near 5 wt.% of the overall mix.

The setting bond was reached using phosphoric acid (PH) or monoaluminum phosphate (MAP). The castables were called PC and MC if PH or MAP were used, respectively. These castables were prepared by mixing the aggregate grains, the matrix powder and the corresponding amount of PH or MAP and water previous to casting in metallic moulds (probes 25.25.150 mm^3), by vibration. A third castable (TC) with the same aggregates but using talc, alumina and clay instead of MgO, silica and alumina in the matrix fraction was prepared. These castables were mixed with water and cast in a plaster mould.

Dilatometric analysis was recorded using Netzsch equipment.

Flexural strength was determined in a J.J. Instrument model T22K.

Crystalline phases were analyzed by X-ray diffraction (XRD) with a Philips PW 3710 equipment using Cu-Kα radiation.

The open porosity was determined by the water immersion method (ASTM C-20).

RESULTS and DISCUSSION

Figure 1 shows the diffractograms of the cordierite-mullite aggregates and the solid dry castable, without PH or MAP addition.

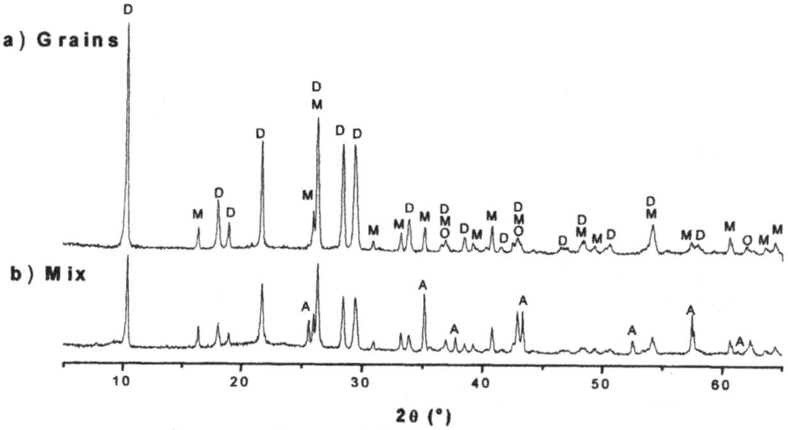

Figure 1. XRD of a) cordierite-mullite grains and b) mix prepared: aggregate plus matrix powder.
Phases: D=cordierite, M=mullite, O=magnesium oxide, A=alumina

The acidic solution required for PC and MC castables was near 25-28 wt.% on the solid basis to obtain a good consistency. Tixotropic properties were not observed because the acidic medium causes a large flocculation of the paste. The paste obtained had some viscosity that favored its application as a plastic or trowellable refractory. The setting time began at twenty minutes and finished within five hours.

The setting time, as it was previously reported, can be controlled by retarders. When ammonium phosphate is employed the incorporation of retarders was found to increase the overall compressive strength of the composition [9]. No good explanation for the behavior of many phosphate systems is provided, however the delay in setting time can be related to the pH of the solution, because the setting reaction is an acid-base reaction, so, the rate can be controlled also by the acidity of the phosphates in the solution.

The reaction between magnesium oxide and phosphoric acid or MAP is rather violent and no hard material is obtained (not set) if the reaction conditions are not controlled.

In this investigation a good setting bond was approached mainly using an adequate amount of MgO with low surface reactivity. Large amounts of magnesium hydroxide or magnesium carbonate on the MgO surface cause a rather violent reaction without good hardening properties, so the MgO was calcined at 1100°C for 1 h. Also MgO/phosphate ratio is an important parameter to take into account.

The setting rate depends on room temperature as the higher the temperature, the faster the reaction. In this case the probes were cast at temperatures between 18 and 24°C.

The amount of liquid necessary to reach a good consistency of the paste (near 25 wt.%) is due mainly to the high porosity of the mullite-cordierite grains and it is enough to form magnesium phosphate hydrates. Aluminum phosphates coming from the MAP or from the reaction between the PH and the alumina of the matrix would be a good thermal setting-bond agent.

To detect the phases formed during setting, a portion of the matrix (without the cordierite-mullite aggregate) was treated with PH or MAP solutions using the same ratio as in the castable. These pastes were hardened at room temperature and then analyzed by XRD.

The XRD analysis made on the hardened matrix (at room temperature) showed the presence of $MgHPO_4 \cdot 3H_2O$ (newberyite) and other low reflections that could be assigned to $MgAl_2(PO_4)_2(OH)_2 \cdot 8H_2O$ (gordonite) as new crystalline phases. Other low intensity peaks appeared and could be assigned to aluminum phosphate phases, but their identification was very difficult. In the materials in which the setting agent was PH, the crystalline phases were more readily developed and identified than when MAP was employed.

The chemical setting takes place through the reaction of phosphoric acid anions on the magnesia particles. Then the magnesium phosphates form hydrate compounds that set like a hydraulic cement. When PH is used a more intense reaction with magnesium oxide would be expected than with MAP, but this fact was not reflected in an increase of the mechanical strength (in the air set materials). This fact evidences that an air setting effect during drying and the presence (or the formation) of aluminum phosphates could contribute to harden the system.

The TC was cast in a plaster mould using approximately 10 wt.% of water. After drying and calcining at different temperatures its physicochemical properties were evaluated and compared with the PC and MC castables.

All the probes remained in air for 24 hours at room temperature, then, they were removed from the mold and thermally treated at different temperatures in an electric furnace for two hours.

Dilatometry of the green materials is shown in Figure 2. For the PC and MC castables no linear changes were observed up to 1100°C. At this temperature a two-step contraction began. The PC material had a smaller contraction at 1200°C than the MC one, and this fact can only be attributed to the initial aluminum phosphate content that generated more liquid phases with temperature increase.

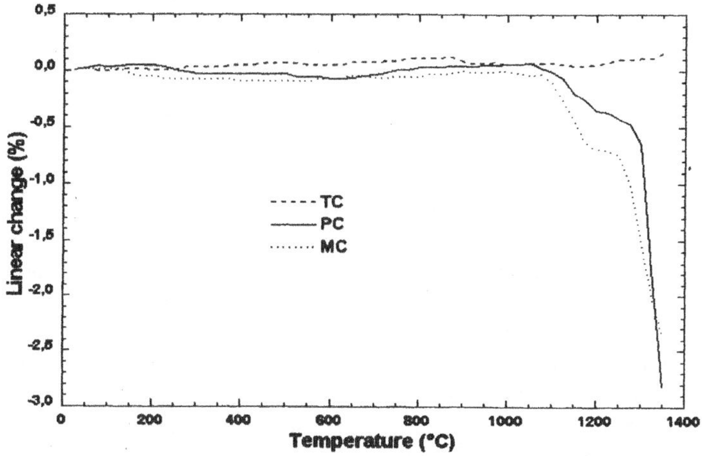

Figure 2. Dilatometry of the green materials.

The most probable phosphate liquid phases formed are: Mg-metaphosphate with a melting point (mp) of 1165°C and compounds of the SiO_2-P_2O_5 system. Silica does not react with PH at ordinary temperatures, and does not produce a phosphorous compound having bonding property. At high temperatures,

$SiO_2{\cdot}P_2O_5$ and $2SiO_2{\cdot}P_2O_5$ are formed, and a liquid phase is produced between 1100°C and 1300°C. Therefore $SiO_2{\cdot}P_2O_5$ compounds are not desirable for refractories.

The TC material showed expansion during heating up to 1350°C. In this material no phosphate compounds were present, so a minor amount of liquid phases was formed.

In Table I the linear change of the probes is shown. At 1350°C the linear changes (contraction) were 1.09 % and 1.57 % for PC and MC, respectively, considered from demolded dimensions. The TC material expanded from its original length as it was observed with dilatometry.

Table I. Linear change of the materials after thermal treatment at different temperatures

Linear change (%)	PC	MC	TC
Demolded – air	0.00	-0.03	---
Demolded - 110°C	-0.22	-0.29	---
110 - 600°C	-0.02	-0.12	---
110 – 1100°C	-0.91	-0.66	0.00
110 – 1350°C	-0.88	-1.19	0.55
Demolded - 1350°C	-1.09	-1.57	---

XRD analyses of PC and MC materials calcined for 2 h at 1350°C showed that the main crystalline phases were cordierite ($2MgO{\cdot}2Al_2O_3{\cdot}5SiO_2$) and mullite ($3Al_2O_3{\cdot}2SiO_2$). Figure 3 (a and b) shows the diffractograms of PC and MC at room temperature, dried at 110°C and calcined 2 h at 1100 and 1350°C. In PC at room temperature (20°C), mullite and cordierite peaks of the aggregates and alumina were observed as principal phases. Also low peaks of magnesia (MgO) were detected. The MgO present was lower than the original composition formulated, showing that it reacted to form non crystalline phosphates. Small amounts of newberyite ($MgHPO_4{\cdot}3H_2O$) and variscite ($AlPO_4{\cdot}2H_2O$) were detected but both phases disappeared by heating. At low temperatures, berlinite (b-$AlPO_4$) appeared as a minor phase. The diffractogram at 1100°C showed a large peak of d=0.408 nm corresponding to the cristoballite form of the aluminum orthophosphate (c-$AlPO_4$). Tridymite form (t-$AlPO_4$) was also present. At 1350°C the phases cordierite and mullite increased owing to the matrix transformation and the alumina was totally consumed. Light excess of MgO and traces of $AlPO_4$ were detected. In MC castables the same phases were observed as in PC, but berlinite was only detected at room temperature.

In TC material at 1350°C, cordierite and mullite as main phases and spinel ($MgO{\cdot}Al_2O_3$), alumina (Al_2O_3) and quartz (SiO_2) as minor additional phases appeared.

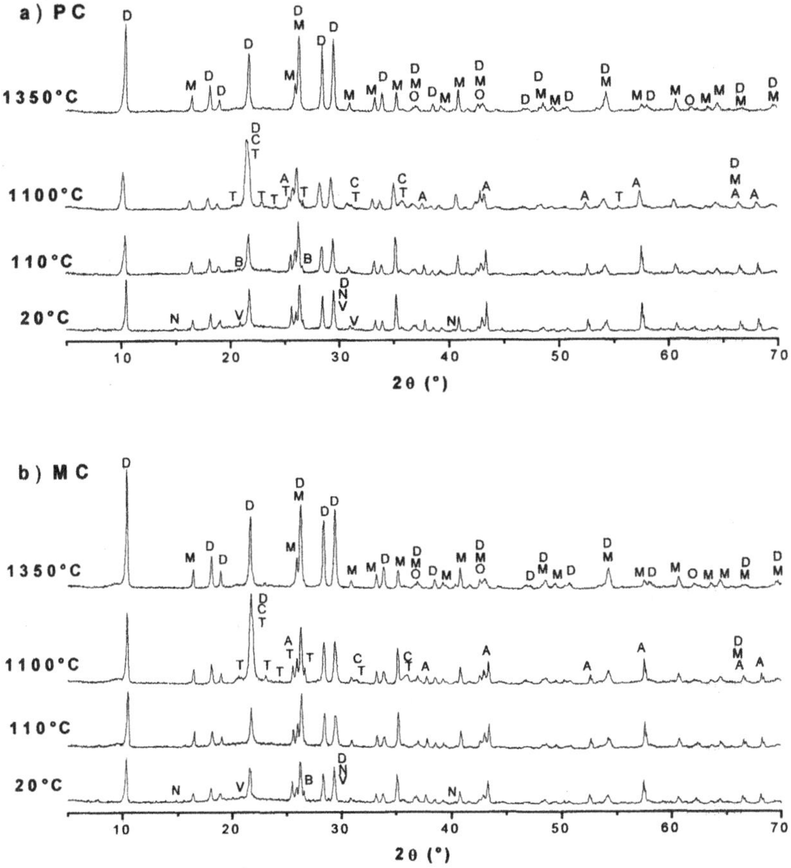

Figure 3. XRD at different temperatures of a) PC and b) MC castables. Phases: D=cordierite, M=mullite, O=MgO, A=Al_2O_3, B=b-$AlPO_4$, C=c-$AlPO_4$, T=t-$AlPO_4$, N=$MgHPO_4 \cdot 3H_2O$, V=$AlPO_4 \cdot 2H_2O$

Some properties of the materials at 1350°C are shown in Table II. Castables calcined for 2 h at 1350°C had an apparent density near 1.70 g/cm^3 and an open porosity of about 30 %. The low density permits to consider them as lightweight refractories with good insulating properties. PC and MC showed very similar and lower thermal expansion coefficients (α) than the TC material. Values near 1.0 10^{-6} (°C^{-1}) of thermal expansion indicated that the materials behave as a pure cordierite instead of a cordierite-mullite composite.

Table II. Density, porosity and thermal expansion coefficient of the materials

	Apparent density (g/cm^3)	Open porosity (%)	α 25–1000°C (°C^{-1})
PC	1.74	29.40	0.90×10^{-6}
MC	1.77	29.35	0.97×10^{-6}
TC	1.81	32.40	1.29×10^{-6}

The mechanical strength (MOR) behavior was similar for PC and MC castables. The flexural strengths of these materials at room temperature, after chemical setting were near 2.6 MPa. Figure 4 shows the MOR values obtained for the castables calcined at different temperatures. Strengths increased with temperature, reaching a maximum near 10 MPa at 1350°C.

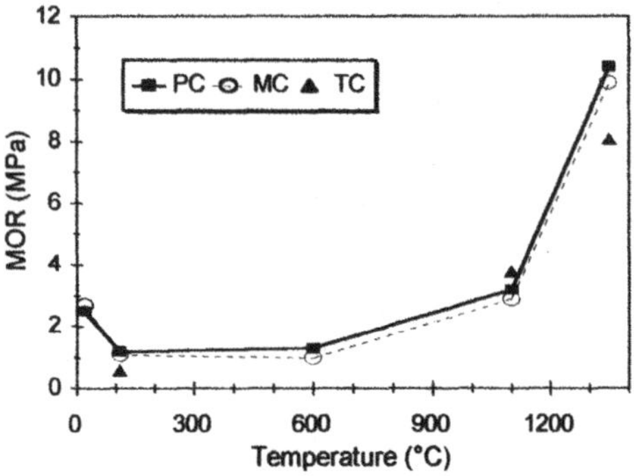

Figure 4. Flexural strength (MOR) of PC and MC castables treated at different temperatures

For both materials, a MOR decrease was observed from the air set material to that treated at 110°C. The analysis of the matrix at these temperatures, using XRD, showed that the peaks of crystalline phosphate hydrate compounds ($MgHPO_4 \cdot 3H_2O$ newberyite and $MgAl_2(PO_4)_2(OH)_2 \cdot 8H_2O$ gordonite) disappeared at 110°C. This fact proved that these compounds are responsible for the chemical setting of the castables giving an important contribution to strength at room temperature.

Increasing the temperature a heat setting of the material occurs through the formation of stable phosphate compounds (mainly $AlPO_4$) with an increasing of the MOR values as is observed for commercial heat setting refractories.

The TC castable had a MOR of 0.5 MPa at 110°C, then it increased and reached near 8 MPa at 1350°C.

The MOR values obtained for these materials would be related to the low density and high porosity of the aggregate grains (cordierite-mullite) employed. If low porosity or dense aggregates are used, the mechanical strength will increase. Most of the water used to reach consistency for application is employed to fill the pores and causes a decrease in mechanical strength.

The materials fired for 2 h at 1350°C were evaluated with a thermal shock resistance test. The method employed is based on the loss of flexural strength after a thermal shock as follows: heating of the probes at 1000°C and then a rapid cooling by water immersion at 20°C ($\Delta T \cong 1000°C$). This operation can be repeated on the probe as heating-cooling cycles. After these operations the modulus of rupture is measured on the probes. The report of this test is usually the ratio between the MOR of the original material without thermal shock (MOR_i) and the MOR after (MOR_f). Probes of PC, MC and TC castables obtained at 1350°C were submitted to three cycles of heating-cooling operations described above. In Table III the results of this test are shown.

Table III. Thermal shock test on the castables obtained at 1350°C.

	MOR_i (MPa)	MOR_f (MPa)	Ratio MOR_i / MOR_f
PC	10.4	2.8	3.7
MC	9.9	3.5	2.8
TC	8.1	5.4	1.5

The castables retained considerable strength after the thermal cycles. The TC castable showed the best thermal shock resistance compared with PC and MC. TC had a low original modulus, high porosity and lower glassy phases than PC and MC. The phosphate materials at 1350°C had lower thermal expansion coefficients but high glass content, indicating that a calcination temperature of 1350°C seems to be too high for this kind of materials.

CONCLUSIONS

Cold setting cordierite castables were developed employing the reaction between PH and MAP with magnesium oxide. The phases responsible for the chemical setting are mainly magnesium phosphate hydrates.

These castables have a plastic or trowellable application consistency and harden within five hours. Low linear change was observed during heating.

The matrix forms cordierite upon thermal treatment at 1350°C. At this treatment temperature the thermal expansion coefficient of the materials is near $1.0\ 10^{-6}$ ($°C^{-1}$). No significant differences on the mechanical and thermal properties were observed among materials performed with PH or MAP.

The flexural strengths increase with temperature, reaching a maximum near 10 MPa at 1350°C and having an acceptable thermal shock resistance.

REFERENCES

1] E.F. Osborn and A. Muan, "Ternary Systems"; pp.115 in *Introduction to Phase Equilibria in Ceramics*, 2nd ed. Edited by C.G. Bergeron and S.H. Risbud. The American Ceramic Society, Inc. Columbus, Ohio, 1984.

2] K.A. Gebler and H.R. Wisely, "Dense Cordierite Bodies", *Journal of the American Ceramic Society*, **32** [5] 163-165 (1949).

3] C.A. Sorrel, "Reaction Sequence and Structural Changes in Cordierite Refractories", *Journal of the American Ceramic Society*, **43** [7] 337-343 (1960).

4] J.L. Mendoza, S. Widjaja and R.E. Moore, "Evaluation of Cordierite-Mullite-Bonded Kiln Furniture Refractory Saggers"; pp. 184-204 in *Ceramics Transactions*, vol.4. Edited by R.E. Fisher. The American Ceramic Society, Inc. Westerville, Ohio, 1989.

5] Z. Chen, "Cordierite Self Flow Castables for Kiln Furniture", *China's Refractories*, **8** [2] 9-13 (1999).

6] W.D. Kingery, "Fundamental Study of Phosphate Bonding in Refractories: I-III", *Journal of the American Ceramic Society*, **33** [8] 239-250 (1950).

7] J.E. Cassidy, "Phosphate Bonding Then and Now", *Ceramic Bulletin*, **56** [7] 640-643 (1977).

8] F.J. Gonzalez and J.W. Halloran, "Reaction of Orthophosphoric Acid with Several Forms of Aluminum Oxide", *Ceramic Bulletin*, **59** [7] 727-731 (1980).

9] A.K. Sarkar, "Phosphate Cement-Based Fast-Setting Binders", *Ceramic Bulletin*, **69** [2] 234-238 (1990).

10] S.S. Seehra, (Ms.) Saroj.Gupta and Satander Kumar, "Rapid Setting Magnesium Phosphate Cement for Quick Repair of Concrete Pavements. Characterisation and Durability Aspects", *Cement and Concrete Research*, **23** [2] 254-266 (1993).

11] B.E.I. Abdelrazig, J.H. Sharp and B. El-Jazairi, "The Chemical Composition of Mortars Made From Magnesia-Phosphate Cement", *Cement and Concrete Research*, **18** [3] 415-425 (1988).

12] B.E.I. Abdelrazig, J.H. Sharp and B. El-Jazairi, "The Microstructure and Mechanical Properties of Mortars Made From Magnesia-Phosphate Cement", *Cement and Concrete Research*, **19** [2] 247-258 (1989).

13] T. Sugama and L.E. Kukacka, "Characteristics of Magnesium Polyphosphate Cements Derived From Ammonium Polyphosphate Solutions", *Cement and Concrete Research*, **13** [4] 499-506 (1983).

14] T. Sugama and L.E. Kukacka, "Magnesium Monophosphate Cements Derived From Diammonium Phosphate Solutions", *Cement and Concrete Research*, **13** [3] 407-416 (1983).

DIFFERENT TYPES OF *IN SITU* REFRACTORIES

W E Lee, S Zhang and H Sarpoolaky
University of Sheffield,
Dept. of Engineering Materials,
Mappin St, Sheffield, S1 3JD, UK.

ABSTRACT

The use of refractories formed *in situ* to resist corrosion is widespread throughout the materials processing industries. This paper aims to define exactly what is meant by the term "*in situ* refractories" and to describe the various types arising in practice.

INTRODUCTION

Refractories invariably undergo some changes in use due to the high service temperatures pushing them towards equilibrium. This is particularly true when refractories are in contact with liquids and vapours as occurs in most materials processing, such as iron and steel making and glass production, and in waste incineration. A surprisingly large proportion of refractories also contain liquid and vapour phases formed under service conditions. Consequently, what is used to line a furnace is not what is present during use. These changes, which occur once the refractory is installed, lead to production of what have come to be termed *in situ* refractories. Since refractories are designed to give the longest campaign possible it is important that the in service changes are understood and controlled. A recent review[1] described the evolution of *in situ* refractories in the twentieth century and suggested that several types of *in situ* refractories can form. The purpose of this paper is to give a clear definition of what is meant by the term *in situ* refractories and to specify possible different types of *in situ* refractories.

DEFINITION

In situ refractories may be defined as the in use product(s) of reaction within a refractory system or between the refractory and furnace contents leading to improved refractory behaviour.

DIFFERENT TYPES OF *IN SITU* REFRACTORIES

For simplicity 4 types of *in situ* refractories will be described but, as shall become apparent, there is the possibility of extensive overlap between these groups. Also, there may be additional types of which the authors are unaware.

Type I

Type I *in situ* refractories are those arising due to reactions solely within the components of the brick or monolith without any external contribution (Figure 1).

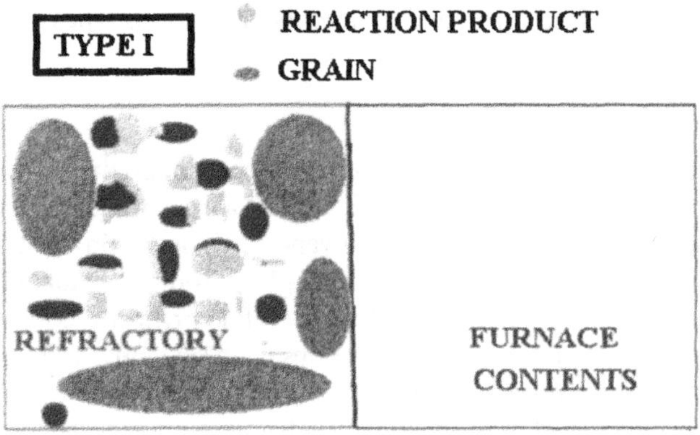

Figure 1. Schematic of Type I *in situ* refractories.

These arise from reactions between refractories components due to the high temperature and so may occur e.g. during heat up of a refractory system. Included in Type I are those *in situ* refractories in which useful phases are generated throughout the matrix often during production/installation. Fine powders are often used in the batch which (due to their high surface area) react to form desired phases. The most common bond phases formed in Type I *in situ* refractories are spinels (of various compositions), forsterite and mullite[2]. Often their formation leads to a direct bond between the individual grains of the refractory aggregate system such as in $MgCr_2O_4$ spinel bonded mag-chrome bricks[3]. Critical features of the reactions leading to Type I *in situ* refractories are the volume changes associated with the reactions[2] and the competition between this volume change and other temperature-induced microstructural changes such as thermal expansions of other phases present and shrinkage contraction arising from liquid phase or solid state sintering. Beneficial behaviour may result if these lead to a tightening of the texture so improving resistance to liquid penetration and strength. The morphology of the product is also important. E.g. formation of acicular or tabular mullite or calcium hexaluminate (CA_6) may open up a microstructure and have an adverse affect on thermo-mechanical properties.

A good example of a Type I refractory is self-formed spinel in castables arising from reaction of fine, matrix additions of alumina and magnesia on firing often in the presence of silica and calcium aluminate cement[4]. The volume increase associated with the spinel formation may tighten the refractories texture helping to resist liquid penetration. Calcium aluminates such as CA_6 also form in the matrix of such systems from reaction of the cement and alumina components. The tabular CA_6 crystals formed are believed to link other phases (such as tabular grain and spinel), interlocking the microstructure and improving strength[5,6].

Type II

Type II includes those *in situ* refractories in which reactions occur within the refractory but which may be assisted by reaction with the (liquid or vapour) furnace contents (Figure 2).

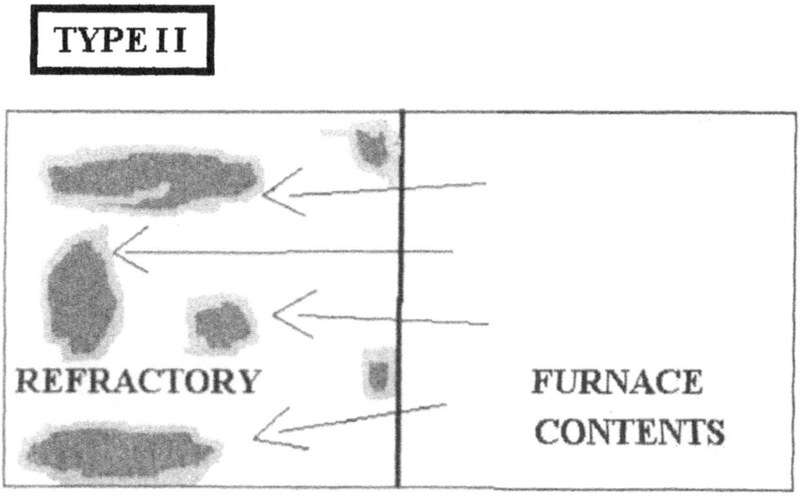

Figure 2. Schematic of Type II *in situ* refractories.

These include formation of dense layers within the refractory adjacent the contents (such as fine MgO in BOS vessels via the MgO-C reaction aided by availability of oxygen from the slag or atmosphere first seen in 1967[7]). Also, included in this category are the MgO-C bricks containing ceramic additives such as B_4C which react to form ceramic/glass phases at high temperature improving oxidation resistance[8]. Similarly, MgO-C bricks with metal additives such as Al, Si, Mg and alloys[e.g.9,10] can be included here since they may react with gases from the atmosphere to form ceramic phases which beneficially getter oxygen and interlink other phases so improving hot strength. The ceramic phases which form are a strong function of the local conditions at the metal location in the brick. The morphologies of the phases formed often indicate their formation mechanism. In Figure 3 two forms of SiO_2 occur. Silica from direct oxidation of silicon metal is present as a shell on the original metal particle. Silica formed indirectly via SiO vapour e.g.

$$Si_s + C_s \longrightarrow SiC_s\ (+ CO_v) \longrightarrow SiO_v + CO_v \longrightarrow SiO_{2s} + C_s$$
$$Si_s + CO_v \longrightarrow SiC_s\ (+ CO_v) \longrightarrow SiO_v + CO_v \longrightarrow SiO_{2s} + C_s$$

occurs as a rim around the original location of the Si particle.

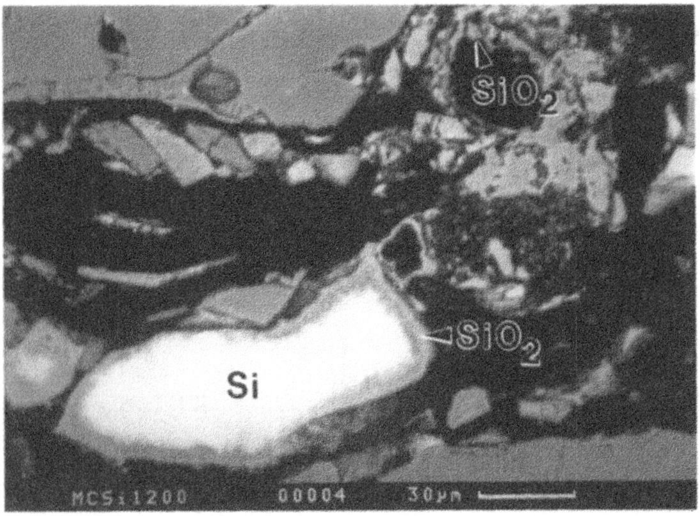

Figure 3. Direct and indirect silica formed in MgO-C brick with added Si at 1200°C.

As with Type I the critical features of Type II *in situ* refractories include the properties of the phases formed. Consequently, while oxidation of additive B_4C to give B_2O_3 rich glass improves oxidation resistance it leads to poorer high-temperature strength since the glass becomes fluid. Often the phase formed covers the grain phase leading to protective coatings which may e.g. be spinel or forsterite (Figure 4).

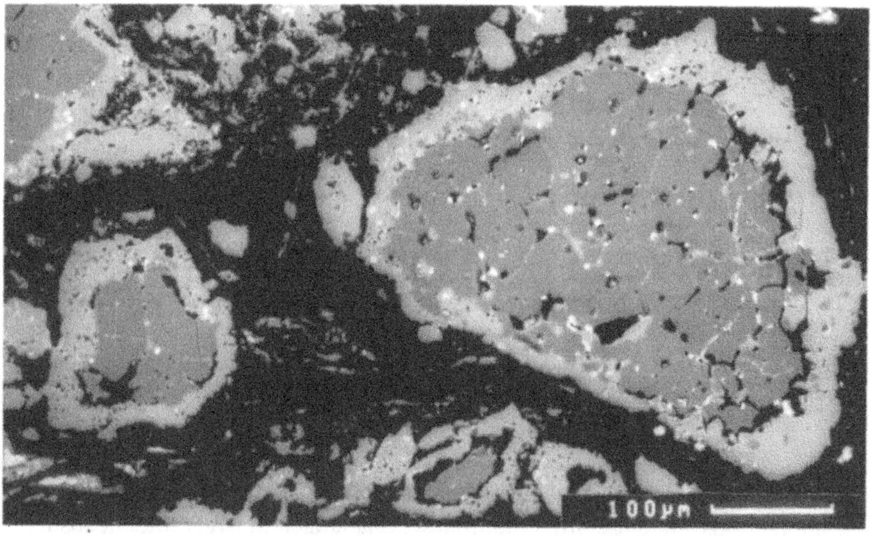

Figure 4. Forsterite (M_2S) coating on MgO grain in MgO-C brick with added Si at 1600°C (courtesy R. Artir, University of Sakarya, Turkey)

Type III

Type III includes those refractories which react with the furnace contents generating a protective interlayer between the refractory and furnace contents (Figure 4).

Often the interlayer forms as a result of indirect corrosion of the refractory solid by penetrating liquid[11]. If the interlayer passivates further attack it is a Type II *in situ* refractory. Commercial examples of such Type II interlayer formation include clinker coatings in the burning zones of cement kilns and viscous glass layers adjacent Al_2O_3-ZrO_2-SiO_2 blocks in glass tanks.

Significant properties of liquid interlayers formed in Type III *in situ* refractories include their viscosity and the solubility of adjacent solid phases in them. In AZS refractories the liquid formed is rich in Al_2O_3 which makes it viscous and so it acts as a barrier between the refractory and the fluid soda-lime-silica glass in the tank. The solid ZrO_2 in adjacent refractory is also slow to dissolve in the alumina-rich glass interlayer.

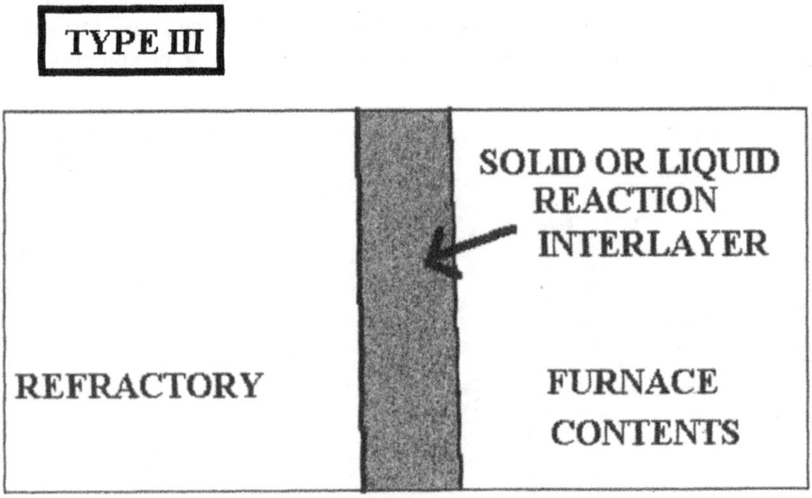

Figure 5. Schematic of Type III *in situ* refractories.

Solid *in situ* refractory interlayers are observed to form in many refractory-slag interactions[11] and are often multiphase. Slag attack of MgO often leads to formation of a magnesiowustite boundary layer in Fe-rich slags[12] and a spinel layer in alumina or chromia rich slags. Complex spinel and calcium aluminate phases (CA_2, CA_6 and gehlenite, C_2AS) often form between alumina and slag[13]. Critical features of such interlayers include whether they are complete and fully cover the (often) aggregate or grain phases being attacked. Incomplete coverage leaves gaps through which aggressive liquid can penetrate. The morphology of the phases in the interlayer is also important. Acicular or tabular phases such as CA_6 or mullite may grow into and penetrate the grain phase, opening up the interface to further attack. Thermal expansion mismatch between the interlayer and other solid phases present may lead to stress and crack formation although the presence of liquid may often alleviate this problem.

The mechanisms by which solid Type III interlayers may passivate further attack vary. Open crystal structures, such as spinel or rock salt (MgO), can accommodate cations such as Fe, Mn, Mg and Cr so removing these cations from the adjacent local liquid rendering it more viscous and less penetrating. Zhang *et al.*[13] observed a change in composition of spinel from Mg aluminate adjacent alumina being corroded to Mg, Al, Fe, Cr, Mn spinel into the slag. This would suggest that the Mg aluminate spinel acted as a nucleus and as the spinel grew into the slag it incorporated the various cations, rendering the local slag more viscous. Another mechanism is when complete layers of refractory phases (such as CA_6) simply act as a prophylactic coating.

A subtle distinction needs to be made here between Type II and Type III *in situ* refractories. As discussed above the formation of an interlayer between slag and solid grain in a crucible-type slag test is Type III. However, if the slag has penetrated the matrix of a refractory brick and then attacks the grain in an identical manner this is Type II, since now reactions occur within the refractory but assisted by reaction with the liquid slag from the furnace contents.

Type IV

Finally, Type IV are those in which the furnace contents are deposited on the refractories to themselves act as the refractory such as slags splashed onto BOS vessel walls.

In BOS vessels the nature of the bond between the slag and refractory is as yet not well defined. If a reaction occurs between the splashed slag and the refractory lining then depending on the nature of the interlayer formed this may introduce some Type II or Type III component. Refractory systems are often designed to freeze liquid either at the refractory surface or at a specific location in the brick at a specified isotherm. Air or water cooling in high thermal conductivity refractories may be used to control the location of the isotherm. In carbon blocks in iron blast furnaces the important isotherm is that at 1150°C (the temperature at which iron + 4% C freezes). In Al reduction cells the 800-850°C isotherms (at which the eutectic bath components freeze) are important. If the slag is frozen at the surface this is a Type IV *in situ* refractory whereas if it is internal to the brick it is Type II.

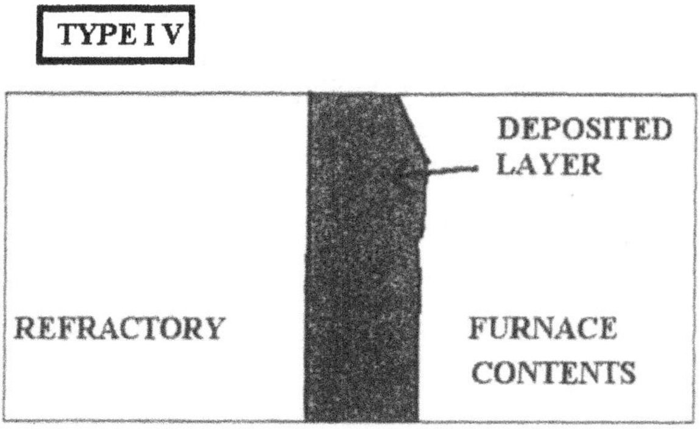

Figure 6. Schematic of Type IV *in situ* refractories.

REFERENCES

[1]WE Lee and RE Moore, "The Evolution of *in situ* Refractories in the 20th Century," J. Am. Ceram. Soc. 81 [6] 1385-1410 (1998).

[2]FN Cunha and RC Bradt, "Reactions of Constituents for *in situ* Bonds of $MgAl_2O_4$, Mg_2SiO_4 and $3Al_2O_3$. $2SiO_2$ in Refractories," pp. 143-152 in proc. 57th Electric Furnace Conference (Pittsburgh, Pa, USA, Iron and Steel Society 1999).

[3]K Goto and WE Lee "The "Direct Bond" in Magnesia Chromite and Magnesia Spinel Refractories". J Am Ceram Soc 78 [7] 1753-60 (1995).

[4]M Fuhrer, A Hey and WE Lee, "Microstructural Evolution in Self-forming Spinel/Calcium Aluminate-Bonded Castable Refractories," J. Euro. Ceram. Soc. 18 [7] 813-820 (1998).

[5]MW Vance, GW Kriechbaum, RA Henrichsen, G MacZura, KJ Moody and S Munding, "Influence of Spinel Additives on High Alumina/spinel Castables," Bull. Am. Ceram. Soc. 73 70-74 (1994).

[6]YC Ko and CF Chan, "Effect of Spinel Content on Hot Strength of Alumina-spinel Castables in the Temperature Range 1000-1500°C," J. Euro. Ceram. Soc. 19 2633-2639 (1999).

[7]RH Herron, CR Beechan and RC Padfield, "Slag Attack on Carbon-Bearing Basic Refractories," Bull. Am. Ceram. Soc. 46 1163-68 (1967).

[8]CF Chan, BB Argent and WE Lee, "Influence of Additives on Slag Resistance of Al_2O_3-SiO_2-SiC-C Refractory Bond Phases Under Reducing Atmosphere," J. Am. Ceram. Soc. 81 [12] 3177-88 (1998).

[9]A Yamaguchi, "Behaviors of SiC and Al Added to Carbon-Containing Refractories," Taikabutsu Overseas 4 [3] 14-18 (1984).

[10]A Watanabe, H Takahashi, S Takanaga, N Goto, K Anan and M Uchida, "Behavior of Different Metals Added to MgO-C Bricks," Taikabutsu Overseas 7 [2] 17-23 (1987).

[11]WE Lee and S Zhang, "Melt Corrosion of Oxide and Oxide-Carbon Refractories," Int. Mater. Reviews 44 [3] 77-105 (1999).

[12]S Zhang, H Sarpoolaky, NJ Marriott and WE Lee, "Penetration and Corrosion of Magnesia Grain by Silicate Slags," Brit. Ceram. Trans. (2000).

[13]S Zhang, HR Rezaie, H Sarpoolaky and WE Lee, "Alumina Dissolution into Silicate Slag," J. Am. Ceram. Soc. 83 [4] (2000).

THE USE OF MODELING IN REFRACTORIES

Charles E. Semler
Semler Materials Services
4213 N. Placita de Susana
Tucson, Arizona 85718

ABSTRACT

The advancement of refractories technology has proceeded continuously since the late 1800's, based mainly on empirical studies until about the 1960's, when the availability of more sophisticated methods began to increase. This paper provides a general overview, and examples, of the increasing use of sophisticated technical methods to enhance the development of new and improved industrial refractories. As computerization advances and the need for better refractories continues, the capabilities and use of technical analysis will certainly increase, with a wide range of practical and technical benefits.

INTRODUCTION

Two general definitions of "modeling" from www.dictionary.com are:

a. The use of a model(s) to promote a product(s).
b. The act/art of sculpting or forming a pliable material such as clay or wax.

But the following more specific definition of "modeling" is an attempt to better describe the meaning of the word, as related to the field of refractories:

Modeling involves the use of theory/equations, computation, and/or simulation to observe, quantify, or predict the effect of a design, an assumption, a change(s), etc., on a hypothesis, a material, a lining, an operating system, etc., to expand the practical and/or technical understanding, and hopefully gain technical, economic, or other benefits.

Modeling studies are not a recent development, although the technology and sophistication have increased markedly in the last ten years. Historically, Winkelman and Schott (1894)[1] developed an equation to evaluate/compare the spalling tendency of glass compositions, based on physical properties, as follows:

$$\text{Spalling Tendency} = t\,(k)^{1/2} \,/\, E\,(p\,c)^{1/2}$$

Where:

t = Tensile Strength
k = Thermal Conductivity
E = Modulus of Elasticity
p = Density
c = Specific Heat

F. H. Norton [2] used the following equation to evaluate the relative thermal shock resistance of various refractory products, based on selected properties, including a measured thermomechanical value:

$$\text{Thermal Shock Factor} = \Delta \,/\, \Phi\, h \times 10^{-7}$$

Where:

Δ = Thermal expansion
Φ = Shear strain in torsion, @ 500C
h = Thermal diffusivity

Thereafter an increasing number of modeling studies followed, although the usage in refractories was slow to develop. A few examples of some early modeling studies on topics in ceramics are listed below, by title:

1955 W.D. Kingery[3] – Factors affecting the thermal stress resistance of ceramics
1956 P. Michaels[4] –Currents in a continuous glass furnace by small-scale models
1961 R.L. Coble[5] – Experimental test of diffusion models in powder compacts
1961 R.E. Carter[6] – Kinetic model for solid-state reactions

Since roughly the 1960's, the use of modeling in refractories has progressively increased, based on the derivation and refinement of the applicable theory and equations that define the fundamental characteristics of refractories. For example, the July 1985 American Ceramic Society Bulletin included five papers on the subject of thermo-mechanical characterization and modeling of refractories. And a 1995 book by C.A. Schacht[7] has provided a thorough discussion of the basic fundamentals involved in the design and analysis of refractory linings, and the development of models to assess the effect of different materials and configurations, including zoning of linings. The number of sessions and papers dealing with modeling and computerized analysis of refractories has progressively increased at the UNITECR (Intl. Refractories Congress) meetings in 1995, 1997, and 1999.

DISCUSSION

The capabilities and use of modeling, for a wide range of refractory applications, will continue to advance, especially because ever faster, more powerful computers, improved software, and better analytical/testing equipment will be available to enhance modeling studies, statistical correlation, finite element analysis, and other techniques. There will be significant benefits for both refractory manufacturers and users.

For example, thermal shock resistance and permeability are two refractory properties, the determination and comparison of which can be expected to change. It is now possible to essentially eliminate thermal shock testing by using property-based calculations to measure and/or compare the thermal shock resistance of refractories, but it will take time to overcome the historical practices and reach this goal. Over the years, a number of thermal stress resistance factors (R factors) have been defined, such as those listed below, which are applicable for various material shapes and boundary conditions:

$$R = \sigma (1 - \mu) / E \alpha$$
$$R' = [k \sigma (1 - \mu)] / E \alpha$$
$$R'' = [\sigma (1 - \mu) / E \alpha] k / \rho c$$
$$R''' = E / \alpha (1 - \mu)$$
$$R'''' = \gamma E / \alpha^2 (1 - \mu)$$
$$R_{st} = (\gamma / \alpha^2 E)^{1/2}$$

Where:

σ = Tensile strength
μ = Poisson's ratio
E = Modulus of elasticity
α = Thermal expansion
k = Thermal conductivity
$k/\rho c$ = Thermal diffusivity
γ = Work of fracture

Numerous studies have been published showing excellent statistical correlation between thermal shock damage of refractories and a pertinent property-based factor. For example, Whittemore and Pavlica[8] reported good correlation between Rst and thermal shock damage, for high alumina refractories, in both the prism and ribbon tests. Their study, and others, have provided experimental evidence that R'''', Rst, or other factors, can be used to evaluate and compare/model the thermal shock characteristics of high alumina refractories, for various conditions, without doing thermal shock tests. And it is likely that the accuracy of property-based evaluation/comparison of thermal shock resistance could be further improved by including findings and data from R-curve (rate of crack growth) testing in the basic theory and applicable equations.

Over the last fifteen years the development and use of castable refractories has increased markedly. Because of the practical and economic importance of curing and drying of castables after placement, their permeability has increased in importance. A recent paper by V. Pandolfelli, et al.[9] indicates that Darcy's Law, which has long been the basis for the measurement of permeability in refractories (e.g. ASTM C-577), actually has only a limited range of applicability. Forchheimer's equation has been found to provide a better fit with experimental data. Based on this finding, and ongoing research at the University of Missouri-Rolla[10], which has included development of a permeameter and modeling of castable dewatering, it is expected that it will be possible to determine a prescription curing and dryout treatment for each castable in the near future, instead of using the generic and arbitrary heat/hold curves that have been used for decades.

The use of finite element analysis (FEA) has become commonplace for refractory manufacturers, users, contractors, and others, to evaluate the materials and design for linings that are in use or proposed for a wide range of applications throughout industry. For example, at UNITECR'95 (Intl. Refractories Congress) in Japan, FEA studies were reported for carbon baking furnace walls, basic oxygen furnace linings, ceramic anchors, and carbon-bonded refractory structures.

It is significant that researchers have recognized the existence and importance of nonlinear mechanical behavior in refractories, whereby there is increased need for valid high temperature property data, i.e., the use of property values that fully account for the effects of, and changes due to exposure (temperature, sintering, pressure, chemical change, time, etc.) instead of general, single- point property values from a materials reference compilation.

Two examples of the many successful FEA studies, from the steel and copper industries, are summarized below:

1. The refractory lining requirements were evaluated and optimized for obround ladles used to process/transport molten steel[11]. Using room- and high-temperature properties for three types of MgO-C bricks, it was determined that:

a. Stiff resin-bonded MgO-C bricks should be used in the elongate section of the slagline, and more flexible pitch/resin-bonded MgO-C bricks should be used in the balance of the ladle lining.
b. The lining should be preheated slowly initially.
c. The lining hot face temperature should be maintained above 1038°C (1900°F), throughout it's life.
d. A vertical retention system is necessary to maintain tight horizontal joints.

2. A sophisticated mathematical modeling study[12] has resulted in increased refractory lining life in the Mitsubishi converting furnace (blister copper). The authors defined three main refractory wear mechanisms: (a) the hearth is affected by top-blown powder injection by lance, (b) splashing of copper/slag droplets contributes to sidewall wear, and (c) the melt line is affected by surface waves in the molten bath. Each of these mechanisms was modeled mathematically, and the information derived therefrom provided materials guidelines that were used in developing improved refractories for each of the regions in the converting furnace lining. And their research is continuing, so there will probably be further optimization of the refractories and increased lining life.

It can be expected that the use, and significance of modeling studies in refractories will continue to increase based on improvement and innovation of pertinent theory, computerization improvements, new and improved tests/instruments, expanded property data base, and further use of developing technologies. A few examples of applicable developing technologies, and their use for refractory/ceramic applications, are briefly discussed below:

Fractal Analysis – Has been used to measure the structure change caused in MgO-Cr_2O_3 refractories by thermal shock cycling, allowing a correlation to be established between pore structure and properties[13]. Further generation of such information will help to clarify and define the factors that need to be controlled to reduce or minimize the damaging effects of thermal shock exposure.

Neural Network – FEA combined with the neural network concept has been used to model inhomogeneity and nonlinearity in refractories[14], due to chemical reactions, sintering, and progressively changing thermal expansion; for example, the changes associated with secondary spinel formation in bricks and castables.

Wavelet and Fourier Transformations – The mechanical properties of refractories depend upon the microstructure, and the crack path through different microstructures has been found to correspond with wavelet and Fourier transformations[15].

Voronoi Construction – A spatially local Voronoi construction was used[16] to identify potential sites for impurity segregation in alumina, and a complementary energetic analysis for various isovalent dopants showed that strain energy associated with the ionic size mismatch provided a driving force for segregation.

A rapidly growing technical field is computational materials science, with new programs at many universities and government agencies, which will further promote and enhance the capabilities and opportunities for advancement of refractories development and technology. For example, NIST (National Inst. for Science and Technology) has a Center for Theoretical/Computational Materials Science, with projects on finite element modeling of composites and models of solidification processes. In addition, NIST has begun to simulate the flow of Portland cement-based concrete using computer modeling. This technique should also be used to evaluate the water requirement/flow characteristics of refractory castables with calcium aluminate or other bonds to provide valuable insight into the critical factors that control this variable.

SUMMARY

Modeling studies have been used for ceramic materials for more than 100 years. And although a correlation/modeling approach was used in the evaluation of thermal shock resistance of refractories in 1925, the improvement and innovation of refractories was based mostly on empirical studies until sometime after roughly the 1960's, when the use of more sophisticated techniques became more common. With the availability and advancement of computerization, the capabilities and applications for industrial refractories have increased markedly. And this trend will certainly continue, with the advancement of the scientific and technical capabilities, and the ongoing need for new and improved refractories that are more cost effective. It will be necessary for the fundamental principles and equations to be reviewed and updated, along with the development of new theories/equations, based on the availability of improved testing and analytical facilities that will support continued advancement. To enhance the significance and use of modeling studies in refractories, it will be important to (a) expand the property database, especially including more high temperature data, and (b) properly interpret the results based on appropriate experience.

REFERENCES

[1] A. Winkelman and O. Schott, Ann. Physik. Chem., 51, 730 (1894)
[2] F.H. Norton, "A General Theory of Spalling", Jour. Amer. Cer. Soc, 8 (1) 29-39 (1925)
[3] W.D. Kingery, "Factors Affecting Thermal Stress Resistance of Ceramic Materials", ibid, 38 (1) 3-15 (1955)
[4] P. Michaels, "Representation of Currents in Continuous Tank Furnaces by Means of Small Scale Models", Jour. Soc. Glass Technol., 40 (196) 470-81 (1956)
[5] R. L. Coble, "Sintering Crystalline Solids: I, Intermediate and Final State Diffusion Models", Jour. Appl. Phys., 32 (5) 793-99 (1961)
[6] R. E. Carter, "Kinetic Model for Solid-State Reactions", Jour. Chem. Phys., 34 (6) 2010-15 (1961)
[7] C.A. Schacht, "Refractory Linings-Thermomechanical Design and Applications", Marcel Dekker, Inc., New York, 1995, 504 p.
[8] D. Whittemore and S. Pavlica, UNITECR Proc., 1995 (Vol. III, p. 323) and 1997 (Vol. II, p. 909)
[9] V. Pandolfelli, et al., "How Accurate is Darcy's Law for Refractories?", Bull. Amer. Cer. Soc., 78 (11) 64 (1999).
[10] M. Velez, et al., "Computer Simulation of the Dewatering of Refractory Concrete Walls", Jour. Tech. Assn. Refr. Japan, 20 (1) 5-9 (2000)
[11] R. Russell, G. Hallum, and E. Chen, "Thermomechanical Studies of Obround Ladles During Preheating and Use", Iron & Steelmaker, June, 37-43 (1993)
[12] T. Tanaka, et al., "Technology for Decreasing Refractory Wear in the Mitsubishi Process: Fundamental Research and It's Application", Proceedings, Vol. VI, Copper'99, Phoenix, AZ, 171-186 (1999)
[13] J. Nakao, et al., "Textural Change of Magnesia-Chromite Refractories after Thermal Shock Testing, and It's Quantitative Evaluation", Taikabutsu Overseas, 16 (2) 11-16 (1996)
[14] M. Sugawara, et al., "Application of a Neural Network Concept to Analyzing the Expansion Behavior of a Monolithic Refractory Lining", Taikabutsu Overseas, 19 (1) 3 (1999)
[15] K. Yasuda, et al., "Wavelet and Fourier Transformations for Crack Paths in Particle-Dispersed Ceramic Composites", Taikabutsu Overseas, 17 (4) 87 (1997)
[16] J. Cho, et al., "Modeling of Grain-Boundary Segregation Behavior in Aluminum Oxide", Jour. Amer. Cer. Soc., 83 (2) 344-52 (2000)

BATH PENETRATION OF BARRIER REFRACTORIES FOR ALUMINUM ELECTROLYTIC CELLS

George Oprea
University of British Columbia
Department of Metals and Materials Engineering
309-6350 Stores Road
Vancouver, BC, Canada V6T 1Z4

ABSTRACT

Cryolitic bath penetration and corrosion of refractories used as bottom lining in aluminum electrolytic cells are complex phenomena involving physical and chemical interactions at the solid-liquid interface. The average pore diameter rather than the open porosity, as texture parameters, and the nature of the bonding phase could be used for assessing the behavior of a dense refractory towards cryolite penetration. This paper correlates the structural characteristics of the refractory material with the molten bath properties and testing conditions. The corrosion mechanisms were also discussed for bricks, castables and dry granular barrier refractories. Although the basic principle is the same for testing them, there are specific properties in each group of products to be considered in evaluating their corrosion behavior when tested with cryolitic baths.

INTRODUCTION

The life of an aluminum electrolytic cell these days is expected to be 6-8 years. If an industrial trial is required in order to qualify refractories for use as barrier linings, the selection process becomes really slow. Laboratory tests [1-8] are available and used in the pre-selection stage, in order to decide what material to be installed in industrial trials. The interpretation of the test results is still a subject of many disputes due to the difficulties in accepting similarities with the real-life conditions of an electrolytic cell.

There are numerous published data regarding the corrosion process involving cryolitic bath, the Norwegian school [9-13] having a tremendous contribution in the last 12 years in clarifying aspects of the corrosion mechanisms.

Our work should be considered a contribution to the study at laboratory scale of the penetration of refractories by molten cryolitic bath and how the ceramic texture characteristics of the tested refractories could be correlated with their corrosion resistance.

PENETRATION AND CORROSION

Both penetration and corrosion of refractories used as bottom cathode lining in aluminum electrolytic cells are complex phenomena involving physical and chemical interactions at the solid-liquid interface. As in any typical static corrosion process, some physical conditions should be present first to enable the chemical reaction to take place at the interface. Consequently, the ceramic texture and the nature of the bonding phase, which is the first to be penetrated by the liquid, are determinant characteristics of the refractory [14]. Properties of the melt, especially viscosity and surface tension, also play an important role in establishing the contact of the two phases. The interfacial tension and the wetting behaviour of the couple refractory-melt determine the penetration of the melt through the open pores before any chemical reaction starts. The open pores and microcracks in the ceramic texture are the capillaries for the initial penetration of the molten bath into a refractory material. The penetration rate *dh/dt* of a liquid into a capillary can be expressed by Poiseuille's law:

$$\frac{dh}{dt} = \frac{d^2 \cdot \Delta p}{32\ \eta \cdot h} \quad (1)$$

where *d* is the capillary diameter, *Δp* the capillary sucking pressure, *η* the dynamic viscosity of the liquid, *h* the liquid penetration depth and *t* is the time. The pressure *Δp* can be expressed versus the liquid surface tension σ_{LV} and the wetting angle *θ*, as follows:

$$\Delta p = 4\sigma_{LV} \cdot \frac{\cos\vartheta}{d} \quad (2)$$

After eliminating Δp from the equations (1) and (2) and integrating, it results:

$$h^2 = \frac{d \cdot \sigma_{LV} \cdot \cos\vartheta}{4\eta} \quad (3)$$

This equation does not consider the temperature as a variable, although the surface tension, the viscosity of the melt and the wetting angle of the refractory-melt decreases with increasing the temperature. As a result, in isothermal conditions, when η, σ_{LV} and $cos\theta$ remain constant ($k = \sigma_{LV} cos\theta / 4\eta$), the depth of penetration depends only on the pore diameter *d* of the refractory material.

$$h^2 = k \cdot d \quad (4)$$

When there is a temperature gradient on the direction of penetration, the increase of the melt viscosity and surface tension will certainly slow down the penetration. This is why, for the bottom cathode lining of aluminum electrolytic cells, it is very important at what level the freezing isotherm is positioned in the refractory lining. A safe design will have it as high as possible, close to the melt. A thermally efficient design will have it far from the penetrating melt, but in this case, the safety of the lining relays mainly on the corrosion resistance of the refractory material used as barrier beneath the cathode.

PERMEABILITY AND POROSITY

The permeability and porosity can be correlated only if the porous medium consists of spherical particles of identical dimensions in an isotropic compact arrangement. In this case, the permeability depends on that particle dimension and implicitly, on the porosity created by that particular arrangement.

Although for the porous ceramics in general, there is no simple correlation between porosity and permeability, such a correlation can be used in simplified models to calculate the pore dimensions. For example, considering the "straight capillaric model" [15] of a ceramic texture, the flow per unit area of a fluid with viscosity *μ*, through a capillary of uniform diameter *d* is given by the law of Hagen-Poiseuille:

$$q = -\frac{n \cdot \pi \cdot d^4}{128\mu} \cdot \frac{dp}{dx} \quad (5)$$

where *dp/dx* is the pressure gradient along *n* capillaries of diameter *d* per unit area of cross section of the model.

The flow can also be expressed by Darcy's law:

$$q = -\frac{k}{\mu} \cdot \frac{dp}{dx} \quad (6)$$

where *k* is the "specific permeability".

It results that:

$$k = n \cdot \pi \cdot d^4 / 128 \tag{7}$$

If the total volume is $V=x \cdot A$, the pores volume V_p of the model will be:

$$V_p = n \cdot \frac{\pi \cdot d^2}{4} \cdot x \cdot A \tag{8}$$

and the porosity becomes:

$$P = \frac{V_p}{V} = \frac{n \cdot \pi \cdot d^2}{4} \tag{9}$$

Eliminating n between equations (7) and (9) yields the equation:

$$k = \frac{P \cdot d^2}{32} \tag{10}$$

If one considers that in a normal porous medium only one third of all the capillaries in the volume unit will participate to the flow on one direction, the permeability is lowered by a factor of 3 and equation (10) becomes:

$$k = \frac{P \cdot d^2}{96} \tag{11}$$

This equation does not correlate correctly the permeability with the porosity and the pore dimensions for all types of ceramic textures.

A simplified texture is also represented by the "serial type model". According to it, the n capillaries per unit area in each dimensional direction of a pore diameter d have the length s longer than the length of the model x, due to a very tortuous shape of each tubular pore.

Defining the tortuosity as the ratio between the length of the flow channel and the length of the porous medium ($T=s/x$) and using the Hagen-Poiseuille and Darcy's equations, the permeability becomes:

$$k = \frac{1}{96} \cdot \frac{P \cdot d^2}{T^2} \tag{12}$$

In other words, at the same pore dimension and porosity, the permeability decreases with increasing the tortuosity.

For manufactured ceramics, the shape of the pores and their dimensions depend on many factors, such as particle size distribution of the initial mix of solids, the shaping or compacting method used to obtain the green ceramic body, drying and firing conditions. The initial particle size certainly decides the average pore dimension, while the compacting method influences both the tortuosity and the pore dimension, for identical drying and firing conditions. One can also presume that two refractory bricks, made from identical raw materials, dried and fired in identical conditions, but one shaped by extrusion and the other one by dry pressing, would have very different permeabilities and porosities and, as a result, different pore dimensions. Extrusion, allowing for a finer particle size distribution, is expected to produce a brick with lower porosity and especially permeability than dry pressing.

On the other hand, for two bricks made from identical raw materials and shaped by the same method, but fired at different temperatures, the porosity and permeability will follow the same trend of variation versus temperature. For particular porous media, more elaborated models [16-18] would better express the flow, hence there will be a better correlation between permeability, porosity and pore dimension.

Kozeny's theory, using the "hydraulic radius" concept, explains the permeability as being conditioned by the geometrical properties of a porous medium, which is an assemblage of channels with various cross-sections, but with the same length. The permeability is expressed in terms of the specific surface S and porosity P of the porous medium and from Kozeny's and Darcy's flow equations it results:

$$k = c \cdot P^3 / S^2 \qquad (13)$$

where c is Kozeny's constant and has different values for different pore shapes (i.e. c=0.5 for circular). Although the specific surface is a measure of a properly defined hydraulic radius, Kozeny's theory, as well as the capillaric model theories, neglects the influence of conical flow in the constrictions and expansions of the flow channels.

For a given group of porous ceramics, such as the dense refractory bricks used as barrier lining in aluminum electrolytic cells, there are similarities in manufacturing and consequently, in their ceramic textures after firing. Based on that, if the permeability is measured by the same method, any correlation according to the above models would allow for a good comparison of their textures.

Correlations between permeability and porosity, using the capillary model equation (10) for two different ranges of pore dimensions, are presented in Figure 1a, and between permeability and pore dimensions at different porosities, in Figure 1b [19]. At pore diameters below 5μm the permeability levels are below 10 centidarcys, for porosities up to 25%. Due to this fact, the method of measuring the permeability plays an important role, in the sense that the detection limit should be in the range of milidarcys, in order to have a good resolution.

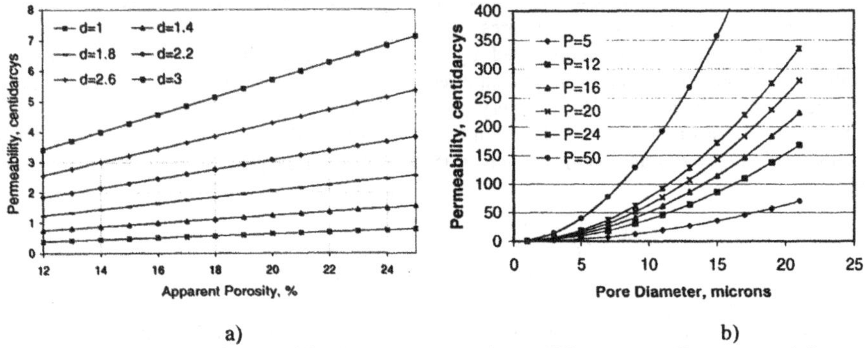

Figure 1. Permeability variation with (a) apparent porosity at different pore diameters and (b) pore diameter at different porosity values.

CRYOLITIC BATH

Chemistry: Being well known that the bath is the "heart of the Hall-Heroult process"[20], impressive improvements of the bath chemistry increased the current efficiency up to 94-96% for cell lines varying from 100 to 300kA. These achievements are the results of the permanent research work done especially by Reynolds, Alcoa and Pechiney and in addition, there are probably other companies that made efforts in developing their own bath chemistry.

There are three main groups of bath compositions in use at the present time:

- traditional or "classical " baths with 3-7% AlF_3
- modified baths with 2-4% AlF_3 , 2-4% LiF and 2-4% MgF_2
- low ratio baths with 8-14% AlF_3

A combination of the last two will give a so-called lithium-modified low ratio (LMLR). Tabereaux [21-22] reported that a 1% lithium fluoride modified low ratio bath would optimise the beneficial aspects and minimise the negative aspects. LiF in the bath reduces the energy consumption, improves the bath stability, but also reduces the current efficiency and increases the cell operational difficulties.

The bath additives added to molten cryolite (Na_3AlF_6) to modify the physico-chemical properties in order to improve the cell performance [23-25] are LiF, AlF_3, CaF_2 and MgF_2.
In order to express the basicity or acidity of the cryolitic bath [26], according to Lewis' theory of acidity and basicity, a cryolite ratio (CR) is used and can be defined as a molar (CMR) or weight (CWR) ratio. With cryolite ($3NaF.AlF_3$) as an example, CMR=3 and CWR=1/2CMR=1.5.
Each additive in the bath has its own contribution to the acidity of the bath [27] and as a result, the bath ratio for a LMLR for example, will be different from the cryolite ratio.
An equivalent weight ratio (EWR) can be calculated as follows [21]:

$$EWR = 1/2EMR = 1/2\frac{(\%_{NaF}/M_{NaF})-(\%_{MgF2}/M_{MgF2})+(\%_{LiF}/M_{LiF})}{(\%_{AlF3}/M_{AlF3})} \qquad (14)$$

where LiF is, like NaF, a Lewis base and MgF_2 is, like AlF_3, a Lewis acid.
It is generally accepted that Al_2O_3 and CaF_2, the other two components of the bath, are neutral or amphoteric, because they do not change the acidity or basicity of the bath, so they are not considered in the bath ratio calculation [28,29].
There are many studies about how all the additives change the physical properties of the bath [30-35]. Without elaborating too much, but enough to understand the bath behaviour in our experiments, some of these properties should be pointed out:
Freezing Point: Baths containing additives [26] show a decrease of the freezing point for bath ratios in the range 1.35-1.50. For a bath composition containing 8.11% AlF_3, 4.81% CaF_2, 1.68% LiF and 5% Al_2O_3, the theoretical freezing point, starting with 1010°C for pure cryolite (Na_3AlF_6), will be 939°C. The measured value [30] was 948°C.
A few other results are presented in Table I.

Table I - Freezing temperatures for three experimental baths in [30]

Mix	Composition, %						Fusion
	Na_3AlF_6	Al_2O_3	AlF_3	LiF	CaF_2	MgF_2	Temperature, °C
I	85.85	4.0	1.75	2.8	2.0	3.6	940
II	80.10	4.0	8.90	0.0	7.0	0.0	956
III	79.0	4.0	10.00	3.0	4.0	0.0	936

Vapour Pressure: Above a liquid bath with additives of metal fluorides the vapor pressure increases when Lewis bases are added (KF, RbF, CsF, BeF_2, MgF_2, AlF_3) and decreases for acids (LiF, NaF, SrF_2, BaF_2).
At the temperature of 1027°C , LiF and NaF have a vapour pressure of 0.43 and respectively 0.32 torr. As a result, the only component above a cryolitic bath, even for a LMLR, will be $NaAlF_4$, which during condensation will decompose into $Na_5Al_3F_{14}$ (s) and AlF_3(s).
Mineralogical Compositions: Solid samples after slow cooling acidic baths in the Na_3AlF_6-AlF_3-Al_2O_3 system with CMR=2.5-2.8 and different additives show three main mineralogical components (Table II): cryolite, alumina and chiolite ($Na_5Al_3F_{14}$).

Table II - Mineralogical compositions of cryolitic baths in [25]

Additives	Phases			
None	Na_3AlF_6	Al_2O_3	$Na_5Al_3F_{14}$	
5% CaF_2	Na_3AlF_6	Al_2O_3	$Na_5Al_3F_{14}$	$NaCaAlF_6$
5% MgF_2	Na_3AlF_6	Al_2O_3	$Na_5Al_3F_{14}$	Na_2MgAlF_7
5% LiF	Na_3AlF_6	Al_2O_3	$Na_5Al_3F_{14}$	Na_2LiAlF_6

If the bath becomes basic (excess of NaF), as usually happens inside or underneath the cathode blocks, the chiolite phase and all of the above complex alumino-fluorides will decompose with the formation of Na_3AlF_6 , $NaMgF_3$ CaF_2 , and Li_3AlF_6 .

Viscosity: For a few particular bath compositions containing 9.35 to19.40% AlF_3 and up to 3.5% LiF [35] the viscosity values showed variations between 2.1 and 2.9 mPa·s at 940°C. A higher concentration in NaF for high ratio baths increases the viscosity above 3.6 mPa·s. The viscosity curves in Figure 2 were extrapolated below the freezing points for an easier reading of the viscosity values at the intersection with the viscosity axis. From the viscosity variations, it results that the increase of AlF_3 in bath decreases the viscosity and the LiF increases it. Although the viscosity is increased by each percent of LIF with 0.13 mPa·s, which is not very much, the slope of the linear variation with the temperature is increased. Based on similarities of Na^+ and Li^+ cations and also on differences between them as glass modifiers, it is expected that NaF in the high ratio bath will increase the viscosity even more.

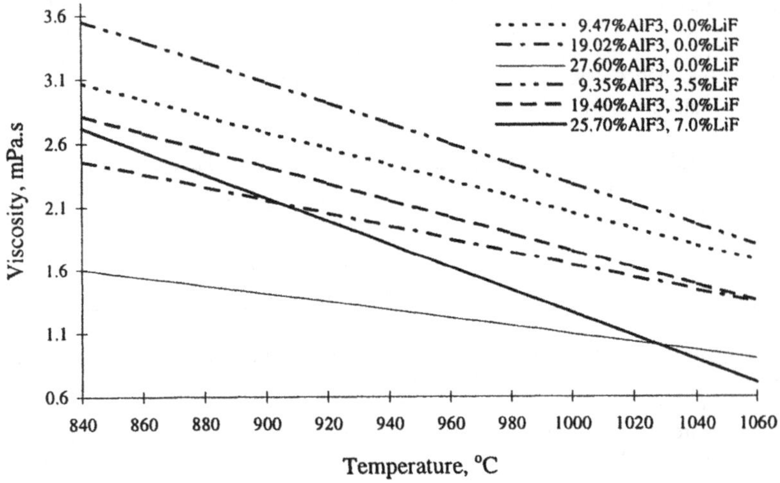

Figure 2 - Bath viscosity versus temperature, from literature data [35]

Bath penetration through the cathode: For a normal operating regime in the reduction cell, the bath will slowly penetrate the cathode through open pores [36-39]. This process is preceded by the sodium vapour attack, which causes the cathode surface to be wet by the cryolite. The moment of reaching total penetration is different for different cathodes (graphite, amorphous carbon, etc.) and can be between four and more than 200 days. Once penetrated, the cathode becomes permeable to the cryolitic bath, which will continue to "flow", due to the liquidostatic pressure above the cathode. The bath will continue to "flow "down through the cathode only if the refractory

underneath cannot stop it. The cold face temperature, on the bottom of the pot can be a good indication of this evolution. From the reported data [40] about the thermal regime of various aluminum pots, the system reaches equilibrium where not too much bath penetration should be expected. In fact, the majority of the reported results on the chemistry of the bath, which impregnates the cathode after more than 2000 days in operation, do not show any bath components. That is a sign that the penetration was stopped maybe when the whole system (impregnated cathode - bath underneath it - reacted refractory) reached equilibrium.
Nevertheless, the gravitational advance of the bath is dependent on the quality of the refractory underneath the cathode. For this regime of operation, probably quite a large range of refractories in the SiO_2 - Al_2O_3 system will perform similarly [41], even for different contents of SiO_2 or Al_2O_3 , densities and porosities. The results of the autopsies reported by different authors in papers presented or published [2, 11, 42-43], generally verify the results of the laboratory tests, even if the principles used to explain the mechanisms of reaction are very different [8, 9, 44-47].

LABORATORY TESTING METHODS

Because in the electrolytic cell during the life in service of any refractory material, the corrosion process is typically static, the classical crucible method could be used.
Although the general principle is the same for most laboratory testing procedures in use, there are differences between them, mainly regarding the following 5 parameters:

- Crucible dimensions at similar geometry or different geometry at comparable crucible volumes
- Bath composition and in particular the cryolite ratio (CR) which expresses the basicity or acidity
- Temperature, which can vary in the range of 900-1000°C
- Duration or holding time at the maximum temperature, which can vary from 12 hours to 16 days.
- Atmosphere, that is regularly reducing because of an assumed high concentration of carbon monoxide beneath the cathode.

It can be considered that after a period of time from the start-up of a cell with a new lining, a melted layer of reacted bath will be formed between the cathode and the barrier refractory. This layer can play a sealing role against penetration of carbon monoxide, metallic sodium and fluorides in vapor state. If we accept this, then we should accept that a testing at low carbon monoxide partial pressure or even in pure air would bring useful information.
In order to simplify the description of the testing procedures, only the differences related to sample preparation procedures were presented below. According to this all the others - bath composition, temperature, duration and atmosphere - were considered variable parameters because they could be different when used by different laboratories.
In order to compare the test results we varied these parameters within the same group of samples, but we maintained them constant within a procedure. As an example, for testing barrier bricks the bath was of two types, low and high ratio, the temperature was 940°C, the atmosphere was reducing and oxidizing for T1 and T3 methods and reducing for T2. Because the preparation procedures were detailed presented in a previous paper [5] only a few differences will be emphasized below.

Sample Preparation

Bricks: Procedure T1 uses a 25.4 mm crucible, 40 mm deep, drilled in a half brick and 40 g of bath.

Procedure T2 uses a cylindrical specimen, 45 mm diameter, core drilled through the whole thickness of a half brick. The specimen, installed tied on the circumference inside of a carbon crucible, will be in contact with 100 g of bath, only at one of the flat ends, in a strong reducing atmosphere.

Procedure T3 uses a 50 mm diameter crucible made of the half brick core drilled for T2, mortared on the top of the other un-drilled half and filled with 100 g of bath.

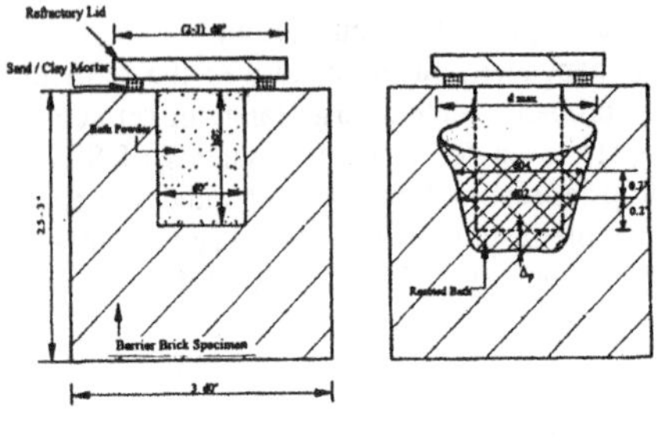

a) b)

Procedure T1

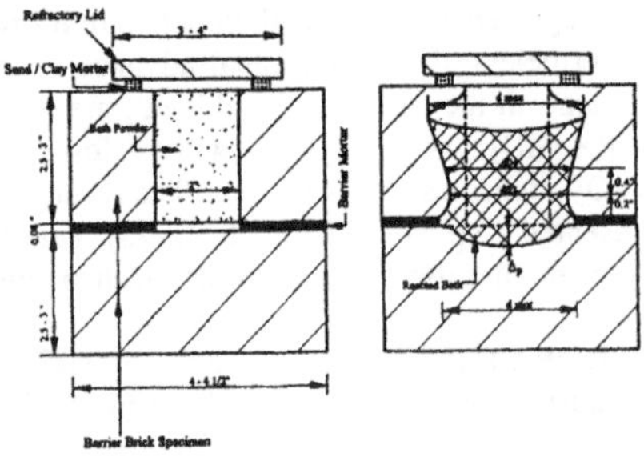

a) b)

Procedure T3

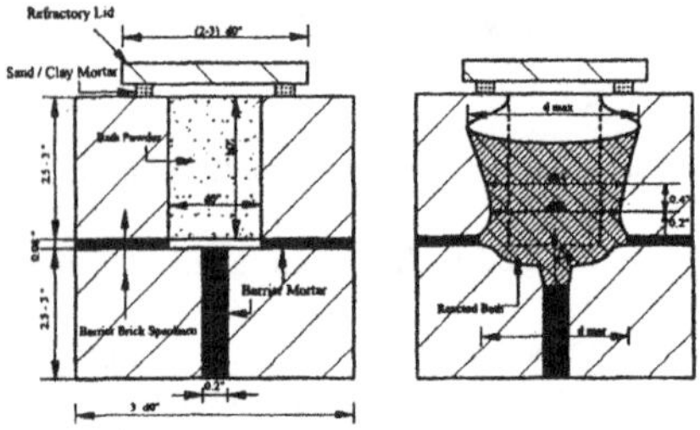

a) b)

Procedure T4

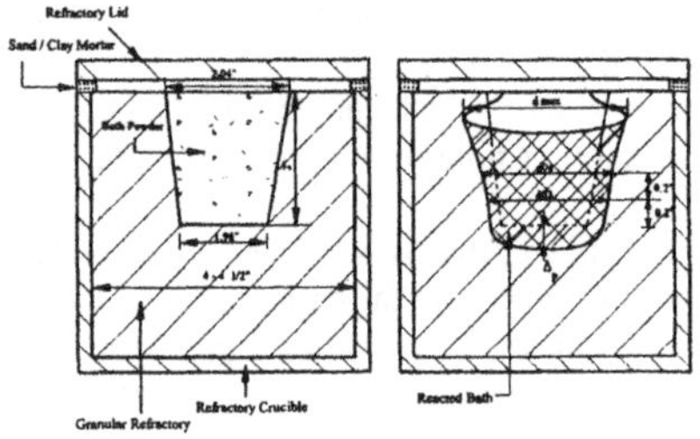

a) b)

Procedure T5

Figure 3. Cross sections through specimens prepared using T1,T3, T4 and T5: (a) before and (b) after test

Mortars: Procedure T3 was described above.

Procedure T4 is similar to T3 except for the bottom undrilled half brick, which in this case is cut in two halves, perpendicular to the 114x114 mm face, and the resulted two quarters are mortared together using the mortar to be tested, with a 5 mm joint.

Powders: Procedure T5 uses a crucible made of rammed powder around a steel molding core and 100 g of bath.

Procedure T6 uses a carbon or graphite crucible, with the powder rammed flat on the bottom and 50 g of bath on the top. Similar to T2, this method can be used only in reducing atmosphere. All the other methods can be used either in oxidizing or reducing.

Testing Conditions

Cryolitic Bath: Two types of bath were used, a low ratio and a high ratio, both well homogenized before starting the whole experimental work. Low Ratio Bath (LRB) was a lithium modified bath of industrial use, having a cryolite ratio (by weight) of 1.22. High Ratio Bath (HRB) was prepared by melting together an industrial bath (also lithium modified, with a cryolite ratio of 1.44) and sodium fluoride, to obtain a final cryolite ratio of 2.55.

Temperature and Atmosphere: Except for T5, at 950°C, all the other tests were performed at 940°C. T2 and T6 use only reducing atmosphere. All the other methods can use either reducing or oxidizing atmosphere.

Duration: The brick specimens prepared using T1, T2 and T3 and the dry barrier powders prepared using T5 were fired at the maximum temperature for 48 hr. A few brick samples, prepared using T3, were also fired 24 hr, in order to compare with the 48hr test results at the mortar level. The mortar specimens prepared using T4 were all fired for 24 hr.

Corrosion Criteria

In order to have the effect of the corrosion process expressed in different ways, measurements at various levels of the corroded areas were taken on the cross-sections of each specimen after testing (Figure 3). These measurements were used to calculate and define better the "corrosion criteria". As an example, the measurements after testing according to method T1 were used to define six individual corrosion parameters:

- *diameter d_1* at 1 cm above the initial bottom of the crucible
- *diameter d_2* at 2 cm above the initial bottom of the crucible
- *depth of penetration (Δp)* in the gravitational direction
- *corroded area*, related to the initial rectangular shape of an imaginary cross-section of the cavity
- *reacted bath*, related to the initial bath volume (determined on quenched specimen)
- *corroded volume*, calculated with corroded area values and related to the initial volume of the crucible

All these were expressed as a dimensional increase in percentage or specific units.

The above corrosion parameters can be expressed as ratios between the final volume (x_f) and the initial value (x_i):

$x = x_f / x_i > 1$, where $x_{max} = 2$, corresponding to the 100% linear increase of that dimension.

A."Corrosion Criterion" (CRC) can be defined and it can be calculated based on the geometrical mean of the individual corrosion parameters, as follows:

$$CRC\,(h,d_1,d_2,S_R,V_R)=[(x_1\cdot x_2\cdot x_3\cdot x_4^2\cdot x_5^3)^{1/8}-1]\cdot 100 \qquad (15)$$

where: $x_1 = h_f/h_i = (h_i+\Delta p)/hi$, with h_i = the initial thickness of the molten bath and Δp - the vertical penetration

$x_2 = d_1/d_0$, with d_0 - the initial diameter of the cylindrical cavity

$x_3 = d_2/d_0$

$x_4 = S_R/S_0$, with S_0 the surface area of an imaginary cross section of the initial molten bath in the cylindrical cavity

$x_5 = V_R/V_0$, with V_0 the initial volume of the molten cryolitic bath

Although each corrosion parameters, as calculated above, is dimensionless, its contribution to the order of the radical will be as a cube (x_5^3) or second power (x_4^2), when it expresses a volume or surface increase. The number of corrosion parameters to be taken into consideration when calculating CRC is unlimited and with a higher number, a more objective selection will result. A general formulation will be:

$$CRC = [(x_1 \cdot x_2 \cdot x_3 \cdot \ldots \cdot x_n)^{1/n} - 1] \cdot 100 \qquad (16)$$

When tests are run in different conditions (oxidizing, reducing, LRB, HRB) all the measured parameters could be used in a unique CRC, or can be calculated for each testing specimen, as follows:

$$CRC = \{[(x_1 \cdot x_2 \cdots x_n)^{1/n} \cdot (x_1 \cdot x_2 \cdots x_m)^{1/m} \cdot (x_1 \cdot x_2 \cdots x_p)^{1/p}]^{1/3} - 1\} \cdot 100 \qquad (17)$$

$$CRC_{oL}(\text{oxidizing, LRB}) = [(x_1 \cdot x_2 \cdot \ldots \cdot x_n)^{1/n} - 1] \cdot 100 \qquad (18)$$

$$CRC_{oH}(\text{oxidizing, HRB}) = [(x_1 \cdot x_2 \cdot \ldots \cdot x_m)^{1/m} - 1] \cdot 100 \qquad (19)$$

$$CRC_{rH}(\text{reducing, HRB}) = [(x_1 \cdot x_2 \cdot \ldots \cdot x_p)^{1/p} - 1] \cdot 100 \qquad (20)$$

where: $m, n, p \in N$ are numbers defining the individual corrosion parameters

If an individual corrosion parameter had a value higher than 2, then it was considered equal to 2 in the CRC calculation.

CORROSION DATA

EXAMPLE #1

One of our preliminary studies on granular refractories called dry barrier materials, presented in [48], emphasized how the refractory's properties (chemistry and mineralogy) correlate with bath chemistry (low and high ratio) and testing conditions (temperature, duration, atmosphere). Calcined materials containing 25, 30, 45, 60 and 70% Al_2O_3 were tested with LRB, using both T5 and T6 methods.

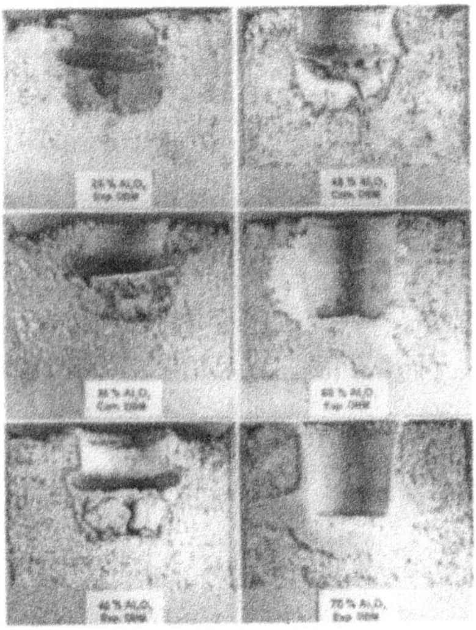

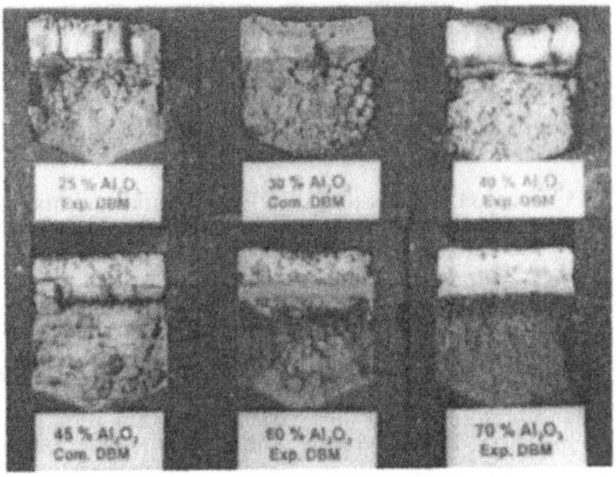

Figure 4 - Experimental and commercial Dry Barrier materials after cryolite cup tests with LRB:

a) (left) using T5 in oxidizing atmosphere;

b) (above) using T6 in reducing atmosphere

Penetration versus atmosphere

One set of tests used these calcines at identical particle size distribution, typical for a commercial dry barrier material. The results, as shown in Figure 4, emphasized the importance of the atmosphere on their behavior when tested with a regular low ration bath (LRB). In an oxidizing atmosphere (Figure 4a), the materials with up to 45% Al_2O_3 had a good capability for holding the bath, with no penetration, while those with 60 and 70% Al_2O_3 were completely penetrated. These differences disappeared when all six dry barrier materials were tested in reducing atmosphere (Figure 4b), using method T6.

Penetration versus pore dimension

The second set of experiments consisted in testing all six calcined materials at four different particle size distributions, corresponding to 4 different fractions obtained by screening: -4+8 mesh, -8+20 mesh, -20+65 mesh and –65 mesh. Theoretically, the largest average pore size belongs to the –4+8 mesh fraction and the smallest to the –65 mesh fraction.

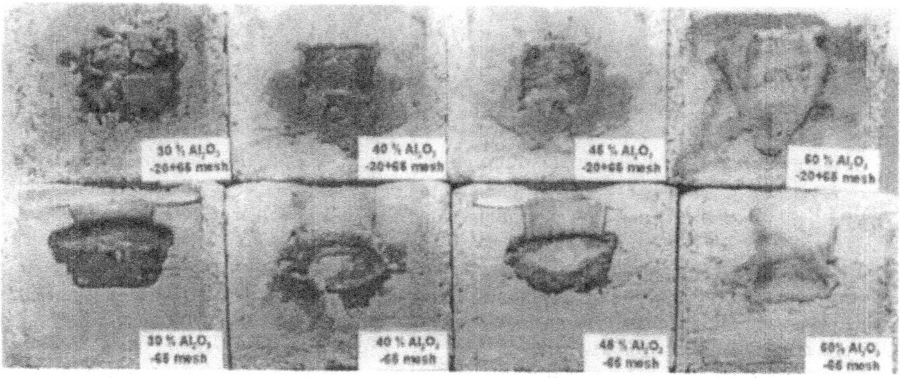

Figure 5 - Cryolite cup tests with LRB using T5 in oxidizing atmosphere on four calcines and two particle size distributions.

All 6 calcines tested with LRB in reducing atmosphere, using T6, showed partial or total penetration for the coarse fractions –4+8 mesh and –8+20 mesh and no penetration for the fine fractions -20+65 mesh and –65 mesh.

When tested in oxidizing atmosphere, even the fine fractions had a different behavior. As seen in Figure 5, the "holding capability" of the -20+65 mesh fraction was maintained only by the 30% Al_2O_3 calcines and all the others, with higher alumina, were penetrated. In the finest fraction group (-65 mesh), the penetration occurred at 60% Al_2O_3.

Penetration versus refractory's reactivity

All test results in an oxidizing atmosphere suggested that a strong correlation exists between the bath penetration and the refractory's chemistry and mineralogy, both included in the term "reactivity". It appears that the tendency to penetration increases with the Al_2O_3 content. In other words, a refractory material richer in silica would perform better when tested with LRB in air.

Penetration versus bath ratio at different pore dimensions

The third set of experiments was performed with bricks and dry barrier materials tested with two different baths – LRB and HRB. In order to verify how the pore dimension would influence the penetration, two different textures of bricks and dry granular materials were used. A fine texture in brick is usually obtained by extrusion and a coarse texture by dry pressing. As a result, a stiff-mud brick (SM), representing the fine texture and a dry pressed brick (DP), of similar chemical composition, were used. In the granular materials group, the fine and coarse versions were obtained from identical calcines. The results after testing in air with both LRB and HRB are presented in Figure 6.

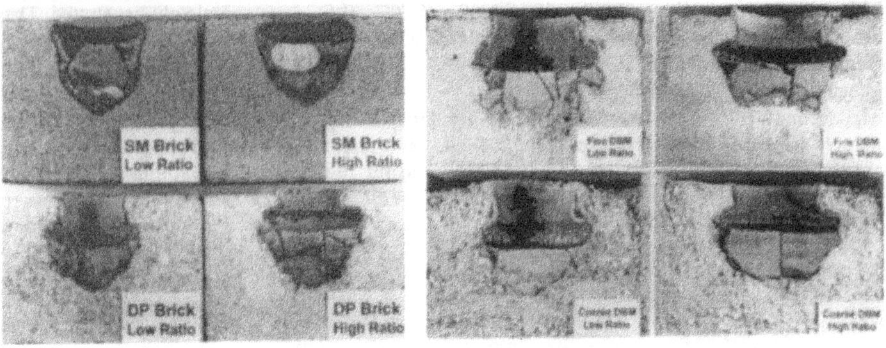

a) b)

Figure 6. Barrier materials after cryolite cup tests with LRB in an oxidizing atmosphere:
a) Bricks (SM - extruded; DP - dry pressed), using T1
b) DB materials made out of high-fired calcines, using T5

The extruded brick (SM) having the smallest pores from all 4 barrier refractories used in this experiment showed no penetration for both LRB and HRB. The dry pressed brick (DP), usually with an average pore dimension one order of magnitude larger than SM, showed penetration for LRB and no penetration for HRB.

At this stage of our experiments, it was very clear that not only the refractories' properties, but also the testing atmosphere would strongly determine their behavior when tested for cryolitic bath penetration.

<u>**Conclusions from example # 1**</u>

- Any refractory dry barrier product, in a very wide range of compositions from 25 to 70% Al_2O_3 or Al_2O_3: SiO_2 ratio from 0.3 to 2.8, could fail in a laboratory test and presumably in the real aluminum electrolytic cell, if it shows a low reactivity when tested with a low ratio bath.
- Less vulnerable to this kind of test and in our opinion, much safer in use, are the dry powder barrier refractories which develop fast a viscous layer of product of reaction. This layer is impermeable to further physical penetration of a very fluid low ratio bath, due to any incidental local failure of the cathode.
- In the case of a dense barrier brick or tile in the Al_2O_3 - SiO_2 system, the main parameters to be considered when assessing the behaviour towards cryolite penetration are open porosity and pore dimensions more than the number of pores.

EXAMPLE #2

The results on three groups of barrier refractories, bricks, mortars and "powders" (simplified term, adopted for the so-called dry barrier materials) were already reported [5, 6, 41, 44] in order to compare the cryolite penetration and corrosion data obtained by using a few of the most used method. The selection of the barrier material in each category was based on our experience in using the testing methods that were to be compared. The objective was to emphasize how different material properties can change the test results, and the correlation that exists between these properties and the corrosion behavior when materials are tested in contact with cryolitic baths.

Table III. Characteristics of the tested refractory materials

No. Crt.	Refractory Material Code	Chemical Composition Al_2O_3 %	SiO_2 %	Fe_2O_3 %	TiO_2 %	Others %	$Al_2O_3 : SiO_2$ Ratio	Bulk Density g/cm^3	Apparent Porosity %
Barrier BRICK									
1.	B1	35.84	54.66	2.73	2.10	4.67	0.65	2.21	18.0
2.	B2	24.94	61.02	7.96	1.34	4.74	0.41	2.22	14.1
3.	B3	29.43	60.43	5.19	1.97	2.98	0.49	2.06	21.9
4.	B4	45.94	47.54	1.74	2.80	1.98	0.96	2.27	15.8
5.	B5	41.78	54.10	1.06	2.57	0.49	0.77	2.27	16.0
6.	B6	27.94	59.30	7.58	1.49	3.99	0.47	2.15	18.2
7.	B7	27.94	59.30	7.58	1.49	3.99	0.47	2.04	21.5
8.	B8	25.55	69.30	2.04	0.87	2.24	0.37	2.06	20.0
9.	B9	25.55	69.30	2.04	0.87	2.24	0.37	1.98	23.4
Barrier MORTARS									
10.	M1	43.38	46.60	1.12	2.50	6.40	0.93	-	-
11.	M2	27.87	62.26	3.79	1.73	4.35	0.45	-	-
12.	M3	27.58	61.60	3.75	1.71	5.36	0.45	-	-
Barrier POWDERS									
13.	P1	27.47	62.02	4.63	1.72	4.16	0.44	1.93	22.7
14.	P2	26.36	67.49	2.60	0.91	2.54	0.39	1.72	18.4
15.	P3	35.77	58.31	2.86	1.01	2.05	0.61	1.89	18.2
16.	P4	38.49	54.25	1.91	1.76	3.59	0.71	1.64	22.4
17.	P5	44.00	51.29	0.85	2.35	1.51	0.86	2.11	18.2
18.	P6	44.45	52.14	1.16	0.94	1.31	0.85	1.93	17.5

Note: * Bond strength according to ASTM C 198-91, on 1.5 mm joint, after drying @ 110 C.

Materials

The properties of interest for the selected barrier materials are presented in Table III.

The first five bricks, B1-B5, from the group of nine, are commercial products and they are successfully used by different smelters. One is extruded (B2), the other four are dry pressed, with an alumina content ranging from 24 to 45%.

Two experimental bricks, B6 and B7, were made of the same raw materials with the same particle size distribution. They were shaped by dry pressing and fired in similar conditions, but at different maximum temperatures, in order to create slightly different textures and mineralogical compositions. It can be assumed that the pores would have a similar shape, distribution and very close dimensions. The B6 brick being fired at a higher temperature than B7, the difference was mainly in the amount of pores, owing to more vitreous phase and partially in the pore size distribution. The experimental bricks B8 and B9 were chemically and mineralogically identical, because they were made from the same raw materials and fired at the same temperature, but they were shaped by different processes. Brick B8 was extruded, similar to the commercial product B2, using a similar particle size distribution of raw materials. The raw materials were different and were fired in different conditions than B2. The other brick, B9, was dry pressed, similar to the

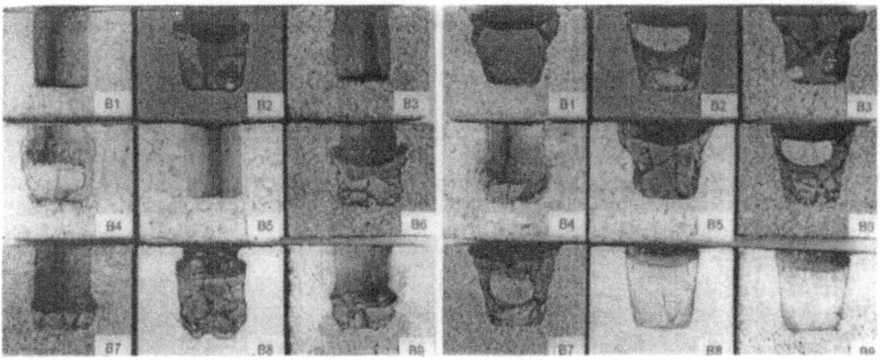

a) b)

Figure 7. Barrier brick samples B1 to B9 after 48 hours test in oxidizing atmosphere, using procedure T1: a) Low ratio bath; b) High ratio bath

product B3. It had the same particle distribution, but used raw materials identical to the ones for B8. It was fired in the same conditions as B8. As a result, the pore size distribution in the B9 brick could be assumed to be very different than for B8, due to the forming process.

As seen in Table III, the porosity values for all these four experimental bricks were similar to B3, the most porous commercial product, rather than to B2 (14% porosity) or to B1 (18%).

Two mortars, M1 and M2, were commercial products in the super-duty range (M1) and medium duty range (M2). The mortar M3 was an experimental mortar similar to M2.

The group of 6 barrier powders (P1 to P6) was chosen to demonstrate that at identical particle size distributions (and as a result, pore diameters), similar chemistries (in groups of two, P1&P2, P3&P4, P5&P6) and levels of Al_2O_3 (aprox. 25, 35 and 45%), the cryolite penetration also depends on the mineralogical composition, hence on the thermal history of all the components of the mix. Three dry barrier materials (P1, P3 and P5) were commercial products, while the other three (P2, P4 and P6) were created for the purpose of this demonstration. The product P1, although a low duty material, has a well-known performance of more than 2800 days as a barrier material against cryolite penetration in aluminum electrolytic cells.

Results on Bricks

Figure 7 presents cross-sections through all the specimens prepared using the procedure T1 and fired for 48 hr in oxidizing atmosphere. In Table IV the corrosion parameters are presented as dimensional increase. From all nine specimens tested with LRB, only two, B2 and B8, were not penetrated at all, while three, B1, B3 and B5, were totally penetrated. The other four, B4, B6, B7 and B9, were partially penetrated. In order to have a correct comparison, none of the measurements on penetrated specimens were used for rating the corrosion effect. For the same reason, the measurements on the specimen B4 with HRB were also not used.

From Table IV, the best corrosion behavior when tested with LRB is easily decided by the fact that the only bricks with no penetration at all are B2 and B8. When tested with HRB, the best were B8 and B9, followed by B6 and B7.

Table IV. Penetration and corrosion data for brick samples B1 to B9, after 48 hr., in oxidizing atmosphere, using T1

BRICK	Dimensional Increase by Corrosion						
	Diameter @ 1 cm above base, %	Diameter @ 2 cm above base, %	Depth of penetration - cm -	Corroded area - cm^2-	Reacted bath - cm^3-	Corroded volume - cm^3-	Corroded volume - % -
High Ratio Bath, 48 hr.							
B1	24.5	35.0	0.28	5.99	47.2	31.1	164.5
B2	22.1	26.3	0.05	4.81	49.7	27.8	147.2
B3	25.5	32.9	0.11	5.38	66.3	27.2	144.0
B4 ••	9.3	46.4	0.29	3.38	9.8	16.3	86.2
B5	27.4	32.7	0.04	4.94	42.0	23.3	123.4
B6	16.8	22.1	0.02	4.06	44.1	21.1	111.6
B7	17.8	25.3	0.12	4.70	43.7	23.4	123.6
B8	19.8	26.1	0.20	3.79	38.0	17.5	92.6
B9	18.4	24.7	0.21	3.96	35.7	17.8	94.1
Low Ratio Bath, 48 hr.							
B1 •••	-	-	-	-	-	-	-
B2	26.3	31.7	0.05	3.83	28.9	21.8	125.3
B3 •••	-	-	-	-	-	-	-
B4 ••	31.1	36.4	0.18	4.54	27.3	23.0	132.4
B5 •••	-	-	-	-	-	-	-
B6 ••	29.8	31.9	0.31	4.40	20.9	21.4	132.4
B7 ••	32.1	30.0	0.23	2.17	2.9	9.6	55.1
B8	32.3	33.3	0.46	4.79	32.8	23.9	137.5
B9 •	29.8	39.2	0.44	3.07	4.5	13.0	74.9

Note: • slight penetration; •• extensive penetration; ••• total penetration
* Base was considered the bottom of the initial 1inch blind hole

For tests using the testing ***method T1*** in an oxidizing atmosphere, a corrosion rating scale based on the individual parameters from Table IV, would look as follows:

	HRB
Corroded Area, cm^2	B8<B9<B6<B7<B2<B5<B3<B1
Reacted bath, cm^3	B9<B8<B5<B7<B6<B1<B2<B3
Corroded volume, cm^3	B8<B9<B6<B5<B7<B3<B2<B1
Diameter @ 1 cm above base, %	B6<B7<B9<B8<B2<B1<B3<B5
Diameter @ 2 cm above base, %	B6<B9<B7<B8<B2<B5<B3<B1

In order to quantify the combined effect of the individual corrosion parameters, the CRC was calculated in 5 different ways as shown in Table V.

The combined contribution of the individual corrosion parameters in CRC ($x_1 \cdot x_2 \cdot x_3 \cdot x_4 \cdot x_5$) should be considered the most complete and the corrosion rating scale would now look as follows:

$$B_9 < B_8 < B_6 < B_7 < B_5 < B_2 < B_3 < B_1 < B_4$$

TABLE V. Corrosion rating using CRC with T3 data.

	$X_1 \cdot X_2$ %	$X_1 \cdot X_2 \cdot X_3$ %	$X_1 \cdot X_2 \cdot X_3 \cdot X_4^2$ %	$X_1 \cdot X_2 \cdot X_5^3$ %	$X_1 \cdot X_2 \cdot X_3 \cdot X_4^2 \cdot X_5^3$ %
B_1	29.6	24.2	48.1	64.0	63.2
B_2	24.2	16.5	36.9	63.6	56.8
B_3	29.1	20.7	42.7	67.9	62.0
B_4	26.5	24.1	50.2	66.5	67.2
B_5	30.0	19.9	40.0	58.9	54.3
B_6	19.4	12.9	30.7	55.6	49.1
B_7	21.5	16.1	36.1	56.3	52.7
B_8	22.9	18.4	33.2	51.4	47.2
B_9	21.5	17.7	33.6	48.3	46.1

The least corroded and in other words the most corrosion resistant bricks are B_9 and B_8, which are dry pressed and respectively extruded. They were made of the same raw materials at a total Al_2O_3 of 25.55% and a Al_2O_3:SiO_2 ratio of 0.37.

The results using the testing ***method T3*** showed similar results with ***method T1*** (Table IV), above. Only brick B5 had a different behavior, showing no penetration when tested with LRB. Other than that, bricks B2 and B8 were also not penetrated by the LRB. Brick B5 was slightly better than B8, although the last one was the least corroded at the top of the bath. With HRB, brick B8 shared the lead with B2, although B9 was next, when considering all four individual corrosion parameters: the diameter at the mortar level (x_1'), the diameter at the bath top level (x_2'), the corroded area (x_4) and the corroded volume (x_6). Using the most complete corrosion criterion, CRC (x_1'. x_2'. x_4^2. x_6^3), the corrosion rating will look as follows:

	B_8 <	B_2 <	B_9 <	B_5 <	B_6 <	B_3
CRC,	16.5	19.2	19.5	19.8	20.1	23.0

All the results on bricks definitely indicate that there are other characteristics, different from the ones presented in Table III, which might decide the behavior of the refractory brick tested for cryolite resistance.

Influence of the Refractory Properties

Chemistry: In order to correlate the test results in Table IV with the brick compositions given in Table III, two different theories could be used. According to one, let's call it "A", a better resistance to cryolite penetration would be expected for refractories with higher alumina and lower silica, and according to the other theory, "B", a better resistance for higher silica and lower alumina.

If we arrange the data in Table III in a different order, it will look as follows:

Al_2O_3	B2<B8=B9<B6=B7<B3<B1<B5<B4
SiO_2	B4<B5<B1<B6=B7<B3<B2<B8=B9
Al_2O_3: SiO_2	B8=B9<B2<B6=B7<B3<B1<B5<B4

According to theory "A", the best corrosion resistance would belong to B4 and B5, followed by B1, while according to theory "B", the most resistant would be B8 and B9, followed by B2.

According to the overall test results in Table II, for HRB bath the best behavior belongs to the bricks B8 and B9, followed by B6 and B7. All four are at the low end of Al_2O_3: SiO_2 ratio and

Al_2O_3 content. The results are in contradiction with theory "A" according to which the expected best would be B4 and B5.

Density and Apparent Porosity: From Table I the results are as follows:

Density	B9<B7<B8=B3<B6<B1<B2<B4<B5
Apparent Porosity	B2<B4<B5<B1<B6<B8<B7<B3<B9

Bricks B8 and B9 followed by B6 and B7 have relatively high apparent porosities and the lowest bulk densities, except for B6. Brick B9 has the highest apparent porosity, the largest pores of all the dry pressed bricks, and the lowest density of all the nine bricks tested.

Chemistry and Mineralogy (Reactivity): The fact that these two groups, B8 and B9 and respectively B6 and B7, are situated together on the "corrosion rating scale", could prove that the chemistry and mineralogy play important roles in the corrosion behavior of any barrier refractory material.

Influence of the Bath Chemistry

A regular LRB used in the electrolytic process showed easier penetration through the open pores than the HRB. This indicates that the penetration depends on both chemical and mineralogical compositions.

Influence of the Atmosphere

Figure 8 presents the cross-sections of the specimens tested according to ***method T2*** in a reducing atmosphere. The data in Table VI are the corrosion parameters calculated from the measured cross sectional values.

TABLE VI. Corrosion rating using CRC with T2 data.

	$(x_3 \cdot x_4^2)^{1/3}$ %		$(x_3 \cdot x_6^3)^{1/4}$ %		$(x_3 \cdot x_4^2 \cdot x_6^3)^{1/6}$ %	
	HRB	LRB	HRB	LRB	HRB	LRB
B_1	10.3	13.1	11.1	14.5	10.2	12.9
B_2	10.1	11.5	12.7	12.0	11.9	11.0
B_3	11.1	13.5	13.2	15.3	12.0	13.6
B_4	9.9	13.5	9.9	15.3	8.9	13.5
B_5	7.2	11.5	5.3	11.5	5.2	10.3
B_6	10.2	14.1	12.1	17.0	11.1	15.3
B_7	11.6	13.4	13.9	15.1	12.6	13.3
B_8	6.1	12.9	4.1	14.6	4.3	13.0
B_9	6.9	13.6	5.1	15.4	5.0	13.5

It is important to note that none of the brick specimens showed cryolite penetration. Unusual corrosion patterns were obtained with HRB for specimens B2, B6 and B7 at the interface with the graphite crucible or petroleum coke powder. The bath corroded deeper on the circumference of the specimen, where three phases were present: carbon, cryolitic bath and refractory. The only explanation could be the "edge effect" at the interface bath-refractory due to direct contact with the carbon as a third phase. This could make these refractories more reactive at that level, due to a combination of the strong reducing effect of the carbon and oxidizing effect of the fluorine. The center side of the cylindrical specimens was less corroded, indicating that the above explanation could be correct.

a) b)

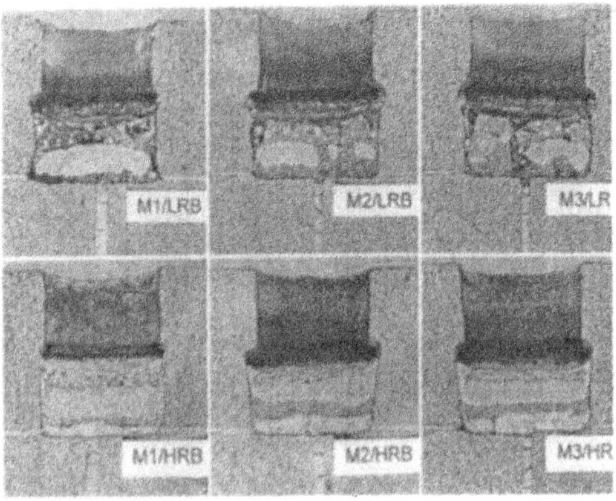

Figure 8 (above). Barrier brick samples B1 to B9 after 48 hours test in reducing atmosphere, using procedure T2:
a) Low ratio bath; b) High ratio bath

Figure 9 (left). Barrier mortars after 24 hours test in air, using method T4, with low ratio (LRB) and high ratio (HRB) baths.

Using the primary data in [5], the CRC was calculated as combinations of three independent corrosion parameters: depth of penetration (x_3), corroded area (x_4) and corroded volume (x_6). As seen in Table VI, according to the results, using the testing ***method T2*** with HRB, the brick B2 should be ruled out for area and volume increase, but it had one of the best behaviors with LRB. If one considered only the depth of penetration, which was measured along the axis of the cylindrical specimen, then B2 appears in a different position in the corrosion rating scale, as follows:

	HRB
Depth of penetration, cm	$B8<B2<B9<B5=B6<B1=B3<B7<B4$
Corroded area, cm^2	$B8<B9\approx B5<B4<B1<B3<B6<B7<B2$
Corroded volume, cm^3	$B8<B9=B5<B4<B1<B6<B3<B7<B2$

Without considering the circumference corrosion, the test results would look different if, for example, the interface refractory-cryolitic bath would be as large as a real electrolytic cell bottom. In other words the "edge effect" does not exist practically underneath the cathode in the real electrolytic cell. Based on the above results the test method T2 should not be considered for testing barrier refractories, but it could be useful for side wall linings.

Results on Mortars

The test results on mortar specimens prepared using T4 are presented in Figure 9. The brick used with these 3 experimental mortars was B2. It can be seen that a slight penetration for both mortars M2 and M3 occurred when tested with the LRB and that no penetration occurred with HRB. Probably at extended holding times, the penetration by the LRB would show an increase when tested in oxidizing atmosphere and will continue to be nonexistent in reducing. The only corrosion parameters to describe the behavior of a mortar would be the diameter of the reacted bath at the mortar level, for T3 and T4, and the depth of penetration for T4.

Results on Powders

Data related to granular barrier refractories tested according to both T5 and T6 methods were presented in previous papers [6, 41, 48]. The common method [5] is similar to T6 and the specimens after testing look similar to the ones presented in Figure 4b. The amount of bath, which decides the gravitational penetration of the cryolitic bath, plays an important role in evaluating the corrosion [7]. When bath used is doubled in a reducing atmosphere according to T5, the test results bring more information than using T6. For this reason and also based on our experience and the database created by developing such new products, a method based on T6 would not bring enough information necessary to select a refractory material. If it is used, then it should be

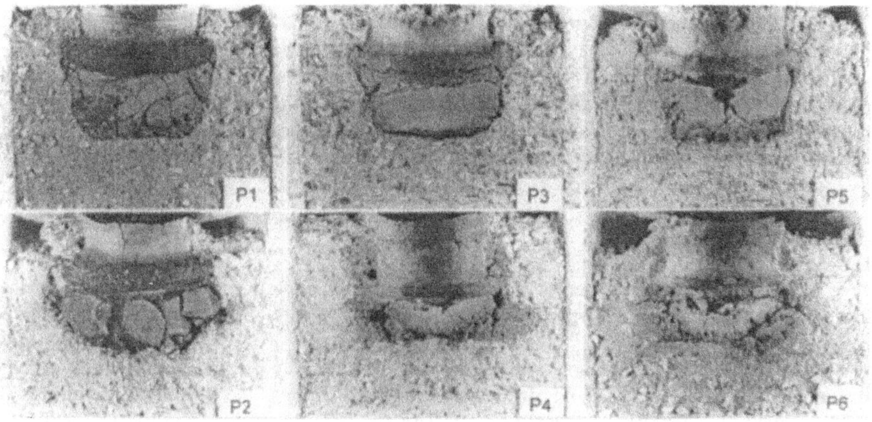

Figure 10. Barrier powder samples (P1-P6) after 48 hours with LRB, in oxidizing atmosphere (T5).

combined with a method based on T5 in an oxidizing atmosphere.

The corrosion parameters calculated from measurements are very similar , although the corrosion patterns look different for the specimens P1, P3 and P5 tested with LRB (Figure 10). That appears to be rather strange, considering the differences between them, as shown in Table III. As seen in Figure 10, similar materials could behave differently when tested in oxidizing atmosphere with

LRB. The penetration appears to increase with increasing the Al_2O_3, being the worst for P6, a superduty type of material. It is to be mentioned that all six materials had identical particle size distributions, with minor differences in particle shapes due to their hardness and shearing behaviour during crushing and grinding. As a result, their permeabilities and pore diameters should be considered identical. In this case, their corrosion behaviour would depend only on other properties of the refractory grains, such as chemical and mineralogical composition, and related to them, the chemical reactivity with the cryolitic bath. Their chemical compositions, as presented in Table I, have some similarities, in the sense that P1 is similar to P2, P3 to P4 and P5 to P6. Their mineralogical compositions though were really different, as generated by different thermal treatments of different natural clays for P1 to P4 and flint for P5 and P6. The penetration observed for P2, P4 and P6 could be firstly due to the higher wetting capability by the LRB and secondly to their inertness or lack of reactivity during the initial contact with the molten bath.

Conclusions for example # 2

- All the testing procedures used for barrier bricks in reducing and oxidizing atmosphere, with high ratio and low ratio baths, emphasized the fact that the extruded bricks (B2 and B8) have a corrosion resistance superior to dry-pressed bricks, at any $Al_2O_3 : SiO_2$ ratio. This behavior is due to the pore size rather than any other characteristics.
- The dry-pressed bricks could achieve good performances when tested with cryolitic bath only if they have very low open porosities at high-fired quality (B5). Also, they could perform well at relatively high porosities (B9, B6 and B7) if the bricks are made of special raw materials and are fired in special conditions to generate a texture extremely resistant to the fluoride attack.
- Comparing the testing procedures for bricks, we consider that all provide useful information. The procedure T1 is the most suitable for oxidizing atmosphere and both T3 and T4 are the most adequate when a mortar has to be tested along with the brick. Due to the "edge effect", the procedure T2 in reducing atmosphere should be used only for side walls lining.
- For testing granular barrier materials, called powders in this paper only for convenience, procedure T5 appears to be the one to be used in both reducing and oxidizing atmosphere.
- The "Cryolite Corrosion Criterion" CRC represents an important laboratory tool for assessing and quantitatively rating the penetration and corrosion of barrier refractories by molten cryolitic baths.

EXAMPLE # 3

Materials

Three groups of refractories were used in our laboratory experiments [19] to correlate their physical, chemical and mineralogical properties with the penetration and corrosion by cryolitic baths.

For a group of five commercial bricks (B1 to B5) manufactured by Clayburn Refractories Ltd. and used as barrier cathode linings in different aluminum smelters, the cryolite penetration was correlated with their texture characteristics: porosity and pore dimension. The Bricks B1, B2 and B4 were made by extrusion while B3 and B5 were made by dry pressing. As shown in Table VII, their chemical compositions are quite similar, the Al_2O_3 ranging from 24.94% (B1) to 29.43% (B3). SiO_2 ranges from 59.30 (B5) to 69.30% (B2). Each type of brick was intentionally used in our testing program at three different levels of porosity (low, regular and high) in order to determine the correlation with permeability and corrosion (Table VIII). Each sample of a certain porosity consisted of 5 specimens, out of which one was used for cryolite corrosion testing and the others to prepare the permeability specimens. All 5 specimens were tested for porosity, absorption and density, with the averages presented in Table VIII. Each brick sample B1 to B5 was represented by 15 brick specimens.

Table VII. Characteristics of the experimental refractory materials

No. Crt.	Refractory Material	Chemical Composition Al_2O_3 %	SiO_2 %	Fe_2O_3 %	TiO_2 %	Others %	$Al_2O_3 : SiO_2$ Ratio	Bulk Density g/cm^3	Apparent Porosity %
Barrier BRICKS									
1.	B1	24.94	61.02	7.96	1.34	4.74	0.41	2.22	14.1
2.	B2	25.55	69.30	2.04	0.87	2.24	0.37	2.06	20.0
3.	B3	29.43	60.43	5.19	1.97	2.98	0.49	2.06	21.9
4.	B4	27.18	60.72	6.57	1.65	3.86	0.45	2.14	18.0
5.	B5	27.94	59.30	7.58	1.49	3.99	0.47	2.15	18.2
Barrier CASTABLES									
6.	M1	33.56	53.71	6.02	1.39	5.31	0.62	2.04	23.1
7.	M2	36.70	53.30	4.08	1.83	4.09	0.69	2.16	17.1
8.	M3	48.31	46.24	1.57	1.46	1.43	1.04	2.34	17.3
Barrier POWDERS									
9.	P1	27.47	62.02	4.63	1.72	4.16	0.44	1.93	22.7
10.	P2	26.36	67.49	2.60	0.91	2.54	0.39	1.72	18.4
11.	P3	35.77	58.31	2.86	1.01	2.05	0.61	1.89	18.2
12.	P4	38.49	54.25	1.91	1.76	3.59	0.71	1.64	22.4
13.	P5	44.00	51.29	0.85	2.35	1.51	0.86	2.11	18.2
14.	P6	44.45	52.14	1.16	0.94	1.31	0.85	1.93	17.5

Table VIII. Porosity, permeability and pore diameter data

Brick	Apparent Porosity	Permeability ASTM	VacuPerm	Aver. Pore Diameter
	- % -	- centidarcys -		- µm -
B1-1	13.45	1.14	0.79	1.60
B1-2	15.27	1.52	1.06	1.64
B1-3	17.46	1.88	1.3	1.70
B2-1	18.43	3.25	1.95	2.10
B2-2	19.83	3.29	1.71	1.97
B2-3	21.27	4.01	3.08	2.29
B3-1	22.01	97	-	11.8
B3-2	23.13	96	-	11.5
B3-3	24.94	119	-	12.3
B4-1	14.29	2.13	0.93	1.81
B4-2	15.34	1.91	1.17	1.78
B4-3	16.69	2.96	1.21	1.95
B5-1	13.64	-	25	7.64
B5-2	14.70	-	25.9	7.49
B5-3	16.69	-	31.7	7.78
C1	23.12	-	0.21	0.54
C2	17.10	-	0.26	0.69
C3	17.31	-	0.29	0.73

The group of low cement castables (C1 to C3) was used for two purposes. The first was to demonstrate that the penetration and corrosion do not depend on the porosity, only on the average pore diameter and/or permeability. In order to demonstrate that, the specimens were cast at a level of water that allowed for the air bubbles to be trapped inside the cast specimen, generating a high open porosity. The second purpose was to demonstrate that for refractories with 25-45% Al_2O_3 cryolite penetration and corrosion depend on the nature and microstructure of the bonding phase and only slightly on the chemistry of the aggregate.

Testing methods

Permeability: Two methods were used to measure the permeability of the experimental refractory bricks and castables. One was the ASTM method C577-96, using nitrogen as a permeating fluid and cubic specimens of dimensions

51×51×51 mm. The other was a method developed by the University of Missouri-Rolla, using a vacuum decay permeameter (VacuPerm) and cylindrical specimens 101.6mm diameter and 25.4 mm thickness. Unfortunately for the experimental materials, both methods were at one of the limits of their operating capabilities. For the ASTM method, which measures up to a few hundred centidarcys, the range 1-10 centidarcys is at the bottom limit of its detection capability. For the VacuPerm method, which has a high accuracy up to 5 milidarcys, the range 10-100 milidarcys will be theoretically outside its detection limit. This is because of the small number of pressure readings during the vacuum decay does not allow for a good statistical data processing. Above 100 milidarcys, it was necessary to modify the specimen dimensions to 47 mm diameter and 63.5 mm thickness in order to make the measurements possible.

Porosity: The ASTM method C20-92 was used for brick specimens and C830-93 for castables.

Cryolite penetration: The method used for measuring the penetration and corrosion by molten cryolitic bath [7] is not standardized, but it is under development by the ASTM Subcommittee C8.05 "Refractories for Aluminum Industry". The method was used in assessing the penetration and corrosion in a few previous papers [5, 6, 41, 44, 49, 51]. A few details from those papers regarding sample preparation and testing conditions are presented below.

Results

Corrosion versus Permeability and Pore Diameter:

The permeability data on the experimental bricks and castables are presented in Table VIII. Most of these data are averages of 4 values for the ASTM method and of 8 values for the VacuPerm method. Comparing the two sets of values, the ASTM method showed higher standard deviation and repeatability than the VacuPerm method at permeabilities below 5 centidarcys. It was necessary to eliminate one or two values out of 4 in order to maintain the rest of a population within a 95% repeatability interval.

The pore diameters, using the equation (10) were calculated from both sets of permeability data. Because some samples were tested only with one method, averages of both methods were calculated for permeability and pore diameter in order to cover the whole group of 15 brick samples.

The measurements of the corrosion parameters, from Table IX-a for LRB and IX-b for HRB, were used to calculate the "Cryolite Corrosion Criterion". Equation (15) was used with the measurements to make possible the penetration and corrosion rating. Averaging the total corrosion values (*CRC* from Table IX) within each type of brick, the following ratings will result:

Corrosion by LRB: $B1<B4<B2<B5<B3$ (21)

Corrosion by HRB: $B2<B1<B4<B5<B3$ (22)

Similar ratings, using permeability and pore diameter data from Table II, would look as follows:

Permeability: $B1<B4<B2<<B5<<<B3$ (23)

Pore Diameter: $B1<B4<B2<B5<B3$ (24)

Regression analyses showed that the corrosion data (*CRC* in Table IX) plotted in terms of permeability or pore diameter (Table VIII) could be expressed by equations of the following type:

$$CRC(k) = a_1+b_1 \ln k \quad (25)$$

and

$$CRC(d) = a_2+b_2 \ln d \quad (26)$$

where k and d are the permeability and respectively the average pore diameter.

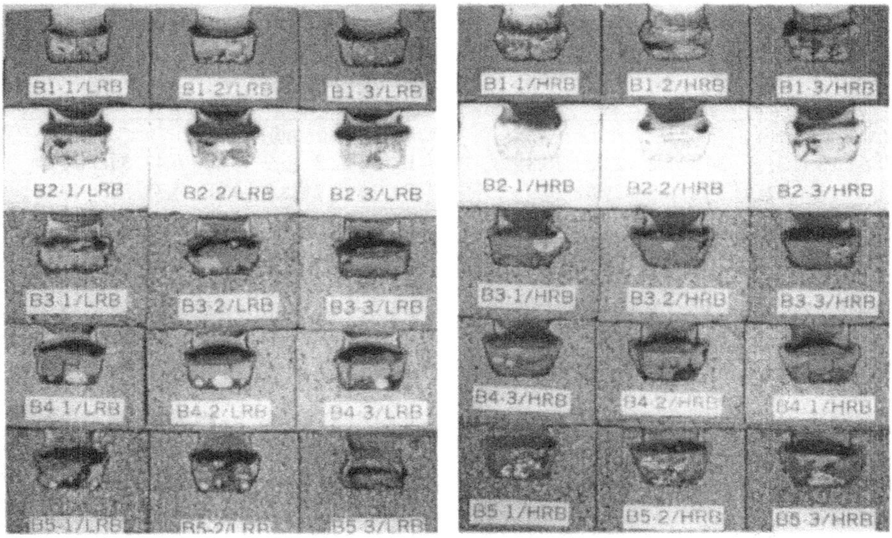

a) b)

Figure 11. Barrier brick samples B1 to B5 after test using T1 in oxidizing atmosphere: (a) LRB and (b) HRB

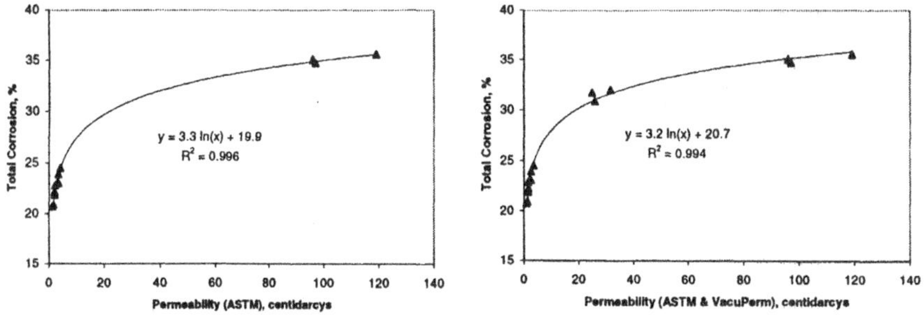

a) b)

Figure 12. Total corrosion versus permeability: a) only ASTM values; b) average of ASTM and VacuPerm values.

Trends of the corrosion variation in terms of permeability for each set of data, ASTM and average of both ASTM and VacuPerm, are shown in figures 12a and respectively 12b. The correlation coefficients were quite acceptable for individual values ASTM or combined ASTM and VacuPerm. The average of the permeability data obtained using both testing methods was used for calculating the pore diameter.

Table IX-a. Corrosion measurements for LRB

Brick	d_{05} %	d_{10} %	d_{max} %	h %	S_{cor} %	CRC %
Low Ratio Bath						
B1-1	18.7	21.3	34.1	5.2	23.6	20.8
B1-2	20.4	22.3	35.0	6.5	21.6	20.9
B1-3	20.9	22.6	36.2	7.3	24.0	22.2
B2-1	21.1	24.0	42.0	6.0	26.6	23.9
B2-2	19.6	23.1	42.5	7.5	24.1	23.1
B2-3	22.7	23.3	43.5	7.5	26.4	24.5
B3-1	29.8	39.8	43.5	10.2	44.5	34.8
B3-2	31.2	38.7	42.5	13.6	43.6	35.1
B3-3	33.0	39.7	45.0	12.5	43.2	35.6
B4-1	20.7	24.3	37.0	5.5	25.9	22.9
B4-2	21.4	22.4	34.3	7.1	23.9	21.9
B4-3	22.6	24.3	37.0	8.1	24.1	23.1
B5-1	24.1	32.7	37.6	4.7	41.9	29.8
B5-2	24.8	33.1	36.8	6.3	44.3	30.9
B5-3	-	-	-	-	-	-
C1	22.1	23.8	52.4	5.2	25.5	25.0
C2	22.0	27.7	38.7	6.3	23.9	23.3
C3	26.4	25.6	40.7	8.3	20.5	23.3

Table IX-b. Corrosion measurements for HRB

Brick	d_{05} %	d_{10} %	d_{max} %	h %	S_{cor} %	CRC %
High Ratio Bath						
B1-1	22.3	36.9	47.0	6.4	35.6	29.9
B1-2	23.7	38.7	45.5	4.7	32.7	29.0
B1-3	23.7	38.7	44.4	7.1	32.8	29.3
B2-1	15.9	29.4	44.9	6.4	33.1	26.5
B2-2	18.0	28.1	46.2	4.7	33.9	26.8
B2-3	19.1	30.6	46.2	7.0	32.3	27.3
B3-1	22.7	35.8	58.2	5.5	49.8	35.7
B3-2	23.7	36.9	57.5	4.7	44.5	34.2
B3-3	25.5	38.7	58.2	5.8	47.9	36.2
B4-1	23.4	35.0	44.5	6.2	28.6	27.2
B4-2	23.4	35.1	42.9	5.4	33.2	28.3
B4-3	24.7	36.1	47.5	7.4	32.2	29.4
B5-1	22.0	36.4	48.7	5.3	36.4	30.1
B5-2	22.1	32.0	45.1	6.1	38.7	29.8
B5-3	24.8	35.0	47.4	6.3	36.2	30.3
C1	24.4	38.3	50.1	3.1	32.9	29.4
C2	19.0	33.4	45.0	6.3	30.1	26.7
C3	22.7	36.4	41.3	3.1	32.4	27.4

For the brick specimens tested with LRB, as shown in Figure 13a, the correlation coefficient R^2 remains above 0.99 for the corrosion variation in terms of these pore diameter values. For the representation of the corrosion in terms of the average permeability data (Figure 12b) the correlation coefficient was also higher than 0.99.

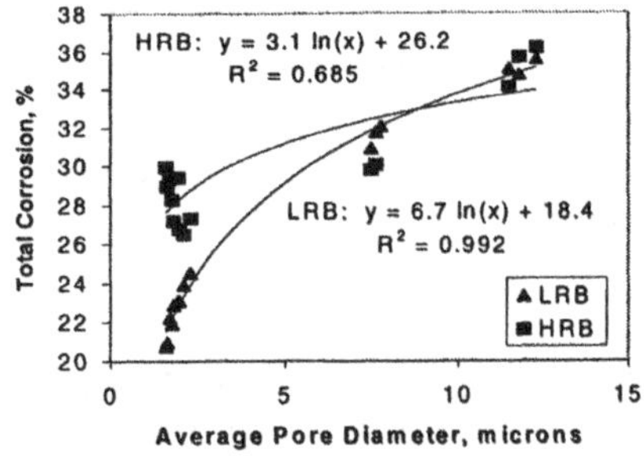

a)

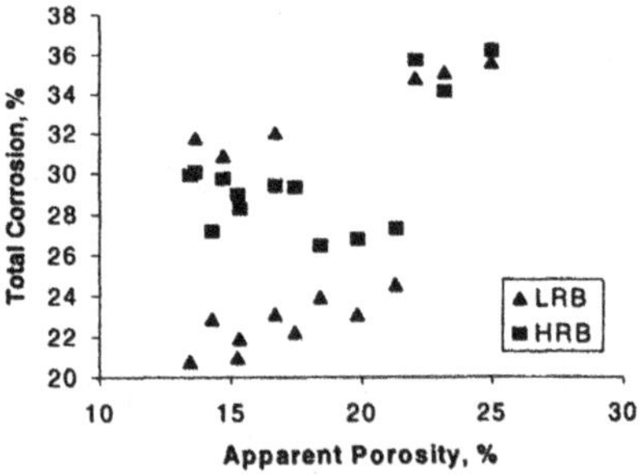

b)

Figure 13. Total corrosion versus (a) average pore diameter and (b) apparent porosity for the bricks B1-B5, tested with LRB and HRB.

Linear and logarithmic correlations (Figure 13a) for the HRB showed coefficients of 0.717 and respectively 0.685, which are relatively low. That explains why the most of already reported data on tests performed using HRB are difficult to interpret and in particular to correlate with the refractory properties.

Corrosion versus Porosity

Our experimental data confirmed the fact that penetration and corrosion, mainly by low ratio baths, usually depend on the permeability or pore dimensions and not on the porosity. Figure 13b shows that there is no correlation between corrosion and porosity for both types of baths, LRB and HRB, used in the experiments. Because many manufacturing conditions, which generally determine the final ceramic texture of the refractory brick, are identical within the same type of bricks, extruded (B1, B2 and B4) or dry pressed (B3 and B5), a linear variation of the corrosion in terms of porosity could eventually be considered.

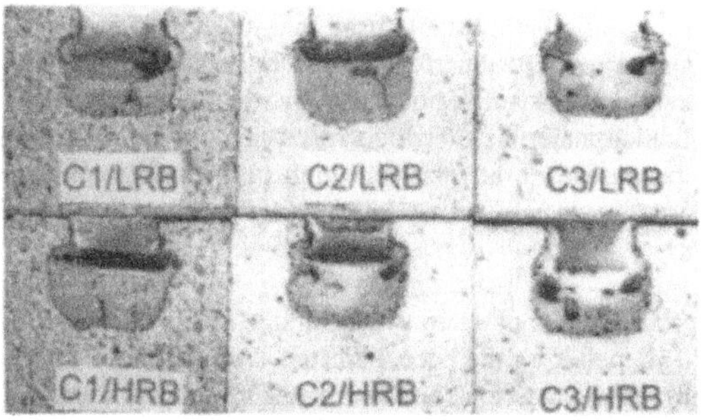

Figure 14. Castable samples C1 to C3 after 48 hours test in oxidizing atmosphere with LRB and HRB.

The fact that there is no correlation between cryolite corrosion and porosity could be better accepted if one tries to correlate the corrosion with porosity for the samples within the castables group. Castable C1, at 23.18% porosity and the lowest Al_2O_3 content and Al_2O_3:SiO_2 ratio, should be the most corroded according to the existing theories. From the data in Table III-a and from the cross-sections of the corroded specimens (Figure 14), it appears that there are no important differences between the 3 castable samples and also between them and the extruded bricks B1, B2 and B4 when tested with LRB. An apparent higher corrosion of C1 by the HRB, as shown in Table III-b, was due to the sectioning through a few large spherical voids, trapped in the castable during preparation.

Corrosion versus Reactivity

For all three castables, the low permeability values and the pore diameters calculated with them explain how important the binding phase is in generating the fine pore structure of the ceramic matrix after firing. The unrealistic part is the fact that each aggregate used in these three castable mixes does not seem to have any contribution to the measured permeability. That could be due to a coating of the porous aggregate grains with an impervious gel-phase of calcium aluminate hydrates and silica fume [50]. Its contribution should not be considered when evaluating the corrosion behaviour because the crucible cavity was drilled into the castable after curing. As a result, each castable has exposed to the molten bath all three components of its texture: the binding phase, the pores and the sectioned refractory aggregate grains.

Three different kinds of refractory aggregates were used: crushed B1 brick in C1, calcined fireclay (similar to that used in B3) in C2 and calcined flint in C3. By comparing the corrosion results, in particular for LRB (Table IX-a), it does not seem to matter what type of refractory aggregate was used in the castable. Because the binding phase was the same in all three castables, one can conclude that the penetration and corrosion depend mainly on the nature and properties of the binding phase rather than on the nature and properties of the aggregate. That could be due to the high reactivity of the silica fume, which would react fast with the molten bath to form the protective layer of viscous glass rich in silica, similar to the bricks B1 to B5.

Conclusions for Example # 3

- When refractory bricks from different manufacturing sources are compared, the permeability data and not the porosity should be used in order to predict their penetration and corrosion behaviour against cryolitic baths, if the refractory material is to be used as a barrier lining in aluminum electrolytic cells.
- The correlation corrosion-permeability and corrosion-pore diameter emphasized the fact that, at identical porosities and similar chemistries, an extruded brick has pores almost an order of magnitude smaller than a dry-pressed brick. As a result, a much better penetration and corrosion resistance and consequently a better performance are to be expected.
- When refractory materials of similar permeability or pore dimensions are compared, their reactivity while in contact with the molten bath should be also considered. Consequently, their resistance to penetration by the regular low ratio bath, usually used in the electrolytic process, should be a deciding factor in assessing the refractory performance. This reactivity is mainly due to the high silica phases of low crystallinity or amorphous, usually difficult to identify by X-ray diffraction techniques. For particular materials, the high reactivity could bring some shadow on the pore dimension issue because even at relatively high permeability and pore diameter values, the penetration and corrosion remain at very low levels.

CORROSION MECHANISM

As mentioned above, the static corrosion process takes place in two distinct steps. The first is the penetration, which according to equation (3) depends on the pore diameter, wetting angle, surface tension and viscosity of the melt at the moment of contact. The second is the chemical interaction, which is usually a dissolution process of the solid in the melt, governed by diffusion (most of the static corrosion processes) or by chemical reaction. For example, if the liquid does not wet the solid, the chemical interaction practically cannot occur. If the liquid wets the solid and the wetting angle is $\theta=0$, then $\cos\theta=1$ and theoretically the "sucking pressure" of the melt into an open pore is maximum. The penetration takes place easily, and the rate would depend mainly on the viscosity of the melt and the pore diameter. Also a perfect wetting ($\theta=0$) gives optimum physical contact between the melt and refractory, allowing for the dissolution to occur. That depends on how the refractory and the melt would interact with one another. A lack of chemical activity from one of the species in contact will allow the penetration to continue further inside the open pores. After a period of time, the dissolution would start producing a thin layer of viscous liquid, known to be an albite glass [1, 12]. This layer stops the penetration from progressing, due to its sealing capability at the interface refractory-liquid bath and would explain how, in example # 2, the samples P2, P4, and P4 were penetrated and P1, P3 and P5 were not. The mineralogical phase responsible for the refractory's behavior is the ceramic matrix for bricks and castables or the fine fraction in the dry barrier refractories.

REFERENCES

[1] O.-J. Siljan, "Studies of the refractory-melt interface in aluminum reduction cells", *Proceedings of the Symposium on Refractories for Aluminum Industry*, Seattle-Bellevue, WA, October 26-28, 1999, 140-159.

[2] F. Brunk, "Corrosion and behavior of fireclay bricks of varying chemical compositions used in the bottom lining of reduction cells, *Light Metals* 1994, 477-482.

[3] C. Allaire, "Electrolysis bath testing of refractories at Alcan", *Journal of the Canadian Ceramic Society*, Vol. 60, No. 2, February 1991, 47-52.

[4] O.-J. Siljan, "Reaction of fireclay refractories in aluminum reduction cells", *Proceedings, UNITECR '95*, Kyoto, Japan, 1995, 280-287.
[5] G. Oprea, "Methods of testing cryolite penetration of potlining refractories for aluminum electrolytic cells", *Proceedings, UNITECR '97*, New Orleans, LA, 1997, 1677-1693.
[6] G. Oprea, "Laboratory characterization of refractory-cryolitic bath interactions", *Light Metals* 1999, Ed. C.E. Eckert, The Minerals, Metals & Materials Society, 1999, 437-444.
[7] G. Oprea, "Standard test method for corrosion resistance of refractories to molten cryolitic bath", presented at the Task Force Meeting of ASTM Subcommittee C8.05 Refractories for Aluminum Industry, San Antonio, TX, 1998
[8] F. Brunk, W. Becker, K. Lepere, "Cryolite influence on refractory bricks: influence of SiO_2 content and furnace atmosphere", *Light Metals* 1993, 315-320.
[9] O.J. Siljan, A. Seltveit, :Chemical reactions in refractory linings of alumina reduction cells", *Proceedings, UNITECR '91*, Aachen, Germany, 1991, 38-42.
[10] K. Grjotheim, H.Kvande, "Penetration barriers in the cathode of Hall-Héroult cells", *Aluminium-68*, January 1992, 64-69
[11] O.J. Siljan, C. Schoning, A. Seltveit, "Investigations of deteriorated alumina reduction cell potlinings", *Proceedings, UNITECR '91*, Aachen, Germany, 1991, 31-36.
[12] T. Grande, J. Rutlin, "Viscosity of oxyfluoride melts relevant to the deterioration of refractory linings in aluminum reduction cells", *Light Metals* 1999, 431-436.
[13] C. Schoning, Grande T., O.J. Siljan, "Cathode refractory materials for aluminum reduction cells", *Light Metals* 1999, 231-237.
[14] W.E. Lee, S. Zhang, "Melt corrosion of oxide and oxide carbon refractories", *International Materials Reviews*, 1999, vol. 44, no. 3, 77-104.
[15] A.E. Scheidegger, *The physics of flow through porous media*, Third Edition, Univ. of Toronto Press, 1974, 127-129.
[16] J.H. Hampton, D.A. Thomas, "Modeling relationships between permeability and cement paste pore microstructures", *Cement and Concrete Research*, vol. 23, 1317-1330, 1993.
[17] J.P. Ollivier, M. Massat, "Permeability and microstructure of concrete: a review of modeling", *Cement and Concrete Research*, Vol. 22, 1992, 503-514.
[18] J. Holly, D. Hampton, M.D.A. Thomas, "Modeling relationships between permeability and cement paste pore microstructures", *Cement and Concrete Research*, Vol. 23, 1993, 1317-1330.
[19] D. Harris and G. Oprea, "Cryolite penetration studies on barrier refractories for aluminum electrolytic cells", *Light Metals* 2000, 419-427.
[20] H. Kvande, "Bath chemistry and aluminum cell performance - facts, fictions and doubts", *J. of Metals*, November 1994, 22-28.
[21] A. T. Tabereaux et al., "Lithium-modified low ratio electrolyte chemistry for improved performance in modern reduction cells, *Light Metals* 1993, 221-226.
[22] A. T. Tabereaux, "Phase and chemical relationships of electrolytes for aluminum reduction cells", *Light Metals* 1985, 753-761.
[23] R. D. Peterson, A. T. Tabereaux, "Effect of bath additives on aluminum metal purity", *Light Metals* 1986, 491-501
[24] K. Grjotheim et al., "Addition of LiF and Mg F_2 to the bath of the Hall-Heroult process", *Light Metals* 1983, 397-411.
[25] A. A. Kostyukov et al., "Physico-chemical study of the system NaF-AlF_3-CaF_2 (MgF_2)", *Light Metals* 1983, 389-396.
[26] N. E. Richards et al., "Further considerations of the acid-base system alkali metal fluoride - aluminum fluoride", *Light Metals* 1983, 3793-89.
[27] E. W. Dewing, "Activities in Li_3AlF_6 - Na_3AlF_6 melts", *Light Metals* 1985, 737-749.

[28] Z. Qiu et al., "Influences of additives on the cryolite ratio of aluminum electrolyte", *Light Metals* 1993, 291-296.
[29] H. Kvande,"The structure of alumina dissolved in cryolite melts" *Light Metals* 1986,451-459
[30] K. Grjotheim et al.,"Low-melting baths in aluminum electrolysis" *Light Metals* 1986,417-23
[31] J. Guzman et al., "The influence of different fluoride additions on the vapor pressure of molten cryolite", *Light Metals* 1986, 425-429.
[32] L. Bullard, D. D. Przybycien, "DTA determination of bath liquidus temperatures: Effect of LiF", *Light Metals* 1986, 437-444.
[33] M. Chrenkova et al., "Density, electrical conductivity and viscosity of low melting baths for aluminum electrolysis", *Light Metals* 1996, 227-232.
[34] G. M. Haarberg et al., "Electrical Conductivity measurements in cryolite alumina melts in the presence of aluminum", *Light Metals* 1996, 221-225.
[35] P. Fellner, A. Silny,"Viscosity of sodium cryolite-aluminum fluoride-lithium fluoride melt mixtures", *Ber. Bunsenges. Phys. Chem.* Vol.. 98, No. 7, 1994, 935-937.
[36] P. Brilloit et al., "Melt penetration and chemical reactions in carbon cathodes during aluminum electrolysis. I. Laboratory experiments", *Light Metals* 1993, 321-330.
[37] L. P. Losius, H. A. Oye, "Melt Penetration and Chemical Reactions in Carbon Cathodes During Aluminum Electrolysis. II. Industrial Cathodes", *Light Metals* 1996, 331-340.
[38] H. Kvande et al., "Penetration of bath into the cathode lining of alumina reduction cells", *Light Metals* 1989, 161-167.
[39] M. Sorlie, H. A. Oye, *Cathodes in aluminum electrolysis*, 2nd ed., Aluminum-Verlag, Dusseldorf, 1994, 115-150.
[40] J. M. Peyneau, "Design of highly reliable pot linings", *Light Metals* 1989, 175-181.
[41] G. Oprea, "Corrosion studies of dry barrier potlining refractories for aluminum reduction cells ", presented at the 8th Symposium on Refractories for the Aluminum Industry, the 47th Pacific Coast Regional Meeting of the ACERS, Seattle, WA, November 1-3,1995.
[42] S. F. Johansson, "Chemical processes in cathode refractories and insulation during pot life", *Light Metals* 1995, 479-492.
[43] S. R. Brandtzaeg et al., "Experiences with anorthite powder-based penetration barrier in 125 kA Sodeberg cell cathodes", *Light Metals* 1993, 309-314.
[44] C. Allaire, "Refractory lining for alumina electrolytic cells", *Journal of the American Ceramic Society*, **75**, [8], 1992, 2308-2311.
[45] G. Oprea, "Corrosion tests for barrier refractories used in aluminum reduction cells ", presented at the Refractory Ceramics Division of the American Ceramic Society Meeting, Huron, OH, October 6-7, 1995.
[46] S. F. Johansson, "An alternative method for evaluation of resistance of pot insulation to bath attack", *Light Metals* 1986, 501-514.
[47] A. T. Tabereaux, M. Windfeld, "Evaluation and Performance of Powder 'Dry-Barrier' Refractories for Use in Aluminum Cell Cathodes", *Light Metals* 1995, 471-477.
[48] G. Oprea, "Wettability and reactivity of silica-alumina refractories in contact with cryolitic baths", *Proceedings of the International Symposium on Advances in Refractories for the Metallurgical Industries II*", Quebec City, Quebec, Canada, August 24-29, 1996, 305-321.
[49] G. Oprea, "Corrosion tests of refractories for aluminum electrolytic cells", *Proceedings of the International Symposium on Advances in Refractories for the Metallurgical Industries III*", Quebec City, Quebec, Canada, August 22-26, 1999, 277-291.
[50] F. Esanu, T. Troczynski, G. Oprea, "Microstructural studies on binding systems of self-flowing refractory castables", *Proceedings, UNITECR '99*, Berlin, Germany, 1999, 26-33.
[51] G. Oprea, "Corrosion tests on refractories for aluminum electrolytic cells", *Proceedings of the Symposium on Refractories for Aluminum Industry*, Seattle-Bellevue, WA, October 26-28, 1999, 189-205.

INTERFACIAL PHENOMENA

Claude Allaire
CIREP-CRNF
Dept. of Eng. Physics & Materials Engng.
Ecole Polytechnique (CRIQ campus)
8475 Christophe Colomb Street
Montreal, Quebec, H2M 2N9

ABSTRACT

According to materials science, interfacial phenomena are the sorts/kinds of reactions between a material or its matrix and its surrounding environment. In case of ceramics, these phenomena can take place during processes such as sintering, grain growth and wetting. In this paper, two important factors (i.e. surface tension and surface energy) influencing the interfacial phenomena are presented. The effect of curved surfaces on the above processes is also shown. Parameters affecting the wetting characteristics of liquids on solid substrates are discussed as well. Emphasis is made to ceramic materials exposed to molten aluminum. Finally, the role of the above phenomena on the direct oxidation of molten aluminum is described.

INTRODUCTION

The corrosion of solid materials by liquids is highly dependent on the characteristics of the solid/liquid interface, which dictates the extension/level of wetting/infiltration of the corrosive liquid on/into the solid substrate. Among the corrosive liquids of interest are the molten aluminum alloys whose effects on the refractory lining of aluminum treatment furnaces have well been discussed elsewhere [1]. It was shown that the direct oxidation of molten aluminum alloys in service is the main cause of the corrosion of the refractories at the metal line in these furnaces. Interfacial phenomena are at the basis of this corrosion process which can lead to damaging the sidewall in the furnaces. Before describing this process, a review of the following topics will be necessary:

- Source of surface tension and surface energy,
- Effect of curved surfaces,
- Methods of measurement,

- Wetting phenomenon,
- Wetting characteristics of aluminum,
- The direct oxidation of molten aluminum.

SOURCE OF SURFACE TENSION AND SURFACE ENERGY

One can suppose that a material is made of spherical atoms having a section dA and a diameter dL. To bring one of these atoms from the inside of the material up to the material/atmosphere interface, a tensile force dF_s is required, as shown in Fig. 1. If dU_s is the variation of the potential energy of the moved atom, thus:

$$\gamma = dU_s / dA = - dF_s / dL = \sigma \quad (1)$$

where γ = Surface energy of the material (Joule/m^2)
σ = Surface tension of the material (N/m)

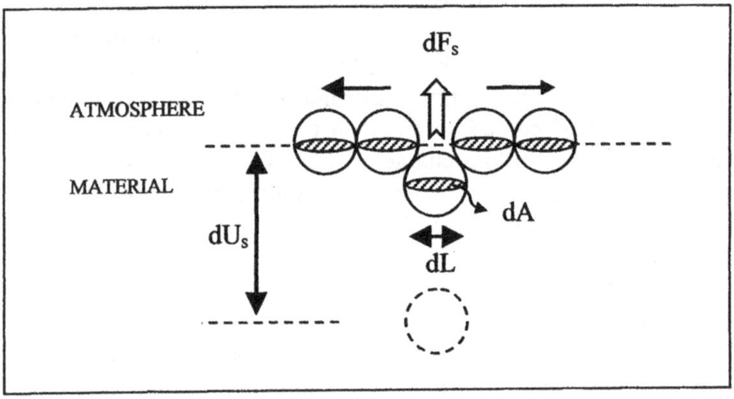

Fig. 1: The basic of surface tension and surface energy.

In equation (1), it is assumed that the atom is not exposed to shear stress when moving from the inside up to the material/atmosphere interface. This is the case of liquids whose surface energy and surface tension have the same value. Since solids can offer a certain resistance to shear stress, unlike liquids, their surface energy is usually higher than their surface tension (see Table I).

The origin of dU_s in equation (1) can be explained as follow. As shown in Fig. 2, for example at the position "1" in a supposed spherical material, an atom is under the influence of its neighbors, which induced an attractive force whose resultant is zero. However, at the surface of the material (e.g. position 2), only half of the neighbors are present and thus a resultant attractive force, F(r), is acting on the atom toward the center of the material. Assuming that not only the first neighbors but also the second, the third and the $n^{ième}$ neighbors have an attractive

Table I. Surface energy of various materials [2]

Material	Temperature (°C)	Surface energy (ergs/cm²)
Water (liquid)	25	72
Lead (liquid)	350	442
Copper (liquid)	1120	1270
Copper (solid)	1080	1430
Silver (liquid)	1000	920
Silver (Solid)	750	1140
Platinum (liquid)	1770	1865
Sodium chloride(liquid)	801	114
NaCl crystal (100)	25	300
Sodium sulfate (liquid)	884	196
Sodium phosphate, $NaPO_3$ (liquid)	620	209
Sodium silicate (liquid)	1000	250
B_2O_3 (liquid)	900	80
FeO (liquid)	1420	585
Al_2O_3 (liquid)	2080	700
Al_2O_3 (solid)	1850	905
0.20 Na_2O-0.80 SiO_2 (liquid)	1350	380
0.13 Na_2O-0.13 CaO-0.74 SiO_2 (liquid)	1350	350
MgO (solid)	25	1000
TiC (solid)	1100	1190
CaF_2 crystal (111)	25	450
$CaCO_3$ crystal (1010)	25	230
LiF crystal (100)	25	340

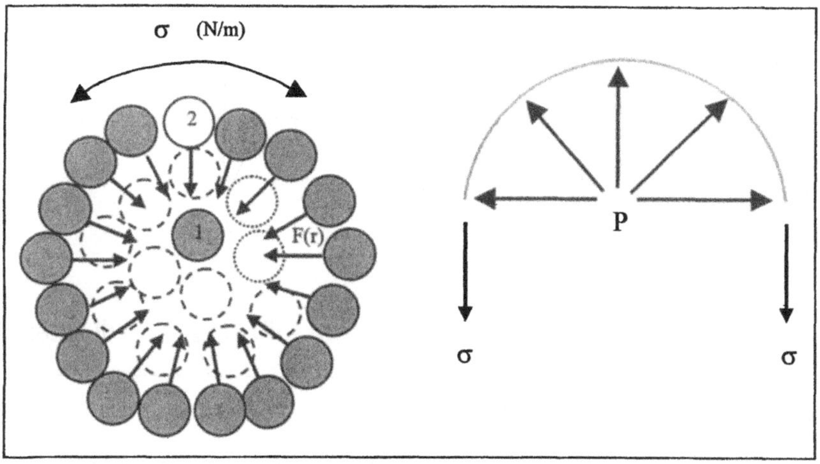

Fig. 2: The role of the attractive bonding force on the surface tension and surface energy.

effect on the atom (which decreases when "n" increases), it may explain why F(r) is continuously decreased as a function of "r". Thus, the variation of "U_s" from positions "1" and "2" (i.e., ΔU_s) is given by the following equation:

$$\Delta U_s = -\int_1^2 F(\dot{r})\,dr \tag{2}$$

To compensate for F(r) in the material, a positive pressure is required. As shown in Fig. 2, such pressure is induced by the equivalent action of the surface tension, σ.

Surface energy (U_s) is, in general, one of the various components of the total energy (U) of an interface, as shown by the following equation (see Fig.3):

$$dU = dU_s + dU_{comp} + dU_{thermal} + dU_{chemical} \tag{3}$$

where: $dU_s = \gamma dA$ (see equation (1))

$$dU_{comp} = -PdV \tag{4}$$

$$dU_{thermal} = TdS \tag{5}$$

$$dU_{chemical} = \Sigma\mu_i dn_i \tag{6}$$

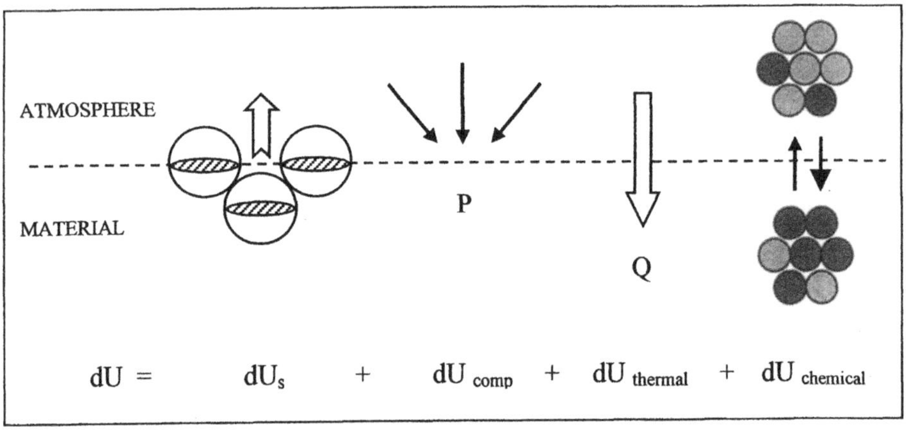

Fig. 3: Interfacial energy components.

The term " dU_{comp} " originates from the variation of volume (dV) at the interface due to the action of an applied pressure (P). While, the term " TdS " originates from the variation of entropy (dS) due to heat flow across the interface. Chemical exchange at the interface is responsible for the term "$dU_{chemical}$", which is given by the summation of the molar fraction of the diffusing species (dn_i) time their chemical potential (μ_i).

According to the thermodynamic, the Heltmoltz free energy (F) of the interface is given by:

$$dF = (\gamma dA - PdV + \Sigma\mu_i dn_i + TdS) - d(TS) \tag{7}$$

This equation may be written as:

$$dF = (\gamma dA - PdV + \Sigma\mu_i dn_i + TdS) - (TdS + SdT) \tag{8}$$

$$= \gamma dA - PdV + \Sigma\mu_i dn_i - SdT \tag{9}$$

For a constant temperature ($dT = 0$) and a constant volume ($dV = 0$), and under an equilibrium condition ($dn_i = 0$), equation (9) becomes:

$$dF = \gamma dA = dU_s \tag{10}$$

According to equation (10), the surface energy. of a material is given by the variation of its Heltmoltz free energy (dF) per unit area (dA).

EFFECT OF CURVED SURFACES

Let us assume a flat semi-infinite material exposed to an atmospheric pressure P_0 (see Fig. 4). To curve the interface toward the inside of the material (concave surface), a compressive force (or stress) should be applied on the atoms existing beneath the material interface. This leads to a pressure increase across the interface. However, to curve the interface in the opposite direction (convex surface), a tensile force (or stress) should be acting on the same atoms, resulting in a pressure decrease along the interface.

As an example, Figure 5 illustrates a deep capillary into a liquid, which is exposed to an atmospheric pressure of P_0. If a soap film at the bottom-end of the capillary is blown-up inside the liquid under the pressure "P" ($P > P_0$), a bubble of radius "r" will be formed. This will create a convex interface resulting in a pressure drop of "ΔP". The energy of that interface, under equilibrium condition, is equal to the energy consumed to blow the bubble. Thus:

$$\Delta P \, dV = \gamma \, dA \tag{11}$$

where: $dV = 4 \Pi r^2 dr$ (12)

$dA = 8 \Pi r\, dr$ (13)

γ = Surface energy of the liquid

Combining the equations (11), (12) and (13), the following Laplace equation is obtained:

$$\Delta P = 2\gamma / r \quad (14)$$

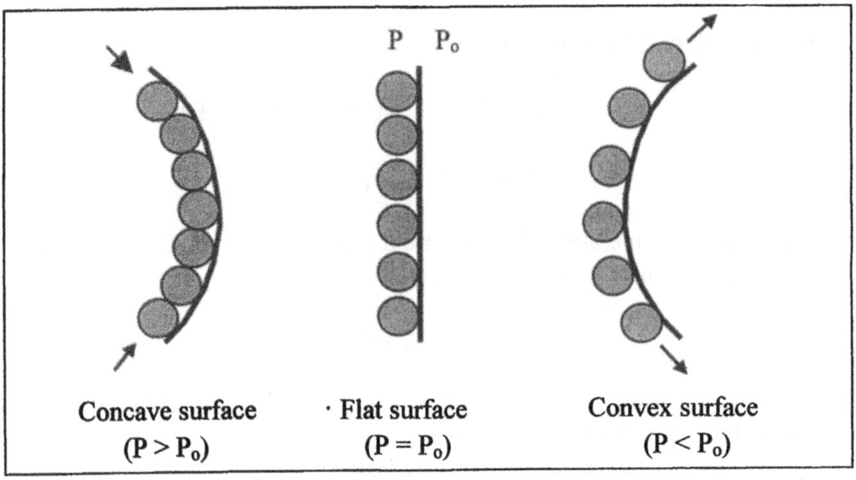

Fig. 4: Effect of the curvature level on the pressure variation across an interface.

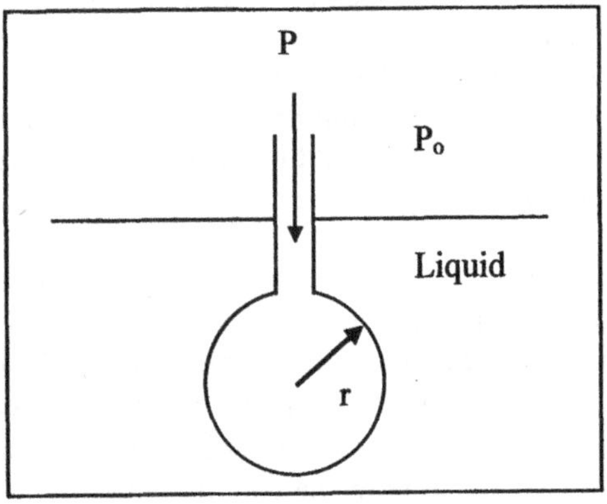

Fig. 5: Blowing of a bubble inside a liquid.

METHODS OF MEASUREMENT

Although there are several techniques/methods to measure the surface energy of materials, two of them are most commonly used in this regard. These are the capillary technique and the sessile drop method.

The capillary technique is based on the Laplace equation and involves the measurement of the deepness of a capillary channel filled with a liquid. If the liquid wets the sidewall of capillary channel (see next section), the pressure drop across the liquid/atmosphere (convex) interface will make the liquid move up inside the channel. Under equilibrium condition, the "capillary force" counterbalances the weight of the liquid column inside the channel, which is above the original liquid level. Thus, according to Fig. 6:

$$\Delta P = 2\gamma / r = 2\gamma\,[\cos(\theta) / R] = \rho g h \tag{15}$$

where ρ = Density of the liquid
g = Gravitational constant
γ = Surface energy of the liquid

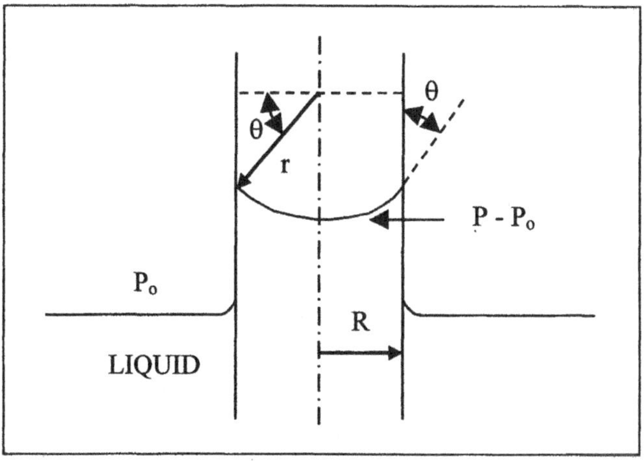

Fig. 6: The capillary technique.

From equation (15) Jurin's law is obtained as:

$$\gamma = R\rho g h / 2\cos(\theta) \tag{16}$$

The sessile drop technique is based on the use of a solid substrate supporting a liquid drop, as shown on Fig. 7. Based on the shape of the drop, after equilibrium, the surface tension of the liquid may be calculated using the following formula:

$$\gamma = \tfrac{1}{2} g Z^2 (1.641\ X/1.641\ X + Z) \tag{17}$$

where X and Z are geometrical parameters (see Fig. 7)

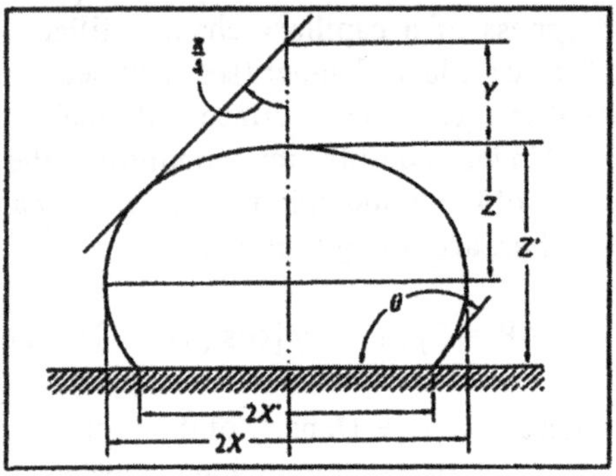

Fig. 7: Geometrical parameters according to the sessile drop technique (after Iida and Guthrie [3]).

This technique is usually performed at high temperature using an apparatus such as shown in Fig. 8. An example of results with this apparatus is presented in Fig. 9.

The sessile drop technique is also commonly used to perform wetting angle measurement.

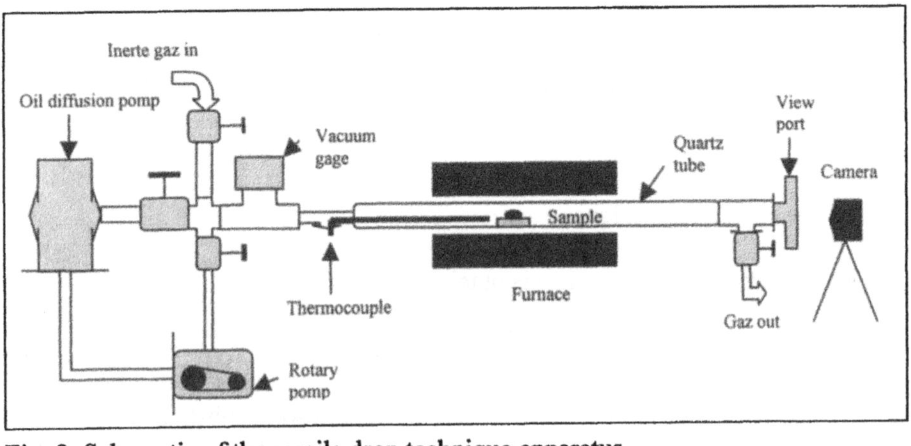

Fig. 8: Schematic of the sessile drop technique apparatus.

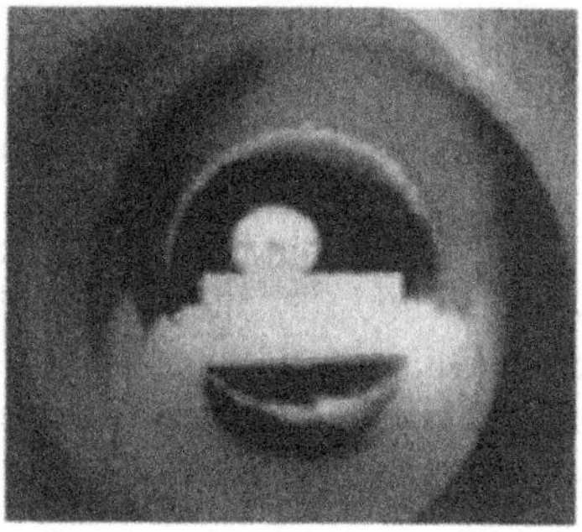

Fig. 9: Appearance of a molten aluminum drop on a refractory substrate.

WETTING PHENOMENON

In Fig. 6, the case of a wetting liquid inside a capillary channel is shown. The curved liquid/atmosphere interface is convex. More generally, the shape of such interface is dictated by the following three vectors (see Fig. 10):

γ_{sv} = "solid-vapor" interfacial energy
γ_{lv} = "liquid-vapor" interfacial energy
γ_{sl} = "solid-liquid" interfacial energy

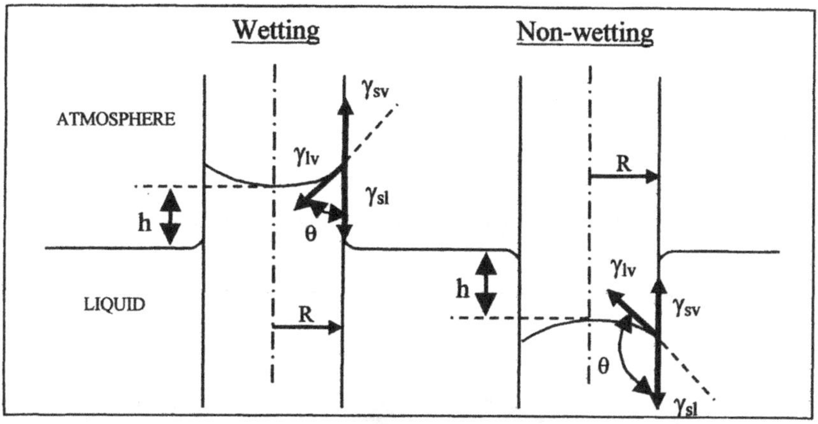

Fig. 10: Wetting and non-wetting liquids inside a capillary channel.

Under equilibrium condition, the sum of these vectors should be zero which fixed the wetting angle (θ), according to the following Young's law:

$$\cos(\theta) = (\gamma_{sv} - \gamma_{sl}) / \gamma_{lv} \quad (18)$$

The case of a non-wetting liquid is also shown in Fig. 10. In this case, the curved liquid/atmosphere interface is concave. This leads to a pressure increase across the interface, which causes the liquid to move downward.

As illustrated in Fig. 11, a wetting liquid lying on a flat solid material produces a wetting angle normally less than 90 °. By contrast, a wetting angle of more than 90 ° is considered as the characteristic of a non-wetting liquid. The extension of spreading of a liquid on a solid substrate can also be characterized by the so-called "work of adhesion", W_{sl}, which is given by the following equation (see Fig. 12):

$$W_{sl} = \gamma_{sv} + \gamma_{lv} - \gamma_{sl} = \gamma_{lv}\ [1 + \cos(\theta)] \qquad (19)$$

The "work of adhesion" is enhanced by an increase in the bonding force between the liquid and the solid.

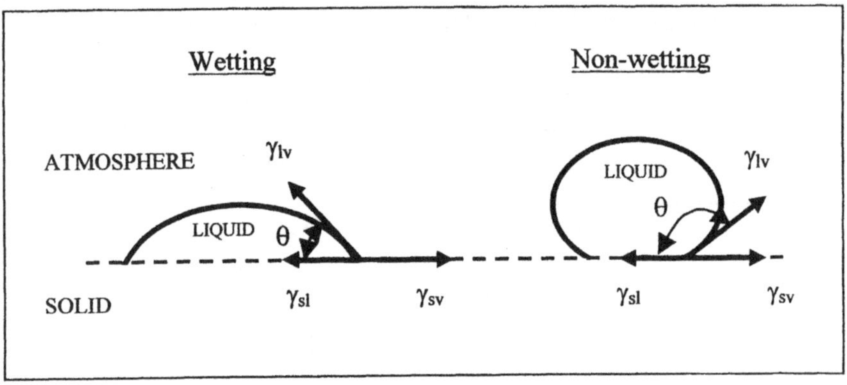

Fig. 11: Wetting and non-wetting liquids on a flat solid substrate.

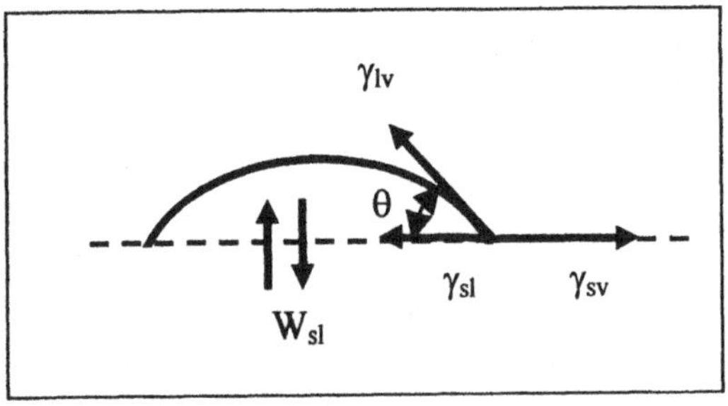

Fig. 12: The work of adhesion (W_{sl}).

WETTING CHARACTERISTICS OF MOLTEN ALUMINUM

Tables II and III give the wetting (or contact) angle of molten aluminum alloys on a sapphire (Al_2O_3) and that of pure molten aluminum on various ceramic materials, respectively. As indicated in Table II, alloying elements such as Se, Te and Ge tend to decrease the wetting of molten aluminum (i.e., increase the contact angle). However, most of the alloying elements in molten aluminum act in opposite way. This is the case, for example, for Cu, Zr, Mg, Si and Zn. The increase of the wetting of molten aluminum containing such elements is related to the decrease of its surface tension, as shown in Fig. 13. Increasing the temperature of the molten metal also tends to decrease its surface tension and thus to promote wetting (see Fig. 14 and Table III).

Table II. Wetting Characteristics of molten aluminum on a sapphire (after Wolf[4])

Increased wetting	
Alloying elements (atomic %)	Contact angle (Degree)
Pure Al	139 ± 5
0.230 Cu	119 ± 5
0.940 Mg	118 ± 5
0.130 Zr	118 ± 5
Decreased wetting	
Alloying elements (atomic %)	Contact angle (Degree)
0.014 Se	155 ± 13
0.017 Te	152 ± 6
0.020 Ge	144 ± 5

The oxygen partial pressure, P_{O_2}, has also a significant effect on the surface tension of molten aluminum. As presented in Fig. 15, an increase in P_{O_2} results in a decrease in the surface tension before equilibrium condition is achieved. The solubility of oxygen in molten aluminum is the main cause of this effect. Increasing P_{O_2} increases the kinetics of oxygen diffusion in the molten metal, which in turn increases its entropy at the liquid/atmosphere interface. Such an increase then lowers the Heltmoltz free energy and thus the surface tension of the molten metal (see equations (1) and (10)).

The detrimental effect of the oxygen partial pressure on the oxidation of Al-5 % Mg alloy is illustrated in Fig. 16. At low P_{O_2} (under nitrogen purging), the oxide layer on top of the molten metal is very thin and contains a low level of aluminum

Table III. Wetting Characteristics of molten aluminum tested on various ceramics (after Geirnaert [5])

Materials	Temperature (°C)	Atmosphere	Contact angle (degree)
MgO	980	vacuum	155
$MgAl_2O_4$	980	vacuum	155
Al_2O_3	900 – 1170	vacuum	118 – 70
	940		170 (0 h)
	970		148 (0 h)
	980	vacuum	152
	1255		85 (0 h)
			77 (0.3 h)
			66 (0.4 h)
			47.5 (0.5 h)
AlN	970 – 1150	vacuum	123 - 84
TiN	980 – 1100	vacuum	84 - 45
B_4C	600 – 670	vacuum	117 - 118
Al_4C_3	1000	vacuum	104
SiC	1100	vacuum	34
TiC	700	Ar	118
	980 – 1060	vacuum	111 - 65
TiB_2	980 - 1110	vacuum	90 - 60

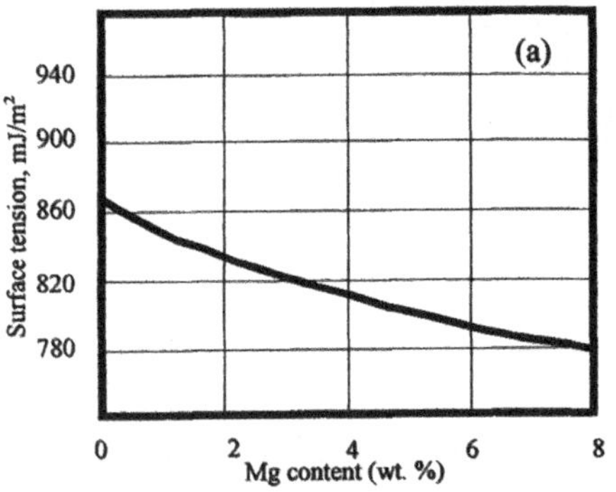

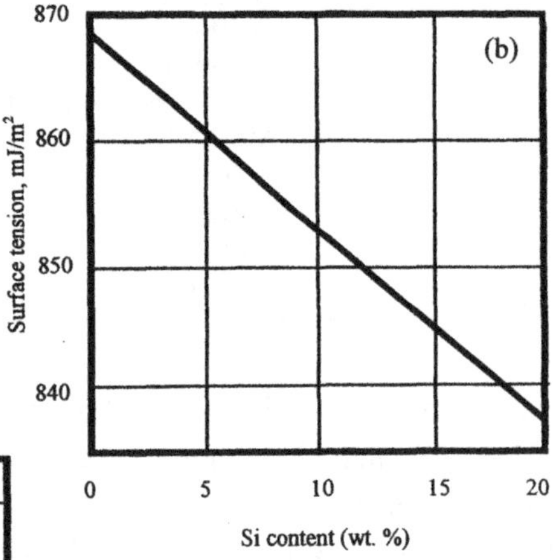

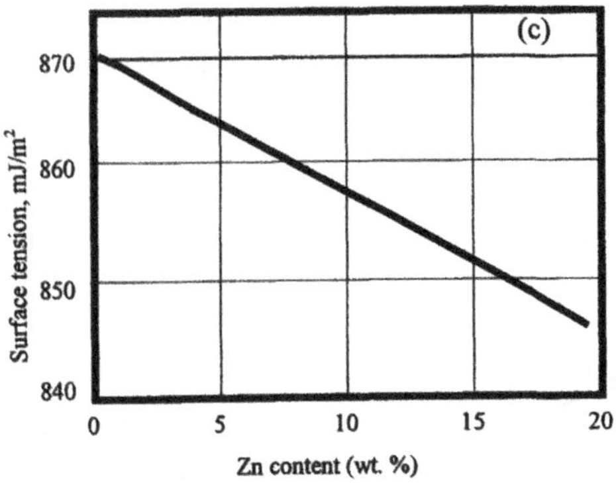

Fig. 13: Effect of the alloying elements on the surface tension of molten aluminum at 973 °K; (a) - Mg (after Garcia-Cordovilla et al. [6]), (b) - Si (after Goicoechea et al. [7]) and (c) - Zn (after Goicoechea et al. [7]).

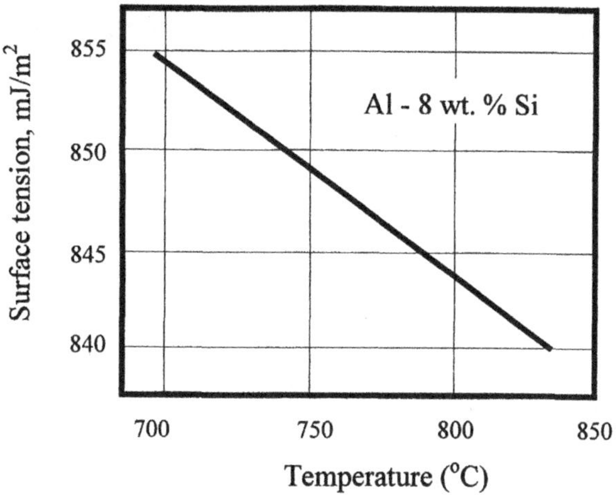

Fig. 14: Effect of the temperature on the surface tension of molten aluminum (after Goicoechea et al.[7]).

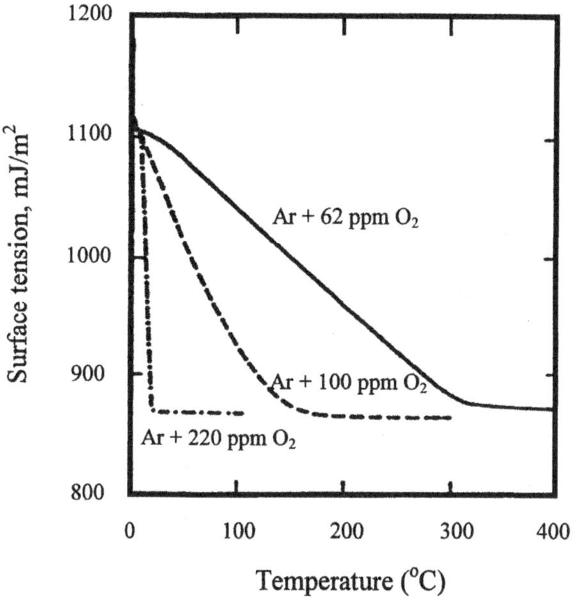

Fig. 15: Effect of the oxygen partial pressure on the surface tension of molten aluminum(after Garcia-Cordovilla et al.[6]).

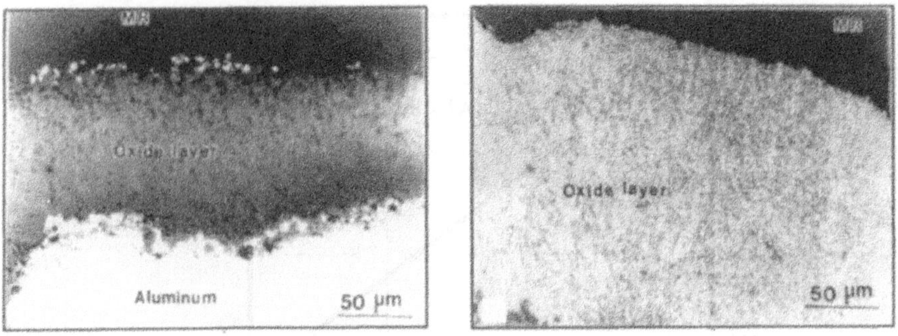

(a) (b)

Fig. 16: Oxide layer on the top of Al-5 wt. % Mg after 24 hours at 815 °C; (a) - under nitrogen purging and (b) - under air.

metal (gray area of the oxide layer under optical microscopy observations). However, at high P_{O_2} (under air), the thickness of the oxide layer becomes much higher, as well as its content in aluminum metal (white region of the oxide layer under optical microscopy observations). Based on the previous discussion, the decrease of the surface tension of the molten metal at high P_{O_2} favors the wetting of the alumina crystals, which forms during the oxidation process. This in turn prevents the formation of alumina-alumina grain boundaries, as shown in Fig. 17. The instability of such boundaries is significantly propagated when:

$$\gamma_b > 2\,\gamma_{sl} \quad \text{and} \quad \gamma_{sl} < \gamma_{sv} \tag{20}$$

where γ_b = Alumina-alumina grain boundary energy

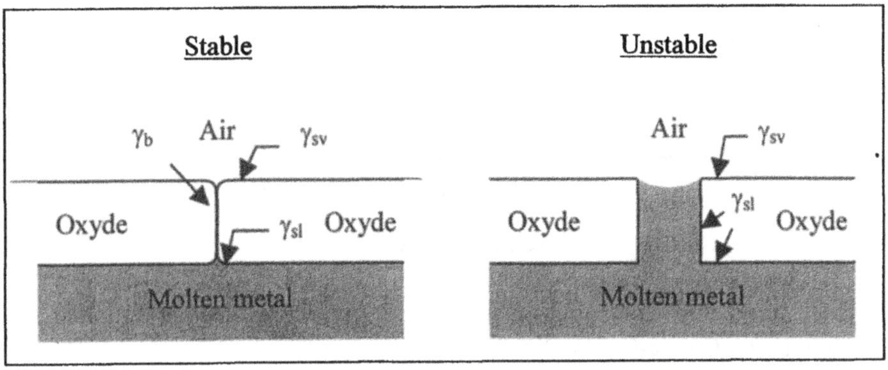

Fig. 17: Stability of oxide grain boundaries (after Newkirk et al. [8]).

Since magnesium decreases the surface tension of molten aluminum, it promotes aluminum oxidation by preventing the formation of alumina-alumina grain boundaries, and thus the formation of a protective oxide layer. This is at the origin of the DIMOX™ process, which is described in the next section.

THE DIRECT OXIDATION OF MOLTEN ALUMINUM

Direct oxidation of molten aluminum alloys above 1050 °C and in the presence of magnesium is shown schematically in Fig. 18. This involves, according to the DIMOX™ process [8], the following steps (see also reference 9):

1. Magnesium evaporation followed by its oxidation, promoting the formation of a thin and continuous magnesia layer (~ 1 μm) at the molten metal/atmosphere interface,

$$Mg\,(g) + 1/2\,O_2\,(g) \text{ ----> } MgO\,(s) \qquad (21)$$

2. Reaction of the magnesia layer in contact with the Al-Mg bulk metal (C_{Bulk}), leading to the formation of a porous primary spinel layer,

$$2\,Al_{(C_{Bulk})} + 4\,MgO\,(s) \text{ ----> } MgAl_2O_4\,(s) + 3\,Mg\,(g) \qquad (22)$$

   This primary spinel layer (~ 1 mm in thickness) contains microchannels because of its higher molar volume as compared to that of the magnesia layer.

3. Infiltration of the bulk metal through the spinel microchannels leading to a reduction of its magnesium content up to the concentration at which the three phases, Liquid (L)-Al_2O_3 (A)-$MgAl_2O_4$ (S), are in equilibrium (C_{LAS}).

4. Reaction of the magnesia layer by the Al-Mg molten metal having the C_{LAS} concentration. This results in the formation of an Al_2O_3/Al composite consisted of columnar alumina crystals growing on top of the primary spinel layer:

$$2\,Al_{(C_{LAS})} + 3\,MgO\,(s) \text{ ---> } 3\,Mg\,(g) + Al_2O_3\,(s) \qquad (23)$$

   (As mentioned before, the presence of magnesium prevents the formation of a continuous alumina layer because it reduces the surface tension of the molten metal.)

   The growth rate of the alumina crystals is controlled by the displacement of the magnesia layer toward the oxygen source. Such displacement is promoted by the action of the magnesium vapor released from reaction (22). The latter is accumulated below the magnesia layer until it reaches some sufficiently high partial pressure causing the oxide to crack. The magnesium vapor fills these microcracks and is oxidized according to the equation (21) to form magnesia. This somehow/somewhat can compensate the magnesia that already consumed during step 2.

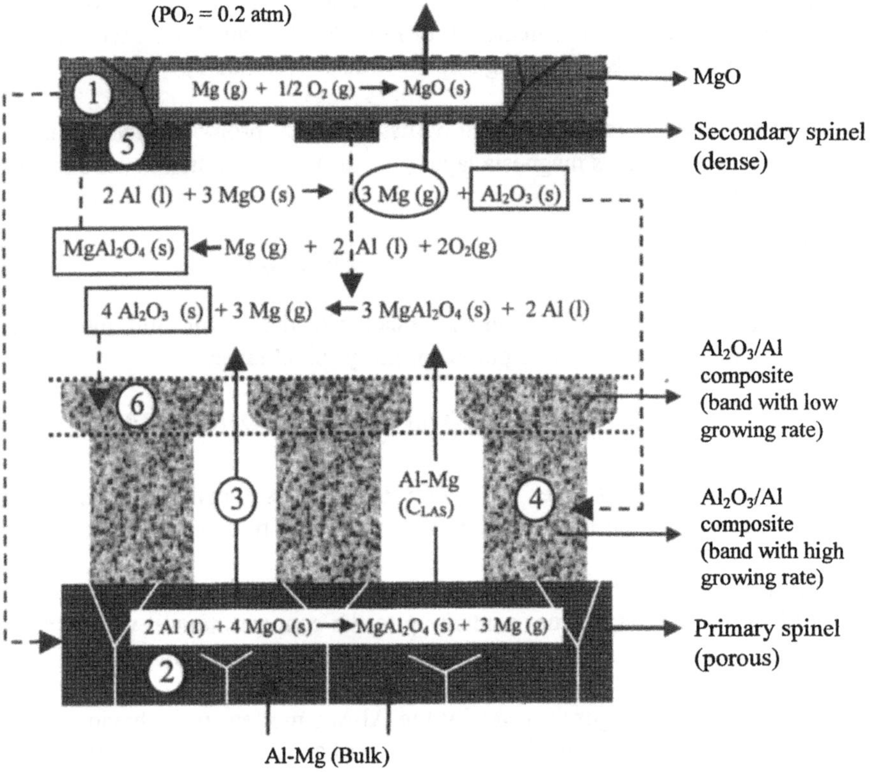

Fig. 18: Schematic representation of the DIMOX^{TM} process [9].

5. Formation of a dense secondary spinel underneath the magnesia layer which prevents the establishment of reaction (23), and thus reduces the growth rate of the alumina crystals. This secondary spinel is formed from the magnesium vapor released from reaction (22):

$$Mg\,(g) + 2\,Al\,(c_{LAS}) + 2\,O_2\,(g) \dashrightarrow MgAl_2O_4\,(s) \qquad (24)$$

The higher density of the secondary spinel, as compared to the primary spinel, could be explained by the fact that only fluid phases participated in its formation.

6. Decomposition of the secondary spinel layer by the action of the molten Al-Mg metal having the C_{LAS} concentration:

$$3\ MgAl_2O_4\ (s) + 2\ Al\ (c_{LAS})\ \text{--->}\ 4\ Al_2O_3\ (s) + 3\ Mg\ (g) \qquad (25)$$

This also leads to the formation of an Al_2O_3/Al composite consisted of columnar alumina crystals, growing on top of the one previously formed. However, because of the slower kinetics of reaction (25), as compared to reaction (23), the rate of corundum growth toward the oxygen source is reduced during this step. Since the molten metal is kept for a long period of time between the alumina crystals, their lateral growth is favored.

During oxidation of the molten metal via the DIMOX™ process, the repetitive formation and decomposition of the secondary spinel layer, according to reactions (24) and (25), respectively, permit the repetition of steps 3 to 6. This leads to the cyclic formation of successive large and thin bands of Al_2O_3/Al composite, having a high and a low growth rate, respectively (see Figures 19 and 20).

CONCLUSIONS

The characteristics of surface tension and surface energy, as well as the effect of curved surfaces were presented in case of ceramics. Two methods for measuring the surface energy and/or the wetting (or contact) angle were introduced: the capillary and the sessile drop techniques. The wetting phenomenon of solid materials by liquids was described. It was shown that most alloying elements decreases the surface tension of molten aluminum and thus favors its wetting property. Increasing the oxygen partial pressure as well as the temperature has a similar effect on the surface tension of molten aluminum. Interfacial phenomena were also shown to be at the origin of the direct oxidation of molten aluminum in the presence of magnesium. The latter prevents the formation of a protective oxide layer since it favors the instability of the alumina-alumina grain boundaries by decreasing the surface tension of the molten metal.

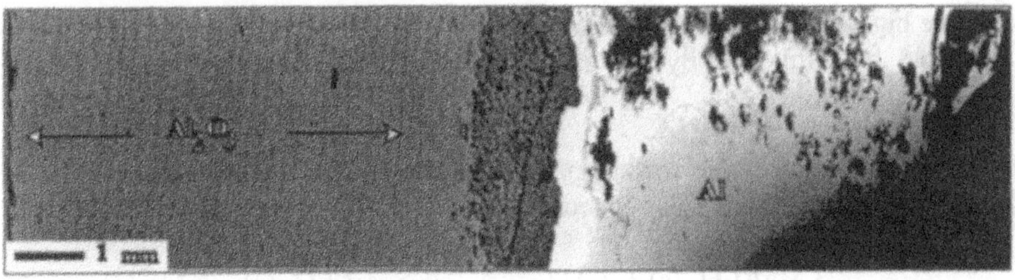

Fig. 19: Microstructure of molten Al-5 wt. % Mg alloy, after oxidation, in air and at 1150 °C. The layer between the alumina and aluminum areas is the primary $MgAl_2O_4$ spinel [10].

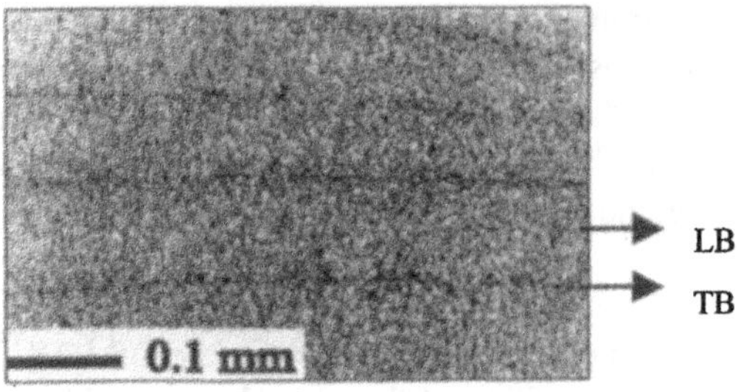

Fig. 20: Microstructure of the Al_2O_3-Al composite produced during the oxidation of molten Al-5 wt. % Mg alloy, at 1150 °C, showing large (LB) and thin (TB) bands [10].

ACKNOWLEDGMENTS

The author is very grateful to Dr. Saïed Afshar and Mrs. Marie-Eve Perron for their contribution to this work.

REFERENCES

1. C. Allaire, and M. Guermazi, "The Corrosion of Furnace Refractories by Molten Aluminum", Aluminum Transactions, Vol. 1, No. 1, pp. 163-170, septembre (1999).
2. W.D. Kingery, H.K. Bowen and D.R. Uhlmann, "Introduction to Ceramics", Second Ed., John Wiley & Sons Inc., U.S.A., © 1960, (1976).
3. T. Iida, and R.I.L. Guthrie, "The Physical Properties of Liquid Metals", © Takamichi Iida and Roderick I.L. Guthrie, Chap. 3 and 5, (1998).
4. S.M. Wolf, "Whisker-metal matrix bonding", Chemical Engineering Progress, Vol. 62, No. 3, pp. 74-78, (1966).

5. G. Geirnaert, "Céramiques – métaux liquides. Compatibilités et angles de mouillages (Ceramics – Liquid Metals. Compatibility and wetting angle)", Bulletin de la Société Française de Céramique, pp. 1-23, (1977).
6. C. Garcia-Cordovilla, E. Louis and A. Pamies, "The surface tension of liquid pure aluminium and aluminium-magnesium alloys", Centro de Inverstigacion y Desarrollo, Empresa Vational del Alumino, SA Apartado 25, 03080 Alicante, Spain © Chapman and Hall Ltd. pp. 2787-2791 (1986).
7. J. Goicoechea, C.G. Cordovilla, E. Louis and A. Pamies, "Surface Tension of Binary and Ternary Aluminum Alloys of the Systems Al-Si-Mg and Al-Zn-Mg", © Chapman and Hall Ltd, pp. 5247-51, (1992).
8. M.S. Newkirk, A.W. Urquhart, H.R. Zwicker and E. Breval, "Formation of Lanxide™ ceramic composite materials", J. Mater. Res., Vol. 1, No. 1, Jan/Feb., 81-89, (1986).
9. C. Allaire, "Mechanism of Corundum Growth in Refractories Exposed to Al-Mg Alloys ", Aluminum transactions, Vol. 3. No. 1, pp. 105-120, (2000).
10. M.E. Perron and C. Allaire, "Effect of Cryolite on the formation of Al_2O_3/Al Composites produced by oxidation of Al-Mg Alloys", British Ceramic Transactions., Manuscrit No. BCT 408, accepted for publication the 16th of November 1999.

Printed and bound by CPI Group (UK) Ltd, Croydon, CR0 4YY

16/04/2025

14658456-0003